Rhythms In Proterozoic Sedimentation : An Example From Peninsular India

Rhythms In Proterozoic Sedimentation : An Example From Peninsular India

Prof. Pradip Kumar Dasgupta
Dr. Ananya Biswas
Department of Geology, University of Calcutta
Department of Geology, Bengal Engineering and Science University, Shibpur (India)

SSPH
SATISH SERIAL PUBLISHING HOUSE
403, Express Tower, Commercial Complex, Azadpur, Delhi - 110033 (INDIA)
Phone : 011-27672852 Fax : 91-11-27672046
E-mail : info@satishserial.com, hkjain1975@yahoo.com
Website : www.satishserial.com

Published by :

SATISH SERIAL PUBLISHING HOUSE
403, Express Tower, Commercial Complex, Azadpur, Delhi - 110033 (INDIA)
Phone : 011-27672852 Fax : 91-11-27672046
E-mail : info@satishserial.com, hkjain1975@yahoo.com

ISBN 81-89304-23-2

Printed at
SALASAR IMAGING SYSTEMS, DELHI-35

Dedicated to the memory of

Prof. Raj Nath

Former Professor of Geology

Banaras Hindu University, Varanasi

and

Prof. Santosh Kumar Ray

Former Professor of Geology

Presidency College, Kolkata

Who shaped our destiny

Preface

I was advised by Professors F.J.Pettijohn, K.S.Valdiya and S. Bhattacharji during early eighties to take up sedimentologic work in the Proterozoic Cuddapah basin, India. I started the work in the basin in 1983 and was joined later by Dr. Ananya Biswas. Present work is the outcome of our collaboration.

Records of sedimentary cycles mostly owing to overprinting by geological events are limited in the Precambrian. Paleo- to Neoproterozoic cyclic deposition is well documented in stratal assemblages of Cuddapah Supergoup, India. Despite identification of some allogenetically controlled periodic sequences, it is however difficult to interpret periodicities in such cyclic stratal assemblages. Cyclic hierarchies based on interplay among various controlling factors are documented from the Cuddapah stratigraphic record. Cyclic bedding, dicyclic bedding, rhythmic bedding, platformal cycles and orbitally forced are some of the salient cyclicities.

The initiation of Cuddapah sedimentation is marked by deep-seated fault-controlled cyclicities of alluvial fan complexes in the Gulcheru Formation in response to extensional basin tectonics. Switching between arid to humid climatic influence is allocyclic and orbitally forced with attendant cyclic arrangements of aeolian- and braided fluvial deposits. The transition facies records the span of time needed for changing depositional styles from the extrabasinal Gulcheru to the intrabasinal Vempallae sedimentation. Cyclic deposits of lagoonal, shallow-marine and associated terrestrial volcaniclastics characterizing the transition facies attest to the oscillatory emergence induced by the pulsatory nature of the basin. Upward shallowing platform carbonate cycles under variable paleoenvironmental dynamics suggest that the cyclic sedimentation in the Vempallae Formation occurred in response to low-amplitude (Milankovich-band) sea-level oscillations.

A trappean-intertrappean sequence with signatures of bimodal volcanism and unmodified mantle xenoliths, rarely reported from the Paleoproterozoic, designates the basal part of the Pulivendla Formation and corroborates reactivation of deep-seated faults. Boulder beds and repetition of couplet facies testify episodic interference by extremely high-energy depositional conditions in the upper part of the formation. This was brought by frequent redistribution of topography of the provenance area and the basin itself in response to regional extension.

Repetitive generation of silicic magma at shallower depths and its intermittent emplacement deposited cyclic ignimbrites, fluvial and related facies in the Tadpatri Formation. This is overlain by cyclic fluvial to shallow-marine sandstone-mudstone facies of the Gandikota Formation.

The Nallamalai Formation demonstrates a cyclic association of turbidites with clastic facies, debrites, felsic to silicic volcaniclastics, seismites, seismoturbidites and tsunamiites owing to variable genetic controls under the influence of extensional tectonics.

The Egalapenta Member of the Srisailam Formation depicts three discrete cyclic deposition types of fluvial to shallow-marine in the lower part, coastal sabkha-aeolian sandsheet in the middle and climbing dunes (draa) to interdunes forming a part of an erg in the upper part. Such cycles, marking the closing phase of the Cuddapah sedimentation owes its origin to eustatic fluctuations of sea-level and/or orbitally controlled climatic variations.

Some salient aspects of Cuddapah sedimentation, its rhythm and basinal evolution have been addressed. These include : an overview of the Cuddapah Supergroup, sedimentation styles of the Cuddapah, possibilities of hydrocarbon reserves, picrite borne mantle xenoliths, Proterozoic erg system, paleoseismicity and cyclic sedimentation in Paleo- to Neoproterozoic Cuddapah Supergroup.

The authors are indebted to many individuals, whose cooperation was essential in presentation of this volume. They gratefully express their thanks in the first place to University Grants Commission, Department of Science and Technology, Govt. of India for partial financial support. Logistic support and valuable cooperation that were essential have been made available to the authors by Y.S.Prakash Reddy, Y.S.Vivekananda Reddy, Jagadiswar Rao, V.V.Krishna Reddy, Mr.Chinaveera Virabhadrudu and Amitabh Bhattacharyya. Prof. Amaljyoti Sengupta, Dept. of Geology, University of Calcutta & Vice Chancellor (Ex), Bengal Engineering and Science University participated in a collaborative work with DST which is duly acknowledged. His valuable guidance and inspiration are instrumental in completion of this book. Dr.Rajesh Mukherjee, a senior worker of our laboratory painstakingly participated in quite a few field and laboratory works. The authors are grateful to Mr. Pratap Dhang, a Research Fellow, for carrying out drafting in connection with this book. We collectively thank all friends, geologists and other workers involved in logistic support. Last but not the least, thanks go to our publishers Mr.Hemant Jain and Mr.Pradip Jain, M/S Satish Serial Publishing House, whose constant involvement has ultimately led to the publication of the work.

Place : Kolkata

Date : 22.11.2005

Pradip Kumar Dasgupta
E-mail : pradipdasgupta@hotmail.com

Ananya Biswas
E-mail : anniegeol@hotmail.com

CONTENTS

Preface v

CHAPTER - I

1. **The Cuddapah Supergroup : An Overview** **1-30**

CHAPTER - II

2. **Cuddapah Sedimentation** **31-157**

CHAPTER - III

3. **Hydrocarbon Possibilities In Frontier Basins : A Case History from the Proterozoic Cuddapah Basin** **159-173**

CHAPTER - IV

4. **Picrite Borne Mantle Xenoliths from Cuddapah Basin** **175-181**

CHAPTER - V

5. **Proterozoic Erg System from the Egalapenta Member, Srisailam Formation** **183-251**

CHAPTER - VI

6. **Paleoseismicity from the Cuddapah Basin** **253-295**

CHAPTER - VII

7. **Cyclic Sedimentation In Palaeo-To Neoproterozoic Cuddapah Supergroup and its Significance in Basinal Evolution** **297-340**

1

The Cuddapah Supergroup : An Overview

Introduction

In the evolutionary history of the earth, the Proterozoic marks a transition from highly mobile and widespread tectonic regimes, reducing atmosphere and acidic oceans of the Archaean to a state of lithospheric rigidity and near-modern air and water characteristic of the Phanerozoic. Sedimentary sequences of the Proterozoic have a high proportion of volcaniclastics and volcanics in the lower part. Banded Iron Formation is a major constituent of the middle Proterozoic. The upper part of Proterozoic sequences is dominated by carbonates. Widespread shallow water sedimentary basins became a dominant feature of the earth surface during the Proterozoic. Proterozoic sedimentary successions are available in mobile belts, epicontinental seas, rifted basins and successor basins (Rogers 1993, Goodwin 1996).

There are reasons to believe that the earth's crustal development has been a time variant self-controlling process with built-in geothermal controls. With falling geothermal gradient the hypersolvus diapirism and growth by underplating during the Archaean changed over to subsolidus ductile flow in the Proterozoic initiating mobile belts sited over areas of mantle upwelling. The exact role of extensional tectonics in the building of Proterozoic volcano-sedimentary prisms adjoining cratonic blocks and the relative contributions from magmatic addition, crustal recycling and crustal redistribution are not well understood.

The Archaean Cratonic Nuclei (ACN) of the Indian Peninsular Shield are thought to have coalesced together through Late Archaean-Palaeoproterozoic accretionary tectonic events. The subsequent

Mesoproterozoic and Neoproterozoic sequences are preserved either in the Proterozoic Basins resting upon with a pronounced "Great Eparchaean Unconformity" upon the eroded surfaces of the intensely deformed rocks grouped under 'Archaean' (Pascoe, 1973) or in the Middle Proterozoic Mobile Belts (MPMB). The latter contains deformed metamorphosed supracrustal sequences and can be ascribed to compressive tectonic regimes. The Proterozoic, in general, represent shallow marine, epicratonic, passive-margin sequences deposited in an extensional tectonic regime.

With reference to the ACN constituting the Indian Peninsular Shield, the major Proterozoic Basins are (Fig.1) :

a) Vindhyanchal (from Bundelkhand Craton)

b) Chattisgarh (from Bhandara Craton)

c) Bastar (from Bhandara Craton)

d) Pranhita-Godavari (between Bhandara and Dharwar Cratons)

e) Cuddapah (from Dharwar Craton)

f) Bhima (from Dharwar Craton)

g) Kaladgi (from Dharwar Craton)

They are pericratonic or epicratonic basins located on the peripheries of the ACN, with the Pranhita-Godavari Basin occupying an intracratonic position. These Proterozoic basins are characterised by specific individuality, which indicates independent evolutionary histories, even while retaining a broad similarity with each other. They have sharp boundaries, often discernible as explicit geophysical anomalies.

Proterozoic sedimentary sequences in India are available in the mobile belts of Singhbhum, Rajasthan and Eastern Ghats as well as in epicontinental/rift basins in the northern fringe of Peninsular India bordering the Indo-Gangetic alluvial trough (Vindhyan Supergroup, Bijawar Group and Chattisgarh basins), along the Pranhita-Godavari valley near the east coast, and further south in the Bhima, Kaladgi and Cuddapah basins. Some of the Proterozoic basins of India, south of 20°N, are covered by the Deccan Traps (Fig.2).

Greatest development of the sedimentary sequences is observable in the Cuddapah basin, giving it the status of strato type for the

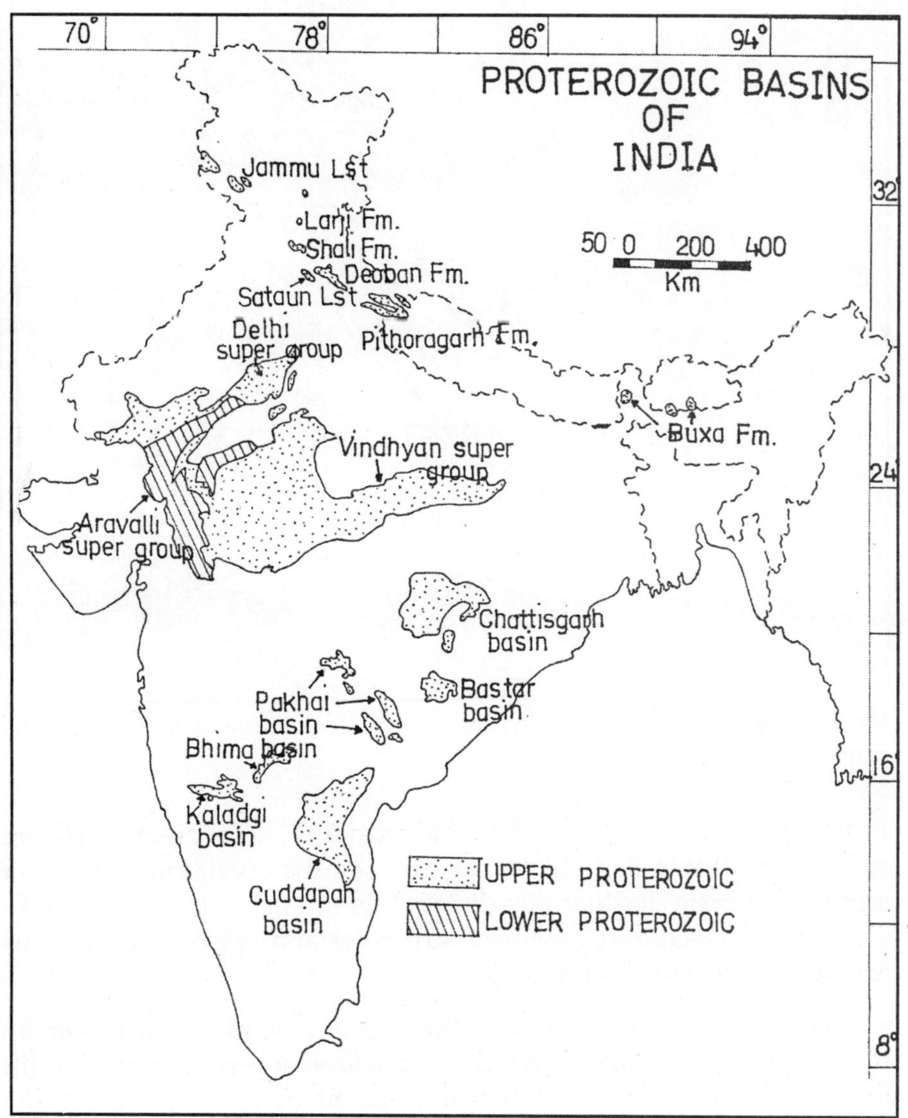

Fig. 1: Distribution of Proterozoic basins of India.

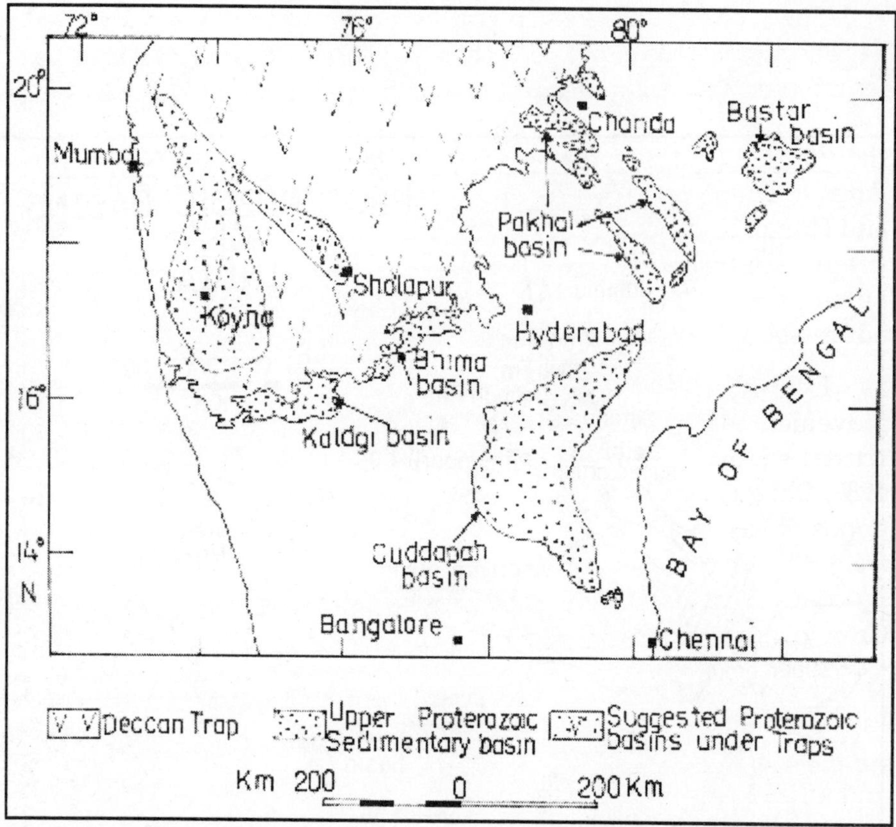

Fig. 2: Proterozoic sedimentary basins below 20°N latitude in Peninsular India and a part of the Deccan Trap.

Proterozoic Eon in India (Karunakaran 1973). Extensive erosion suffered by the Indian shield has exposed the crystalline basement around the basin and permits direct observations of the sedimentary accumulations, contact basement structures and episodes of igneous activity in and around the basin.

The evolutionary history of the Proterozoic platform basins on the Indian shield, some of which experienced several cycles of uplift and subsidence, episodic magmatism prior to and during the sedimentary accumulation, and syn- to post-depositional deformation is extremely complex. The Cuddapah basin of South India is an example of this, but the evidence and role of mantle upwelling and magmatism, lineament reactivation and crustal uplift in basin evolution can be established with some degree of certainty in the case of Cuddapah Basin.

The Cuddapah basin in south-central India covers an area of 44,500sq.km. It has a crescent shape, is convex to the west and has a length of 400km and a maximum width of 145km (Fig.3) (Pichamuthu 1967, Meijerink et al. 1984, Nagaraja Rao et al. 1987). The thickness of the sedimentary pile contained within the basin ranges upto 14km. After the deposition of the Cuddapah, successor basins like Kurnool and Palnad appeared and blanketed the Cuddapah sediments. Today, deeply eroded portions of Kurnool and Palnad expose Cuddapah outcrops. Some of such representative Cuddapah outcrops have been addressed in the present treatise.

Evolution of the Cuddapah basin appears to be related to movement along deep seated faults which were formed in response to crustal stretching (Murthy 1981, Kaila and Tewari 1985, Kaila et al. 1987, Bhattacharji and Singh 1984). There was a progressive shift in depositional areas through time with their subparallel axes trending NE-SW. The time span covered by the evolution of the basin ranges more than 1000m.y. (ca. 1.9m.y. to 600m.y., Jachariah et al. 1999, Crawford and Compston 1973).

The crescent shaped Proterozoic Cuddapah basin of Peninsular India is dominantly arenaceous and argillaceous. Siliciclastic-carbonate suite, basic to acid volcanics, sills in the lower part and siliceous shales and siliciclastics in the upper part is commonly found. Calcareous sediments constitute a small proportion of the total thickness. Volcaniclastics and volcanics make up a considerable proportion of the Cuddapah sequence at different stratigraphic levels, the thickness of the individual units ranging upto a maximum of 3.5km. The basin has a concave eastern margin, 440 km long, with a maximal basinal width of 145km in the middle, covering an area of 44,500 sq.km and containing over 12km of sediments, volcanics and volcaniclastics. The Cuddapah sediments mostly belong to fluvial, aeolian and lagoonal facies, though intercalations of marine sediments are rather common (King 1872, Banerjee 1974, Singh 1980, Rajurkar 1979, Dasgupta 1996, Dasgupta et al. 1990).

One may note in this connection that the Cuddapah Supergroup was recognised as a deposit of shallow cratonic seas for quite a long time. In fact, the Proterozoic siliciclastic and carbonate sequences of the Indian Peninsula are generally known as 'shallow-marine' deposits (Banerjee 1974, Singh 1976, Chanda and Bhattacharyya 1982, Johnson 1965, Chaudhuri 1970, Sreenivasa Rao 1987). This may be due to the

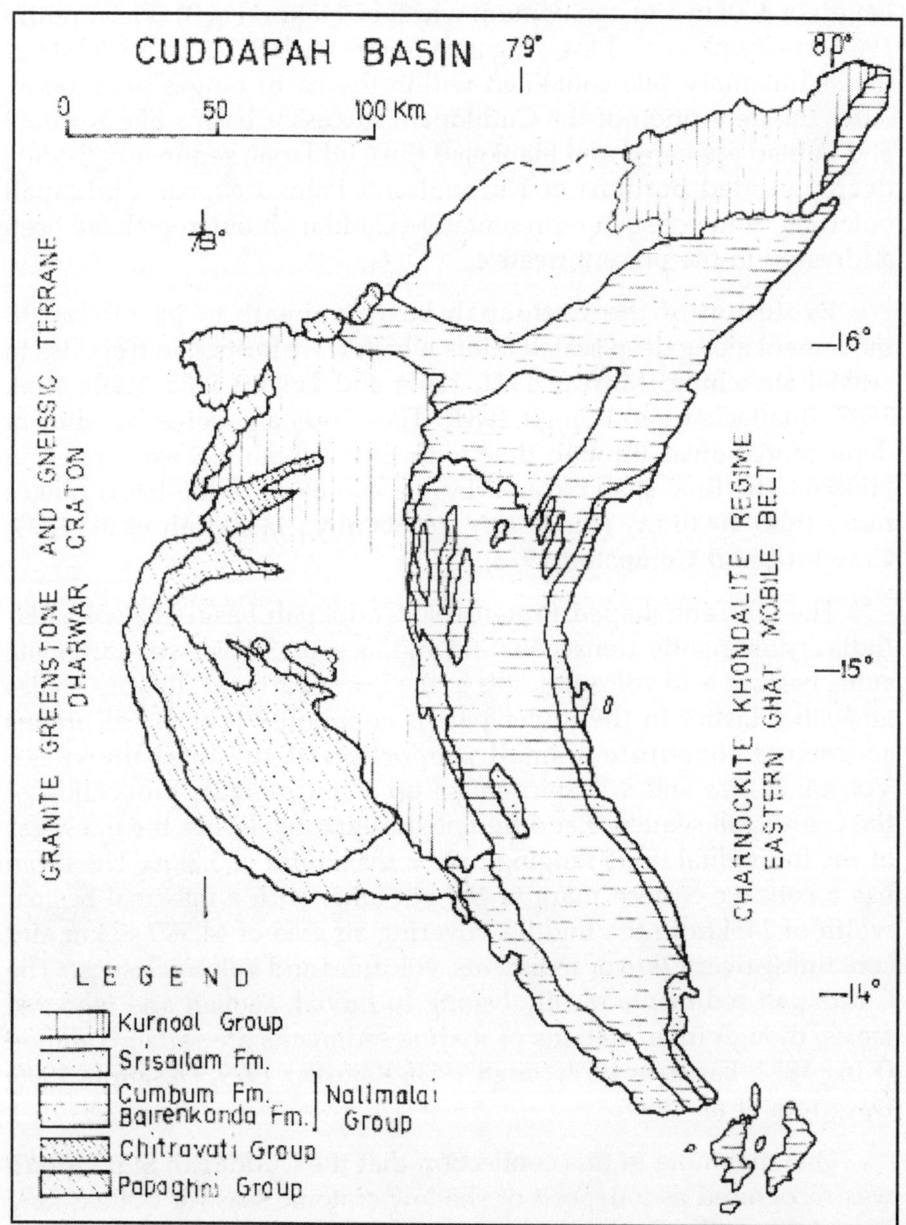

Fig. 3: Geological map of the Cuddapah Supergroup.

influence of the traditional shallow-marine model for the mature cratonic sheet sandstones of the classical carbonate-orthoquartzite suite of the Mississippi Valley, USA, which gained maximum currency during the last few decades (Pettijohn 1975, Sloss 1963, Dott et al. 1986).

The closing phase of Cuddapah sedimentation was marked by deposition of the Srisailam Formation. The three Members of the Srisailam Formation, namely the Kaklet Vagu, Tapasipenta and Egalapenta beds, arranged in ascending order, are characterised by frequent facies changes. The Egalapenta Member at the top, in particular, shows remarkable vertical and spatial facies variations and intertonguing of aeolian, fluviatile and shallow-marine facies.

Stratigraphy

Four major subdivisions of the Cuddapah sequence, namely the Papaghni, Cheyair, Nallamalai and Krishna, arranged in ascending order, were recognised by King (1872). The classification scheme proposed by King and others is given in Table.1. Each subdivision was taken to be laterally continuous throughout the basin. Nearly a century later, an attempt was made to develop a scheme of chronostratigraphic classification for the Cuddapah succession (Sen and Rao,1967). This was immediately replaced by a lithostratigraphic classification which appears to be an improvement on the scheme given by King, 1872; Nagaraja Rao and Ramalingaswamy 1976; Rajurkar and Ramalingaswamy 1975,1978). Extensive mapping coupled with lateral tracing of the lithostratigraphic units for more than two decades resulted in a lithostratigraphic classification which accounts for most of the observed stratigraphic relations (Murthy 1979, vide Table.2).

An Outline of the Geology of the Cuddapah Basin

The Precambrian shield of India consists of a high-grade metamorphic belt of charnockite-khondalite and migmatites, the Dharwar system of schists, greenstone and granites and the Purana group consisting of the Proterozoic Cuddapah and Kurnool sedimentary systems, together with the enclosing Peninsular Gneissic Basement complex. Their evolutionary history is marked primarily by succeeding geosynclinal facies of sediments. Volcanic effusives with associated syntectonic mafic and ultramafic intrusives predominate in the case of charnockite-khondalite and the Dharwars, but platform facies of sediments with some volcanics and mafic-ultramafic intrusives dominate in the Cuddapah system.

Table 1: Stratigraphic Classifications of Cuddapah Supergroup

KING (1872)	SEN AND NARASIMHA RAO (1967)	DUTT (1975)	NARAYANSWAMI (1976)	MEIJERINK, RAO RUPKE (1984)
Kurnool Formation **Khoondair Group** – Nundial Shale; Koilkuntla Limestone **Panem Group** – Pinnacled Quartzite; Plateau Quartzite **Jummula-Mudgoo Group** – Owk Shale; Narjee Limestone **Banganpally Group** – Sandstone and Pebble beds **Cuddapah Formation** **Kistna Group** – Srisailam Quartzite; Kolumnullah Slate; Irlakonda Quartzite **Nallamalai Group** – Byrenkonda Quartzite; Poolampet Shale **Cheyair Group** – Tadpatri Shale; Naggery Quartzite or Pulivendla Quartzite **Papaghni Group** – Vaimpully Slate; Goocheroo Quartzite	**Cuddapah System** **Kistna Series** – Srisailam Quartzite ------Disconformity------ **Nallamalai Series** – Cumbum Shale (Pullampeta Kolamnala Shale); Bairenkonda Quartzite (Nagari Quartzite and Irlakonda Quartzite) ------Disconformity------ **Cheyair Series** – Tadpatri Shale; Pulivendla Quartzite ------Disconformity------ **Papaghni Series** – Vaimpully Shale and dolomites; Gulcheru Quartzite **Archaean**	**Kurnool Supergroup (600m)** **Upper Kurnool** – Nandyal Shale; Koilkuntla Limestone; Panium Shale Quartzite with basal conglomerate ------Disconformity------ **Lower Kurnool** – Auk Shale; Narji Limestone; Banganapalli Quartzite with basal conglomerate ------Disconformity------ Srisailam Quartzite **Nallamalai Group** – Cumbum Shale; Bairenkonda Quartzite ------Disconformity------ **Cheyair Group** – Upper Tadpatri Shale; Lower Pulivendla Quartzite **Papaghni Group** – Vaimpully Dolomite Shale; Guwwacheruvu Quartzite	**Kistna Group** – Srisailam Quartzite Fm.; Kolamnala Shale Fm.; Irlakonda Quartzite Fm. ------Unconformity------ **Nallamalai Group** – Cumbum Shale/Slate; Bairenkonda Quartzite Fm. ------Unconformity------ **Cheyair Group** – Pullampet Shale Fm. with interbedded limestone and dolomites and also volcanogenic tuffs, barytes; Gandikota/Nagari Quartzite Fm. ------Unconformity------ **Chitravati Group** – Tadpatri Shale Fm.; Pulivendla Quartzite **Papaghni Group** – Vaimpully Dolomites and Shale Fm. with volcanic basaltic flows and intrusive doleritic sills; Gulcheru Quartzite and Conglomerate Fm. ------Unconformity------ **Archaean** – Peninsular Gneissic and Granitic Complex and Closepet Granites with basic dyke network	**Kurnool Group** **Khoondair Fm.** – Shale, limestone **Panium Fm.** – Quartzite ------Paraconformity------ **Banganapalle Fm.** – Conglomerate, sandstone, shale, limestone ------Unconformity------ **Nallamalai Group** **Cumbum Subgroup** – Giddalur Fm.; Pullampet Fm.; Srisailam Fm. **Bairenkonda Subgroup** – Dornala Fm.; Iswarakuppam Fm. **Cuddapah Group** **Nagamureru Subgroup** – Tadpatri Fm.; Pulivendla Fm. **Papaghni Subgroup** – Vaimpully Fm.; Gulcheru Fm. **Archaean granites and gneisses**

Table 2: Revised Stratigraphic Classification of Cuddapah Supergroup (Murthy 1979)

		Thickness (in metres)	
Kurnool Group	Nandyal Shale	50-100	500
	Koilkuntla Limestone	15-50	
	Panium Quartzite	10-35	
	Auk Shale	10-35	
	Narji Limestone	100-200	
	Banganapalle Quartzite	10-57	
~~angular unconformlty~~~			
Srisailam Formation	Egalapenta Member	238	600(+)
	Tapasipenta Member	100	
	Kaklet Vagu Member	320	
~~angular unconformity~~~			
Nallamalai Group	Cumbum Formation	2000	2500
	Bairenkonda (Nagari) Formation	1500	
~~angular unconformity~~~			
Chitravati Group	Gandikota Formation	1200	5850
	Tadpatri Formation	4600	
	Pulivendla Formation	1-75	
~~angular uncomformity~~~			
Papaghni Group	Vempalle Formation	1500	6000
	Gulcheru Formation	28-250	
~~nonconformity~~~			
Archaean and Dharwar			

The stratotype Cuddapah basin, convex westward, occupies an area of approximately 44,400 sq.km. extending north-south over a crescent length of 440 km and a maximum width of 150 km east-west. It is situated between 13°15'N to 17°00'N latitudes and 77°45'E to 80°15'E longitude (Fig.3). The thickness of the sedimentary pile is about 6.5 km (Murthy 1979) with varying thickness of the individual formations which is probably due to lateral lithofacies. Variations and each of the major formations are marked by unconformity.

Geology of the Cuddapah basin has been studied by various authors over the last century. King (1872) divided the rocks in the basin into Cuddapah and Kurnool Groups. GSI (1979,1981) have explained the tectonics of the Cuddapah basin on the basis of various geological features.

The Cuddapah basin of South India is one of the major Proterozoic intracratonic basins of India. The basin is bordered on the west by the Peninsular Gneiss showing remnants of greenstone belts. The Nellore Schist Belt and the Eastern Ghats Granulite terrain appear along the eastern margin. The basal formations of the Cuddapah Supergroup lie unconformably on the Peninsular Gneiss along the eastern margin (Fig. 3). The flat lying strata of the basal units have low easterly dips. In contrast, the sedimentary sequences in the eastern part of the basin are steeply dipping and isoclinally folded. The eastern margin of the Cuddapah basin is an easterly dipping thrust along which the sedimentary sequence is tectonically overlain by the Nellore Schist Belt (3.0Ga to 2.6Ga). Deep seismic soundings indicate that the thrust and the related imbrications are steeply inclined and penetrate the base of the crust (Kaila and Tewary 1985, Kaila et al. 1987, Drury and Holt 1983). The basin comprises sediments from late Paleoproterozoic (~1800 Ma, Bhaskar Rao et al.,1995) through Neoproterozoic in age (minimum of ~500 Ma, Ramam and Murthy, 1997) with an estimated cumulative thickness of ~14km.

The widest section of the basin lies just south of the major swing in the structural fabric of the area (Fig.4). Along its eastern margin, the sediments are strongly deformed into a convex to the west oroclinal fold and thrust complex that verges west and northwest. The eastern marginal thrusts, Vellikonda Thrust Front (Fig.5) have brought Archaean gneisses, granulites and gneisses that in most places directly over the Proterozoic sediments that comprise the Cuddapah basin. The intensity of deformation decreases westward to the Nallamalai Thrust Front (Fig. 5) that has brought up older Cuddapah sediments directly over younger Kurnool sediments. Probably in response to an asthenospheric upwelling, extensive episodes of igneous activities (during 2200-3000 Ma) occurred along lineaments. In the Indian Shield intrusion of Closepet-type granites, followed by mafic dyke intrusion caused heating and topographic uplift (and dilation) followed by

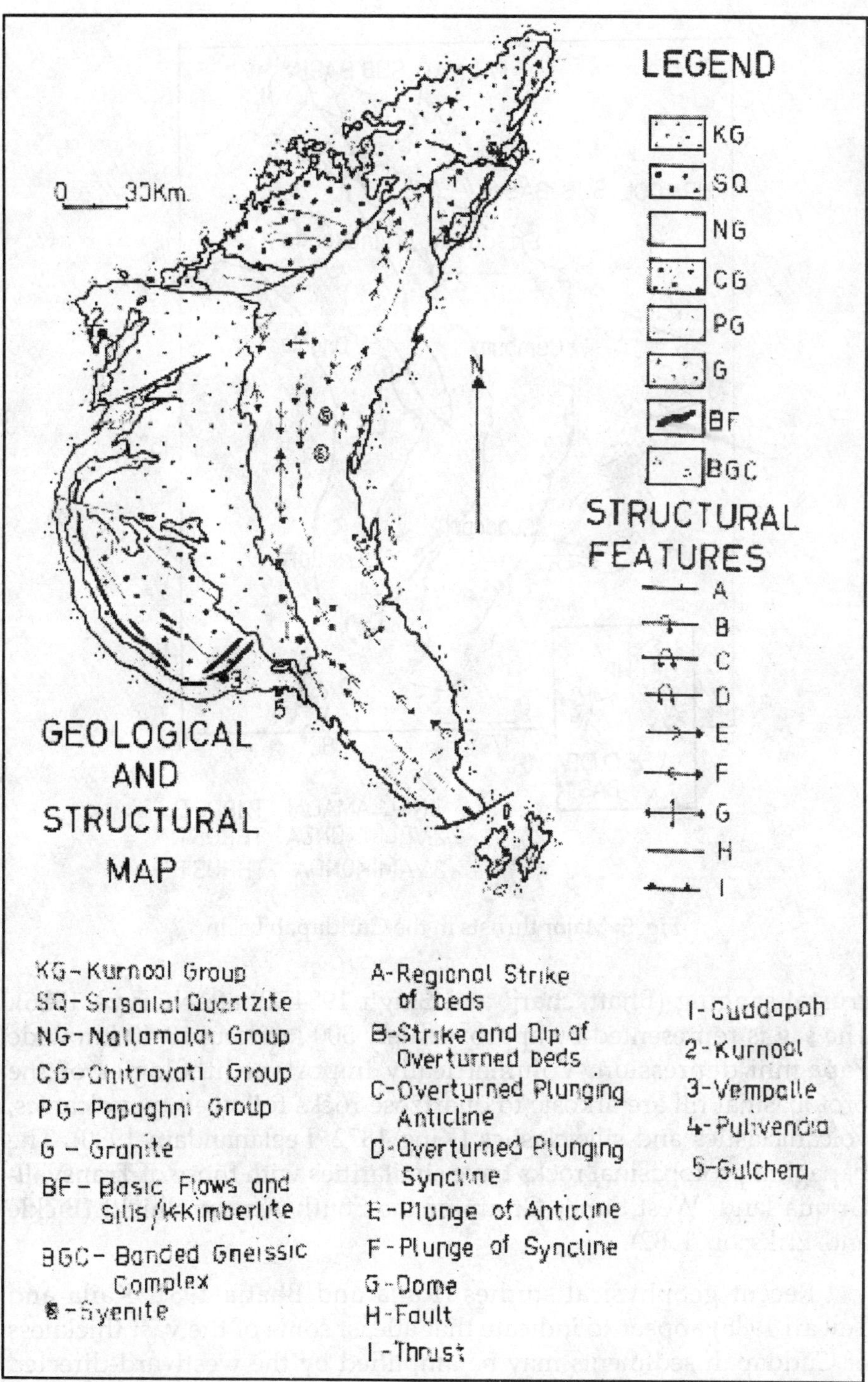

Fig. 4: Geological and structural map of the Cuddapah Supergroup.

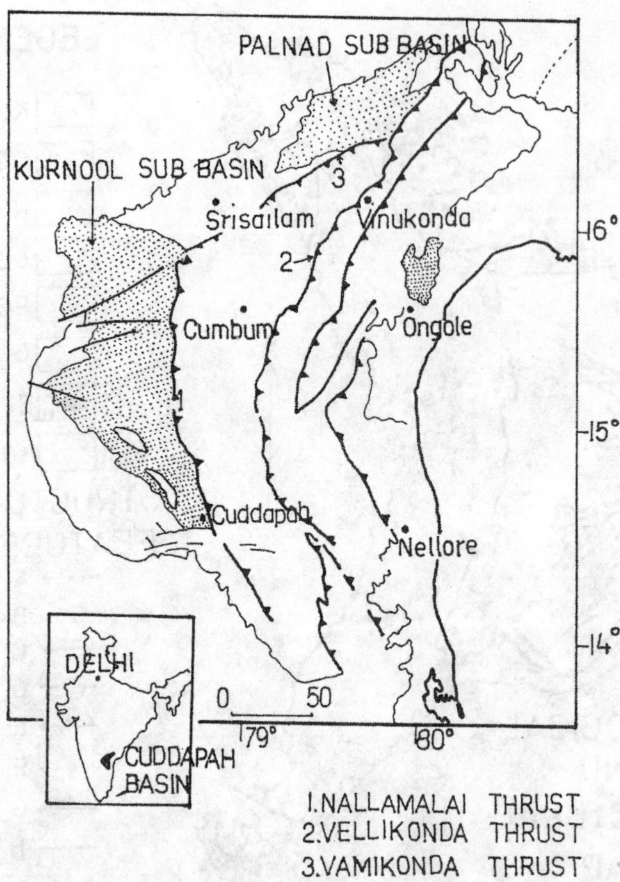

Fig. 5: Major thrusts in the Cuddapah basin.

crustal sagging (Bhattacharji and Singh 1984, Bhattacharyya 1986). The sag is represented by approximately 300 km long 10-40 km wide Papaghni depression. Volumetrically important lithologies of the protobasinal fill are arkosic to quartzose rocks followed by volcanics, volcaniclastics and siliciclastics (King 1872, Leelanandam 1980). The Papaghni protobasinal rocks have similarities with those of Transvall-Griqua land, West Super Group of the South African Shield (Bickle and Eriksson 1982).

Recent geophysical studies (Kaila and Bhatia 1981, Kaila and Tewari 1985) appear to indicate that atleast some of the vast thickness of Cuddapah sediments may be amplified by the westward-directed basement crustal shortening with a component of vertical tectonism.

Narayanaswami (1966), Sen and Narasimha Rao (1967) subdivided the basinal sediments into two major sequences the lower Cuddapah System separated by a profound unconformity from the upper Kurnool system. Since these studies, both systems have been elevated to supergroup and group status respectively (GSI 1981).

Although the Cuddapah basin is presently crescent-shaped, it appears to have originated as a shallow, elongated basin along older NW-SE to NE-SW lineaments and greenstone belt trending NW-SE to N-S within the older granites and gneisses. Early igneous episodes prior to initiation of Cuddapah sedimentation came as periodic pulses. Extensive mafic dykes and dyke-swarms were localised around the southern and southeastern boundaries of the basin. The largest and oldest (2400-2200 Ma) dyke-swarms are found at the southern side of the basin (Balasubramaniyam 1975, Ikramuddin and Stueber 1976). The rock types include mainly dolerite-basalt, gabbro-peridotite and amphibolite. Slightly younger less dense swarms occur on the northwestern side of the basin. The orthoquartzite-carbonate association of the Papaghni sediments of lower Cuddapah (Table. 2) (Bhattacharya 1987) indicates that the basin started shallow with a stable shelf or platform type tectonic environment with a transgressive marine condition. The sedimentary structures indicate that the sediments were derived from the south and west (Murthy 1979).

The Cuddapah rocks are composed of quartzites, graded, pelitic and semi-pelitic turbidites and stromatolitic carbonates. Lateral facies variations, local unconformities of various sorts and a general deepening of the basin to the east are indicated by progressive offlap to successively younger rocks to the east. The younger (upper Proterozoic) Kurnool rocks overstep the featheredge of the older (upper to lower Proterozoic) Cuddapah Supergroup all along the western margin of the Kurnool and Palnad subbasins indicating progressive uplift (and possibly west vergent thrusting) during the Kurnool group foreland basin development. The orthoquartzite-carbonate association of the Papaghni sediments of lower Cuddapah (Table 1) indicates that the basin started shallow with a stable shelf or platform type tectonic environment with a transgressive marine condition. The sedimentary structures indicate that the sediments were derived from the south and west (Murthy 1979). It is indeed interesting that the Kurnool Group is entirely devoid of magmatism. The lower part of the Cuddapah Supergroup is however, characterised by an extensive volcanic and plutonic suite of basalts, gabbros, dolerites and a few felsites.

The youngest post-Cuddapah sills have been dated by Rb-Sr method as 980±110 Ma and an age of 1583±147Ma has been obtained from one of the oldest sills by the same method (Crawford and Comspton 1973). These ages provide respectively a minimum and maximum age for Cuddapah Supergroup sedimentation. Kimberlite dyke that intrudes the Cuddapah sediments was dated by Rb-Sr method as 1225 Ma (Crawford and Comspton 1973). This clearly indicates at least some deep crustal fracturing during basinal events (Kaila et al. 1979).

The Cuddapah Formations show a high degree of consolidation, diagenesis and some crystallisation in denser form. The post depositional history appears to have been more complex in the eastern half of the basin, where severe deformation resulted in large-scale cross-folding, overthrust faulting and culminated in the thrusting of the basement gneisses over the younger rocks in the northeastern margin. The western half of the basin and its sedimentary formations there remained practically unaffected by the post-depositional compressional tectonics from the east.

In the Cuddapah basin, a gentle easterly dip of the sedimentary units of Cuddapah and Kurnool strata is maintained all throughout. There is much domal upwarp in the central part of the basin but doubly plunging structures, cross-culminations and depressions, and overturned isoclinal folding dominate all along the eastern margin (Fig.4). The western margin, however, has remained comparatively undisturbed.

The main structural element of the Cuddapah basin consists of a northerly-plunging, asymmetric synclinorium made up of a gently dipping, practically non-folded western limb and an intensely folded, overturned and thrusted eastern limb. The synclinal nature of the fold in the western margin seems more due to the sinking of the basin floor rather than due to lateral compression (Murthy 1979).

A dominant structural feature is the axes of the folded structures swing from NW-SE to NE-SW in conformity with the general configuration of the basin (Fig.4) starting from an open folded anticline along with the western and central parts of the Nallamalai range, the folding becomes increasingly intense leading to tight isoclinal folds and increasingly overturned isoclinal and recumbent fold. The eastern margin is highly faulted with en-echelon medium to low angle thrust slips, forming a zone of thrust faults, culminating in the thrusting of the basement granites and gneisses over the younger Cuddapah rocks.

Domal upwarps in the middle of the Cuddapah basin are important structural and geomorphic features. Narayanaswami (1966) suggested that these domal upwarps are tectonic impressions developed during the waning phase of the Cuddapah-Kurnool deposition or during some subsequent orogeny and probably the result of some vertical tectonics. Vertical tectonics due to the activation of deep-seated faults by post-depositional compressional deformation is supported by Deep Seismic Sounding (Kaila et al. 1979).

Progressive shifting of sub-basins with their sub-parallel axes trending NE-SW have been recorded (Sen and Rao 1967, Murthy 1981). The following sequence of development of sub-basins has been recognised:

1. Papaghni sub-basin marking the site of deposition of Papaghni and Chitravati Groups;
2. Nallamalai sub-basin marking the site of the Nallamalai Group;
3. Srisailam sub-basin marking the site of the Srisailam Formation;
4. Kurnool/Palnad sub-basin marking the site of the Kurnool Group (Fig.6).

The unconformable relationships between different subbasins (lithostratigraphic units) of the Cuddapah basin suggest the basin is a polyhistory (successor) basin (Kingston et al., 1983; Nilsen and Sylvester, 1995).

The Papaghni Sub-basin

The Gulcheru Formation comes directly over the granitic basement showing clear depositional contact, which is best exposed near the Chitravati Gorge. The basal Gulcheru is a sequence of very coarse to coarse clastics, indicative of a terrestrial fan complex (Dasgupta 1986). Boulders, bouldery and pebbly horizons of polymictic conglomerates within the stacked channels are discernible. Vertically it passes to a channel facies of coarse- to medium-grained clastics. Laterally accreted bars, fining-upward sandstone-mudstone cycles are developed in association with deep to broad asymmetric channels. The facies assemblages with a cyclicity of fining-upward sandstone-mudstone couples suggest passage to braided systems with gradual depletion of energy budget through time.

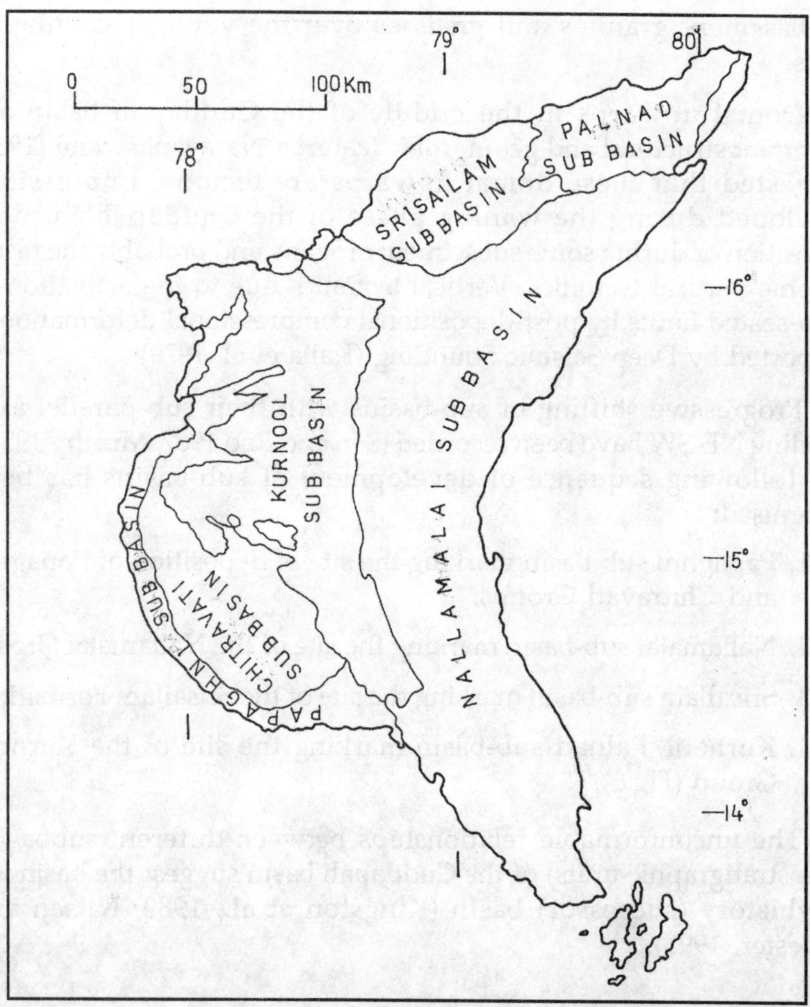

Fig. 6: Different subbasins in the Cuddapah basin.

The Vempalle Formation is separated from the Gulcheru by a transition facies. The Vempalle comprises disorganised lithic complexes of medium- to fine-grained clastics, impure clastics, mixed carbonates and dolomitic carbonates grading into a sequence of well defined carbonate beds. The carbonate beds include sheet bedded limestones, lenticular dolostones with stromatolitic bedding and shales (King 1872). Dolostones with stromatolitic bedding, limestones with their sheet geometry and intervening laminated subordinate thin mudstone units attest to a lacustrine to lagoonal and/ or shallow-marine origin (Murthy 1981, Sen and Rao 1967, Dasgupta et al. 1990).

The Chitravati Group with well-preserved outcrops of the Lower to Middle Proterozoic strata shows unconformable contact over the underlying Vempallae Formation of the Papaghni Group. The Chitravati includes 4.6 km thick Tadpatri Formation comprising ignimbrites and associated volcaniclastics, terrigenous clastics and carbonates and it represents important sequence related to basinal history.

The Chitravati Group is a trappean-intertrappean sequence overlying the Vempalle carbonates with an angular unconformity. Five lithofacies have been recognised within the Group, namely impure carbonates, mudstones with frequent emergence features, subaerial volcanic flows, volcaniclastics at different levels within the sequence and tufa associations (Dasgupta et al. 1990).

The Pulivendla Formation overlies the eroded top of the Vempalle. The Formation represents a channel facies of coarse- to medium-grained clastics. Fining-upward sandstone-mudstone associations constitute the infills of broad, shallow stacked channels. Laterally accreted bars, laminites and penecontemporaneously deformed mudstones are important components of the facies assemblage. The Pulivendla sequence constitutes a braided fluvial system (Dasgupta et al. 1990).

The Pulivendla Formation grades into mudstone/shale-sandstone/carbonate facies complex of the Tadpatri Formation. The upper part of the Formation consists of a thick pile of ignimbrites (3.5km) with occasional intercalations of carbonates and terrigenous clastics. The Tadpatri Formation reveals a lagoonal to shallow-marine milieu alternating with fluvial sedimentation (Murthy 1981, Nagaraja Rao et al., 1987, Dasgupta et al. 1990).

The sandstone-mudstone-sequence of the Chitravati Group overlies the Tadpatri ignimbrites with erosional unconformity. Dominance of channel facies marks the sequence as the product of deposition under braided fluvial environment (Nagaraja Rao et al. 1987, Dasgupta et al. 1990).

The Nallamalai Sub-basin

The Nallamalai Fold Belt, an arcuate fold and thrust belt, about 400km long, occupies the eastern half of the Cuddapah basin. A number of granitic bodies occur along the eastern margin of northern Nallamalai Fold Belt.

Tectonic setting of the Nallamalai is complex and the intensity and stages of deformation increase gradually from northeast to southwest. Multistage deformation (D1, D2, and D3 during the late Paleoproterozoic to Neoproterozoic) is prominent in the Nallamalai Fold Belt (Saha, 2002).

The Thokapallae complex shows two sets of folds : antiform and synform. Axial traces of the antiform and synform trend 035°-215° and truncate abruptly against the fault trending NE-SW. With the increase of intensity of deformation towards southwest in Vontimitta region, the Nallamalai strata show seven folds belonging to polyphase deformation. Fold styles vary from reclined through series of antiforms and synforms with the axial traces nearly parallel trending NNW-SSE, to right superposed folds.

Some amount of folding of the Nallamalai rocks took place before the onset of the succeeding basin namely the Srisailam Formation. Further folding in the eastern part was later followed by major thrusting at the eastern margin and also some Nallamalai rocks westwards over the Kurnool Formation. Folding of the Nallamalai appears to have been accompanied by doming upwards involving or due to local fluidization of basement or diapiric granitic intrusion.

The Bairenkonda Formation of the Nallamalai subbasin is basically an arenaceous facies. The lithologic units include arenaceous and argillaceous sediments. The Bairenkonda assemblage originated in a fluvial and/or shallow-marine condition (Nagaraja Rao et al. 1987).

The Cumbum Formation overlying the Bairenkonda is mudstone-sandstone facies assemblage with intercalations of shales and dolomites at various levels. The Cumbum Formation in the lower part is characterised by volcaniclastic turbidite systems such as T_a, T_{abc}, T_c, $T_{c/c/c}$, T_{bcd} and T_{bc} (Dasgupta et al. 1990). The upper part was deposited in fluvial, lagoonal and/or tidal-subtidal environment (Rajurkar and Ramlingaswamy 1975).

The Srisailam Sub-basin

The Srisailam Formation and the succeeding Kurnool/Palnad rocks are near-horizontal sheets formed in western, northern and northeastern parts of the Cuddapah basin. This overlaps the lower Cuddapah strata as well as the crystalline basement rocks.

According to King (1872), the Srisailam Formation, the topmost unit of the Cuddapah sequence, belongs to the Krishna Group and is underlain successively by Kolumnala Shale and Irlakonda Quartzite. Later studies revealed that the Cumbum Formation, the uppermost unit of the Nallamalai Group, laterally grades into the Irlakonda Quartzite towards west and merge with Kolumnala Shale in the east. Kolumnala Shale is thus the uppermost horizon of the Cumbum Formation (Murthy 1979). The Srisailam Formation rests on the Kolumnala Shale and the Irlakonda Quartzite with an angular unconformity. The angular unconformity is displayed clearly in the north of the valley near Madapuchelima for nearly 10km (Rajurkar and Ramalingaswamy 1975). The regional overlap is further emphasised by the presence of conglomerates at the junction between Srisailam Quartzite and Kolumnala Shale (Murthy 1979). Outliers of Srisailam Quartzite, directly overlying the basement granite, are seen in the northwestern part of the basin. On the other hand the Kurnool Group overlies the Srisailam Formation with an angular unconformity near Nagarjuna Sagar-Macherla area.

Of the three Members recognised in the Srisailam Formation, namely the Kaklet Vagu, the Tapasipenta and the Egalapenta arranged in ascending order, the uppermost Member covers the entire area under study. The lower and the middle Members are exposed only along the valley of the Krishna River.

The Kaklet Vagu, the lowermost Member of the Formation comprises siltstones and mudstones belonging to fluvial siliciclastic facies. Ignimbrites and rhyolites are often intercalated (Dasgupta et al. 1990).

The Kaklet Vagu fluvial sandstone and volcaniclastics are onlapped by argillaceous Tapasipenta beds comprising mudstone, siltstone and fine-grained glauconitic sandstone of shallow-marine origin.

The Egalapenta Member comprising reddish brown sandstone, glauconitic sandstone, mudstone-siltstone and impure carbonates has been ascribed to a shallow-marine origin (King 1872, Murthy 1981, Rajurkar and Ramlingaswamy 1975).

The Kurnool/Palnad Sub-basin

The Srisailam sub-basin is covered by a blanket of the Kurnool sediments deposited in the Kurnool/Palnad sub-basin. The facies associations of the Kurnool Group include conglomerate-sandstone, carbonate, mud-sand and mixed mud-carbonate facies.

The Banganapalle Formation includes coarse clastics, sandstone and shale. It represents beach gravel formed by high energy wave action in transgressive seas (Nagarajarao et al. 1987).

The Narji Limestones consist of well-bedded micritic limestone. It was deposited in a deep water carbonate platform (Nagarajarao et al. 1987). However, interbedded pyrite-bearing sequence probably stands for shallow-marine reducing environment (Vijayam and Kamal 1978).

The Auk Shale comprises shale with thin sandy interbeds. It originated under subtidal condition. The Paniam Quartzite is an arenaceous sequence and represents a near shore environment (Vijayam and Kamal 1978).

The Koilkuntla Limestone, is a facies mosaic of calcareous, coarse arenaceous to argillaceous sediments. The facies complex indicates both shallow-marine as well as deep water environments (Nagarajarao et al. 1987).

The Nandyal Shale is broadly a calc-argillaceous facies. A subtidal condition is attested to its origin (Vijayam and Kamal 1978).

Evolution of the Cuddapah basin took place in four phases :

- Lower Cuddapah subbasins developed with time gaps and unconformities in between.
- Followed by a time gap, development of a linear NS oriented Nallamalai basin to the east took place, which appears to have been thrust over the rigid lower Cuddapah strata.
- This was followed by a shallow spreading type of basin on the eroded part of the Nallamalai, in the northern part represented by the Srisailam plateau.
- Finally, with another time gap, the Kurnool-Palnad basins developed after completion of Cuddapah sedimentation as successor basins.

Structure of the Basin

The Cuddapah basin comprises an undeformed western half and a deformed eastern half called the Nallamalai Fold Belt. The western margin is marked by a profound non-conformity of the Cuddapah sediments resting on the Archaean Peninsular gneissic complex enclosing the greenstone belts of Kadiri, Veligallu and Gadwall areas. The western half is devoid of any major folding, whereas the eastern half is relatively complex with the formation of low amplitude folds in the western part, gradually changing over to tight isoclinal folds towards east. The eastern margin of the basin represents an overthrust. The strata in the western part are nearly horizontal with low radial dips defining an elliptical Papaghni subbasin. Beds are faulted to the east. The granitic basement has been raised along series of minor reverse faults that is prominent to the north of Guvvalacheruvu area. Undeformed Srisailam strata are exposed on the Srisailam plateau, which was deposited in the Srisailam subbasin.

The Cuddapah Supergroup is developed in the Papaghni- and Srisailam subbasins and the Nallamalai Fold Belt in the east, which has resulted from easterly compression and thrusting related to the Eastern Ghats Orogeny.

The Nallamalai Fold Belt

Three phases of deformation is prominent in the subbasin. Folding increases in intensity from west to east in the Nallamalai belt with low amplitude open folds progressively becoming tight isoclinal folds. The fold belt commences where cleavage is noticed in the slates in the middle of the Cuddapah basin. D1 structures are represented by tight to isoclinal folds, a slaty cleavage and local development of mylonites. D2 structures include NE trending tight to open folds (F2) with variable plunge indicating control of large domal structures and a steep crenulation cleavage developed in phyllites and other schistose rocks. These structures are overprinted by E-W trending D3 folds and cleavage, which affect Mesoproterozoic rocks of the Srisailam Formation in the E-W trending Vami Konda range and metasediments of the Palnad area, thought to be equivalent to the Neoproterozoic Kurnool Group. The granite gneisses and schists of the Nellore Schist Belt, bordering the Nallamalai Fold Belt are affected by both early and later events and thrust over rocks of the Nallamalai Group.

The Bairenkonda quartzite occurs in the core of NS doubly plunging anticline, south of Cumbum at Mokshagundam and Taticherla areas. These outcrops can be structurally connected to Cumbum tank ridge. Tight folds with NNE-SSW trending axis is found to the NNE of the Iswarakuppam dome in the Vami Konda range. This has no impact on the Srisailam siliciclastics when traced across the Srisailam lineament. The lineament acted as a barrier to the forces and consequently could not affect the Srisailam Formation. The

Nallamalai Fold Belt occurs in the collisional zone between two cratonic masses, of which one was stable on the western side and the other moved from the east (Murthy, 1981).

Nagari-, Tirumaia and Tirupati Outliers

Position of the Nagari-, Tirumala and Tirupati outliers in the southern tip of the Cuddapah basin is stratigraphically controversial. Their deformation appear to have been caused by the drag effect of the ENE-WSW trending Karkambadi fault noticed at the southern tip of the main basin and northern part of the outliers. They show a curved fold axis trending NNE to NNW which is not consistent with the other fold axial trends in the southern part of the basin.

Iswarakuppam Dome

This is a major domal structure with radial dips in the basin and constitutes quartzite at the core associated with alternating sequence of quartzite and shale/phyllite at stratigraphically higher levels. N-S trending elongated axis of the core cuts through the dome. The Iswarakuppam dome has been considered as a depositional structure and basement high (King, 1872). The dome exposes mainly the Nallamalai sequence instead of basal Cuddapah succession. Unlike the NNE-SSW axial trends of the folds in this part of the basin, the dome has a N-S trending axis. Although there are a lot of controversies regarding the mechanism of formation of the dome, it is presumed that the doming was probably due to uprise of granitic magma.

There are two to three domes with a granitic core, which is either younger juvenile material or remobilized basement intruded into the basement in the northeastern corner of the basin.

Faults

Ramallakota-Gani-Kalva Fault : The Cuddapah-, and Kurnool Formations are affected by ENE-WSW trending fault, which crosses the limits of the Cuddapah basin. It is a wrench fault with variable throws at different places and has caused a steep monocline called the 'Kalva wall'. It has been interpreted as a basement-rooted fault, which got reactivated in post-Kurnool times. The boundary of the Cuddapah rocks, in the southern part of the basin, is marked by a similar wrench fault, called the Karkambadi fault.

Trends of major faults, thrusts and lineaments crisscross the basin as well as the basement (Fig. 7).

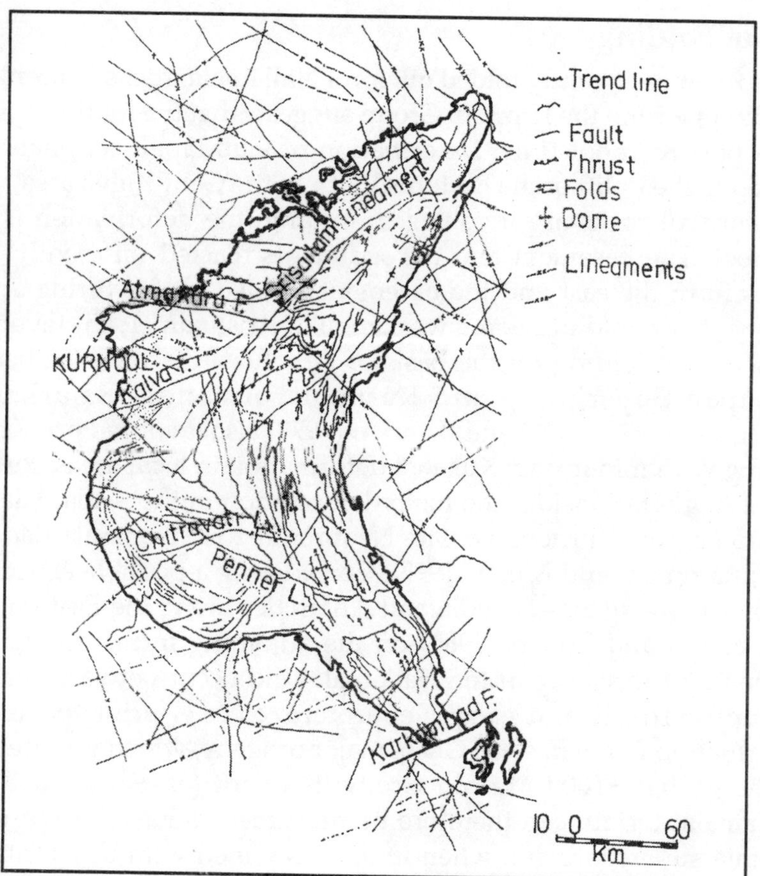

Fig. 7: Major lineaments and faults in and around the Cuddapah basin.

Thrusting

A major dislocation feature in the Cuddapah basin is the eastern boundary thrust. Granitic basement overrides the Nallamalai sediments along this thrust fault to the east of the basin. The en echelon nature of the thrust occasionally extends into the basin. At the contact with the Nallamalai sedimentary wedge, the thrust dips at a moderate angle of 30°-40° towards east. Deep seismic sounding studies support that the sediments extend further eastward beneath the granitic basement (Kaila et al., 1979).

At Khajipet area, the Cumbum Formation of the Nallamalai Group is thrust over the Koilkuntla Limestones of the Kurnool Group. The outliers of the Cuddapah rocks in the eastern part of the basin are thrust slices moving from east.

Age of Folding

Existence of gently folded inliers of Nallamalai rocks (Bairenkonda quartzite) within the Kurnool Group suggests that the folding episodes were post-Kurnool time. Deformation of Cuddapah sediments was related to the uplift of the Eastern Ghats. In the Gani-Kalva area, where the Kurnool rocks are involved in folding, the deformation is post-Kurnool. The basement and the sediments reacted differently to the forces from the east and the basement yielded by fracturing and the sediments by folding. Some workers like Krishnan (1965) favoured a pre-Kurnool folding on the basis of structural discordance between Cuddpah Supergroup and Kurnool Formation. Rajurkar and Ramlingaswamy (1978) and Rajurkar (1979) also favoured pre-Kurnool folding with minor post-Kurnool deformation in local areas. But most of the workers consider the main deformation in the Nallamalai Fold Belt to be post-Kurnool, i.e. late Neoproterozoic (Nagaraja Rao et al., 1987; Rammam and Murty, 1997). Considering a possible effect of the tectonic convergence of India and Antarctica along the Eastern Ghats (e.g. Naqvi and Rogers, 1987) on the adjoining intracratonic basin, folding and thrusting in the Nallamalai did occur between the latest Mesoproterozoic and early Neoproterozoic, i.e. prior to the main deformation in the Eastern Ghats major orogeny, which is correlatable to the global ~1000 Ma old Grenville event (cf. Saha, 2000). The Nallamalai terrain can therefore be presumed to have been a part of Rodinia supercontinent, when India was joined with East Antarctica (Biswas, 2005).

Earlier Works

The first reference to the basin is found in Travernier's writings on his travels during 1639-67 (Ball 1925). The early references to the local geology of the area has been summarised by King (1872). The pioneering geological work on the rocks of the basin was carried out by Oldham, King and Bruce Foote using Atlas Sheets of the Survey of India on 1:4 mile scale.

The geological data and the map of the basin was critically synthesised by King (1872) in a classic memoir. His classification remained unchallenged and was accepted by all for more than a century. No detailed work was carried out by any agency till 1930.

Coulson (1934) carried out detailed geological mapping of parts of the Vempalle belt in the western part of the basin.

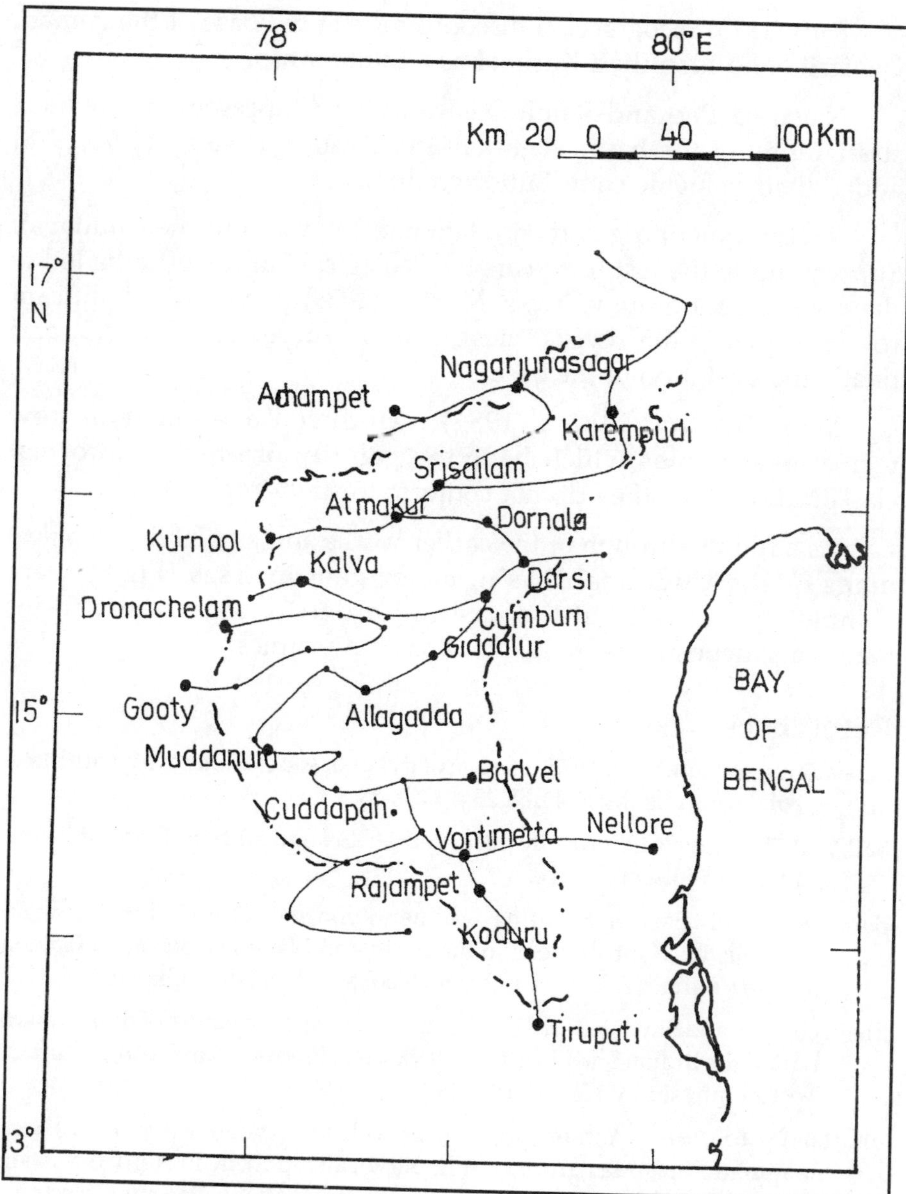

Fig. 8: Major geotraverses completed in the Cuddapah basin.

Narayanswami (1966) has presented the tectonic framework and has proposed a revised five-fold classification (Narayanswami 1976).

Sen and Rao (1967) proposed a thought-provoking model, which stood as a basis for much later work.

Dutt (1975) recognised a disconformity at the base of the Paniam Quartzite and identified Kurnool as a Supergroup.

Nagaraja Rao and Ramlingaswamy (1976) presented a revised stratigraphy of the basin. Rajurkar and Ramlingaswamy (1975,1978) added their valuable contribution to the work.

Revision mapping and correlation of the rocks of the Cuddapah Supergroup in the last few years by Geological Survey of India led to the revised stratigraphy. Y.G.K.Murthy (1979) played his significant role in drawing the revised classification, succession (Table.2) and finally the evolution of the basin.

Meijerink and others (1984) introduced a number of new formational names which have the validity of only an informal classification since they do not conform to stratigraphic codes.

As a follow through of the earlier works and available geological maps of the Cuddapah basin, major geotraverses (Fig.8) were completed through out the basin in order to address the facies variations, depositional milieu and basin dynamics.

References

Balasubramaniyam,M.N.,1975. The age of dykes of south Canara, Mysore State. Geol.Surv.India.,Misc.,Publ.,23,pp.236-39.

Ball,V.,1925. 'Travels in India' by J.B.Tavernier, Second edition. Translated and edited by William Crooke.

Banerjee,I.,1974. Barrier coastline sedimentation model and the Vindhyan example. In: Contributions to the Earth and Planetary Sciences, Golden Jubilee Volume (A.Dey ed.). Quart.J.Geol.Min.Met.Soc. India, pp.101-127.

Bhattacharji,S. and Singh,R.N.,1984. Thermo-mechanical structure of the southern part of the Indian Shield and its relevance to Precambrian Basin evolution. Tectonophysics, v.105, pp.103-120.

Bhattacharji,S.,1986. Asthenospheric upwelling, lineament reactivation, magmatic episodes and ore mineral localization in Proterozoic Basin Evolution on the Archaean Indian Shield. International Basement Tectonic Association, Publ. No. S. 187-200

Bhattacharji,S.,1987. Lineaments and igneous episodes in the evolution of Intracratonic Proterozoic basin on the Indian Shield. S. Sen. Comm. V. Hindusthan Pub. Co., New Delhi, India, 13, 1-15.

Bickle,M.J. and Eriksson,K.A.,1982. Evolution and subsidence of early Precambrian sedimentary basins. Phil.Trans.R.Soc.Lond. A 305, pp.225-247.

Biswas, A., 2005. Mesoproterozoic Nallamalai Sedimentation In The Cuddapah Basin : A Link Between India And East Antarctica In Rodinia Supercontinent. Proceedings of Supercontinents and Earth Evolution Symposium, Freemantle, Perth, Australia.

Chanda,S.K. and Bhattacharyya,A.,1982. Vindhyan sedimentation and paleogeography: Post Auden development. In: Geology of the Vindhyachal (K.S.Valdya, S.B.Bhatia and V.K.Gour, eds.). Hindusthan Publishing Corporation, India., pp.88-101.

Chaudhuri,A. K.,1970. Precambrian Stromatolites in the Pranhita-Godavari Valley (South India). Paleogeogr.Palaeoclim.Palaeoecol., v.7, pp.309-340.

Coulson, A.L.,1934. Asbestos in ceded districts of Madras Presidency with notes on its occurrence in other parts of India. Mem.Geo.Surv.Ind., v.64(2).

Crawford,A.R. and Compston,W.,1973. The age of the Cuddapah and Kurnool Systems, South India. J.Geol.Soc.Aust., v.19(4), pp.453-464.

Dasgupta,P.K.,1986. Picrite borne mantle xenoliths from Cuddapah Basin, South India. Ind. J. Earth Sciences 13(4), 333-338.

Dasgupta,P.K.,1996. Cyclicity in the Proterozoic Cuddapah Supergroup and its bearing on Basin Evolution, India. Session, 3-6, (Sedimentation in Intracratonic basins) 3-6-5, 01214, 0925 p.198, Abs. Vol. 2 of 3, 30th IGC, Beijing, P.R.China.

Dasgupta,P.K., Dey,S. and De,S.,1990. Facies variations and different stages of basin evolution in the Proterozoic Cuddapah Basin. India, Abstract volume on Evolution of Precambrian Crust in India, Geological Survey of India, pp.5-6.

Dott,R.H.Jr., Byers,C.W., Fielder,G.W., Stenzel,S.R. and Winfree,K.E.,1986. Aeolian to marine transition in Cambro-Ordovician cratonic sheet sandstone of the northern Mississippi Valley, USA. Sedimentology, v.33, pp.345-367.

Drury,S.A and Holt,R.W.,1983. The tectonic framework of the South Indian Craton: a reconnaissance involving Landsat imagery. In: Structure and Tectonics of Precambrian Rocks in India (S.Sinha-Roy, ed.). Hindusthan Publishing Corp. Press, New Delhi, pp.178-185.

Dutt,N.V.B.S.,1975. Geology and mineral resources of Andhra Pradesh. Ramesh Publications, Hyderabad, 205p.

Goodwin,A.M.,1996. Principles of Precambrian Geology. Academic Press, London. 327p.

Geological Survey of India,1979. Geological map of the Calcutta University basin (revised), salient features. 3rd workshop on Status, Problems and Programmes in Indian Peninsular Shield. Inst.Indian Peninsular Geology., pp.1-21.

Geological Survey of India,1981. Explanatory Brochure on Geology and Mineral Map of the Cuddapah Basin, 1st ed. (May,1981). Published by Geological Survey of India, Calcutta, 22p.

Ikramuddin,M. and Stueber,A.M., 1976.Rb-Sr ages of Precambrian dolerite and alkaline dykes, southeast Mysore state, India.Lithos 9, pp.235-41.

Jachariah,J.K., Bhaskar Rao,Y.G., Srinivasan,R. and Gopalan,K.,1999. Rb-Sr-Nd isotopic systematics of Uranium mineralised stromatolitic dolomite from Proterozoic Cuddapah Supergroup, South India and constraints on age and provenance. Chem.Geol., v.162, pp.49-64.

Johnson,P.R.,1965. Structure and stratigraphy of part of the upper Pranhita-Godavari Valley with special reference to the Pre-Gondwana. Unpublished Ph.D. thesis, University of Calcutta, 144p.

Kaila,K.L. and Bhatia,S.C.,1981. Gravity study along Kavali-Udipi deep seismic sounding profile in the Indian Peninsular Shield . Some inferences about origin of anorthosites and Eastern Ghat orogeny. Tectonophysics, v.79, pp.129-143.

Kaila,K.L. and Tewari,H.C.,1985. Structural trends in the Cuddapah Basin from deep seismic soundings (DSS) and their tectonic implications. Tectonophysics, v.115, pp.69-86.

Kaila,K.L., Roy Chowdhury,K., Reddy,P.R., Krishna,V.G., Narain,H., Subbotin,S.I., Sallogub,B.B., Chekunov,A.V., Kharetchko,G.E., Lazarenko,M.A. and Ilchenko,T.V.,1979. Crustal structure along the Kavali-Udipi Profile in the Indian Peninsular Shield from Deep Seismic Sounding : J.Geol.Soc.India, v.20, pp.307-333.

Kaila,K.L., Tewari,H.C., Roy Chowdhury,K., Rao,V.K., Sridhar,A.R. and Mall,D.M.,1987. Crustal structure of the Proterozoic Cuddapah Basin of India from deep seismi soundings and gravity data. In : S. Asano and W.D.Mooney (editors), Seismic studies of the continental lithosphere. Tectonophysics, v.140, pp.1-12.

Karunakaran,C.,1973. Barytes, Indo-Soviet symposium on recent trends in exploration of minerals, oil & groundwater, India. Nat.Science. Acad, New Delhi.

King,W.,1872. The Kadapah and Karnul Formations in the Madras Presidency. Mem.Geol.Serv.India, v.8, Pt.1, p.346.

Kingston,D.R., Dishroon,C.P. and Williams,P.A., 1983. Global basinal classification system. Amer. Assoc.Petrol.Geol.Bull.,v.67,pp.2175-2193.

Krishnan,M.S., 1965. Presidential address. Post Cuddapah movements in the Peninsular India. Jour.Geo.Sci.Ass.v.5.,pp.1-8.

Leelanandam,C., 1980. Some observations on the alkaline provinces in Andhra Pradesh. Current Science, v.50,pp.799-802.

Meijerink,A.M.J., Rao,D.P. and Rupke,J.,1984. Stratigraphic and structural development of the Precambrian Cuddapah basin, S.E. India. Precam.Res., v.26, pp.57-104.

Murthy,Y.G.K.,1979. Salient features of the revised geological map of the Cuddapah basin. 3rd Workshop on Status, Problems and Programmes in Indian Peninsular Shield. Institute of Indian Peninsular Geology, Hyderabad, pp.1-21.

Murthy,Y.G.K.,1981. The Cuddapah Basin – A review of basin development and basement framework relations: 4th workshop on status problem in the Cuddapah Basin. Institute of Indian Peninsular Geology, Hyderabad, pp.51-72.

Nagaraja Rao,B.K. and Ramlingaswamy,G.,1976. Some new thoughts on the stratigraphy of Cuddapah Supergroup. Seminar on Kaladgi-Badami, Bhima and Cuddapah Supergroup, Mysore. 17-20 abstracts.

Nagaraja Rao,B.K., Rajurkar,S.T., Ramalingaswamy,G. and Ravindra Babu,B.,1987. Stratigraphy, structure and evolution of Cuddapah Basin. In: Purana Basins of Peninsular India (B.P.Radhakrishna, ed.). Bull.Geol.Soc.Ind. Bangalore, Mem.6, pp.33-86.

Naqvi,S.M. and Rogers,J.G., 1987. Precambrian Geology of India. Oxford University Press. New York,223p.

Narayanaswami,S.,1966. Tectonics of the Cuddapah Basin. J.Geol. Soc.Ind., v.7, pp.33-50.

Narayanswami,S.,1976. Reinterpretation of the lithostratigraphy and tectonics of the Cuddapah Supergroup in Cuddapah basin,A.P. In the seminar of Kaladgi, Badami, Bhima and Cuddapah Supergroup. Mysore-Mineralogical Society of India. (abstracts) 5-8.

Nilsen,T.H. and Sylvester,A.G. 1995. Strike slip basins. In : Busby,C.J. and Ingersoll,R.V. (Eds.), Tectonics of sedimentary basins, Blackwell Science, Massachusetts,pp.425-457.

Pascoe,E.H.,1973. A manual of the geology of India and Burma, v.1(3rd edn.)Geol.Surv.Ind.,Cal.,485pp.

Pettijohn,F.J.,1975. Sedimentary Rocks. 3rd ed. Harper and Row, New York. p.628.

Pichamuthu,C.S.,1967. The Precambrian of India. In: The Precambrian (K.Rankama, ed.). Wiley-Interscience, New York. v.3, pp.1-96.

Rajurkar,S.T.,1979. Pre-Kurnool age of the Cuddapah Orogeny. Indian Mineralogist, v.18, pp.32-39.

Rajurkar,S.T. and Ramlingaswamy,G.,1975. Facies variation within the upper Cuddapah strata in the northern part of Cuddapah basin. Geol.Surv.India Misc.Publ.No.23, Part 1, pp.157-164.

Rajurkar,S.T. and Ramlingaswamy,G.,1978. Atmakur fault and its bearing on the evolution of Kurnool basin. Proceedings of the Symposium Purana Formations of Peninsular India. Univ.Saugar, Sagar, pp.242-245.

Ramam,P.K. and Murty,V.N., 1997. Geology of Andhra Pradesh. Geol.Soc.India.,Bangalore.,245p.

Rogers,J.J.W.,1993. A History of the Earth. Cambridge Univ.Press, 312p.

Saha,D. 2002. Multi-stage deformation in the Nalamalai Fold Belt, Cuddapah basin, South India-Implications for Mesoproterozoic tectonism along southeastern margin of India.

Sen,S.N. and Narasima Rao,CH.,1967. Igneous activity in Cuddapah Basin and adjacent areas and suggestion on the palaeogeography of the basin. In: Proc.Symp.UMP GRB and NGRI.Publ. Hyderabad, A.P., India, 8-261-285.

Singh,I.B.,1976. Depositional environment of the Upper Vindhyan sediments in the Satna-Maihar area, Madhya Pradesh, and its baring on the evolution of Vindhyan Sedimentation Basin. J.Palaeontol.Soc.India, v.19, pp.48-70.

Singh,I.B.,1980. Precambrian Sedimentary sequences of India: their peculiarities and comparison with modern sediments. Precambrian Res., v.12, pp.411-436.

Sloss,LL,1963. Sequences in the cratonic interior of North America. Geol.Soc.Am.Bull., v.74, pp.93-113.

Sreenivasa Rao,T.,1987. The Pakhal Basin – A Perspective. In: Purana Basins of Peninsular India (B.P.Radhakrishna, ed.). Bull.Geol.Soc.Ind. Bangalore, Mem.6, pp.161-187.

Vijayam,B.E. and Kamal,M.Y.,1978. Starved basinal conditions in the Kurnool System, Cuddapah Basin. Symposium on Purana Formations at Sagar University, pp.236-241.

2

Cuddapah Sedimentation

Sedimentation styles and their controls in the Proterozoic intracratonic platformal basins are complex. The cause of initiation and evolution of such basins are some of the unresolved questions of geodynamics.

The Indian shield is attracting the attention of various workers owing to its antiquity and rich endowment of geological events. Various types of sedimentary basins have evolved on the Archaean crust throughout much of geological time; many contain important accumulations of coal and gas; many more are rich in metallic and non-metallic deposits. Not all of these have been studied by various geological, geophysical and geochemical techniques. However, the Proterozoic Cuddapah basin has been studied todate in fair details. Role of lineaments and deep-seated faults reaching upto upper mantle appear to have controlled many igneous episodes and development of the Cuddapah basin. The chapter emphasizes on the sedimentation style and its significance in the basinal evolution of the Cuddapah.

The Papaghni Group

The Cuddapah sedimentation starts with the deposition of the Gulcheru siliciclastics unconformably on the Archaean granitic basement (Fig.1). The Gulcheru is in turn, overlain by the Vempallae Formation, mainly comprising carbonates, dolomites and minor sandy beds. The Vempallae is unconformably overlain by the Pulivendla Formation of the Chitravati Group.

Fig. 1: Unconformity between basal Gulcheru and Archaean basement in the western margin of the Cuddapah basin, Chitravati Gorge.

The Gulcheru Formation

The basal Gulcheru Formation, included approximately in a 14km thick sedimentary pile of the Cuddapah Supergroup, overlies the basement complex. The Gulcheru outcrops in detached manner are discernible along the western margin of the basin from Gavalcheruvuu to Kurnool. Some of the good sections are identifiable along Chitravati gorge, in and around Dadithota and Parnapallae (Fig.2).

Fig. 2: Spatio-vertically continuous basal Gulcheru outcrops in the western margin of the Cuddapah basin.

Facies Analysis

The Gulcheru Formation of the Papaghni Group, the lowest formation of the Cuddapah Supergroup, comprising mainly of conglomerate/grit, sandstone with shale intercalations is conformably overlain by the Vempalle Formation which is mainly a calcareous unit. This clastic sequence develops broadly on undulated erosion surfaces of Precambrian Gneissic Basement (Eparchaean Gap). On the basis of detailed tectono-sedimentological work around Parnapalle, Anantapur District, Andhra Pradesh the following facies have been identified (Table 1).

Table 1: Facies coding scheme (modified after Miall 1977)

Facies Code	Description	Interpretation
1. GMS	Massive clast to matrix suported gravel	Debris flow deposits
2. GSR	Gravel-Sand "couplets", stratified but no internal stratification	Sheet flood deposits
3. BGS	Graded,bimodal granule and pebbly sandstone (granule and pebble) sheets	Suspension deposits
4. St+Sp	Trough- and planar cross-stratified coarse-grained sandstone	Channel-fill element in braided stream set-up
5. MSL	Massive, thinly bedded mudstone with siltstone	Playa-like set-up
6. Sh	Horizontal stratified sand	Upper flow regime

FACIES 1 : (GMS)

This facies consists of lobate, laterally continuous, extremely poorly sorted, polymodal clast to matrix-supported boulder conglomerate with a granule and coarse sand matrix (Fig.3). Angular to rounded fragments of vein quartz, chert, BIF are the main types of clasts. The basal contact with the underlying unit (Facies 2) is abrupt and non-erosive, and stratification is rare, however, at places crude layering may be observed defined by weakly oriented clasts. The amount of finer sized material in the matrix is normally less than a few percent. The maximum particle size (the mean of ten largest clasts per bed) is from 2cm to 32cm and always shows a significant correlation with bed thickness. This facies lacks sedimentary structures such as cross-bedding or pebble imbrication but occasionally shows a fabric in which rod or blade-shaped pebbles consistently lie with their longest ones parallel to the bedding. Sequences commonly coarsen upward as large boulders become concentrated near

Fig. 3: Erosively based channels with poorly sorted, polymodal clast to matrix-supported boulder conglomerate in a granule and coarse sand matrix, basal Gulcheru succession.

the top. This facies is often characterised by vertical orientation of clasts and clustering of clasts in groups of three to five.

Nesting of clasts is common in the conglomerates. Granular sandstone exhibiting channel-form and trough cross-bedding commonly caps each unit of conglomerate; sandstone thickness generally increases with a decreasing maximum particle size in the accompanying (underlying) conglomerate. Grading is fairly present but wherever it is seen, it is inversely graded. Protrusions of clasts at the upper bounding surface of each unit are occasionally seen.

The lobate nature of the deposits with extremely poor sorting, clast to matrix supported conglomerate, with some preferred orientation of the larger clasts and reverse size grading at the base with chaotic fabrics along with protrusion of clasts at the upper bounding surface of each unit suggests that the facies represents cohesionless debris flows. Similar facies have been reported extensively from alluvial fan deposits (Pierson 1981, Blair and Mcpherson 1994, Blair 1987, Beaty 1990, Decelles et al. 1991, Brierley, 1991).

The cohesionless laminated debris-flows (may be called density-modified grainflows, Lowe 1982) deposited this facies. All the mechanisms that act in varying proportions to support clustering of clasts in nests of upto five indicate accumulation due to obstruction by a larger clast or tendency for clasts to migrate towards regions of least internal shear. The sandstone member of the debris flow unit is interpreted in terms of the high stages of flow-water known to follow debris flows on recent alluvial fans (Blair and Mcpherson 1994). It suggests an accumulation of sandstone over the areas representing the relatively distal regions of the fans.

FACIES 2 : GSR

The facies consists of horizontally-stratified gravel-sand couplets of 15cm to 40cm thick (Fig. 4). The sand component of each couplet overlies the coarser fraction. The clasts are pebbles and cobbles of spheroidal, discoidal and cuboidal chert, BIF, quartzite. The clasts are subrounded to rounded and no distinct imbricatlon is observed. In the sand units pebbles are less abundant and cobbles are totally absent. The sands are discontinuous in some cases and individual units bifurcate laterally. The gravels are unsorted, polymodal and lack internal stratification. The gravel sand units occur as up to ten vertical stacked cycles.

Fig. 4: Horizontally-stratified bouldery gravel-sand couplets, Gulcheru succession.

Facies attributes suggest that the gravel-sand couplets resulted from sheet-flood processes (Blair and Mcpherson 1994). The deposition of sand subsequent to gravel deposition implies a suspension fall out. The couplets also suggest low turbulence in the flows that enabled the step-wise deposition of gravel and sand with fall in the flow capacity.

Similar couplets have been described by Blair and Mcpherson (1994), who interpreted their genesis as due to changes in flow hydraulics initiated by flow expansion and decrease in slope gradient.

FACIES 3 : BGS

This facies is characterised predominantly by bimodal granule and pebble sandstones interbedded with rippled and trough cross-stratified coarse sandstone unit (Fig.5). This facies is normally sheet-like with lateral continuity and no significant thickness change over distances of

Fig. 5: Bimodal granule and pebble sandstones interbedded with rippled and trough cross- stratified coarse sandstone, Gulcheru Formation.

upto 100m. They are usually sharp-based and in many instances the basal few centimetres are wavy, possibly rippled/antiduned and lack the granule mode. A remarkable feature of most beds is that the granule mode is relatively well-sorted but is abruptly cut off above a certain maximum grain size. Both inversely and normal graded beds are characterised by an upward increasing maximum particle size (in small pebble/granule grade), although there is often a fining in the topmost few centimetres of the bed. The fine sandstone matrix occupies 5% to 60% of the bed, so the granules are commonly floating, especially in the lower parts of the bed. In some beds the type of grading is such that the coarsest fraction remains constant in size through most of the bed, but the volume of fine matrix significantly decreases upwards. In other cases the coarseness of the coarsest fraction varies continuously. Granules and pebbles generally have their long axes aligned parallel to bedding (Fig. 6), although cases of imbrication, vertical long axes of a chaotic fabric are not uncommon. Some inversely graded beds have been traced into normally graded beds.

Fig. 6: Granules and pebbles generally have their long axes aligned parallel to bedding, Gulcheru Formation.

These graded granule sandstone beds were evidently deposited from suspension. The highly variable fine matrix content together with the range of grading type (or thickness of graded zones), however, suggest that sediment concentration in the dispersion varied considerably, though was never sufficiently high to prevent free movement of clast with respect to each other, at least in the upper and lower parts of the bed.

FACIES 4 : St-Sp

This facies is characterised by medium to coarse-grained sandstone with abundant trough- and multidirectional planar cross beds in sets 10cm to 30cm thick and ripple-drift cross-stratification (Fig. 7). Grain size ranges from coarse sand to granules, with local lenticular to massive bodies of clast-supported pebbles and cobbles. One of the noteworthy features of this facies is the prolific development of erosion surfaces. The cross-laminae filling the troughs are often deformed. Horizontal bedding is subordinate but commonly forms laterally extensive horizons 50cm to 75cm thick. This facies is commonly interbedded with facies 2 and sometimes with facies 1. Channels occur upto 3m deep and 12m wide. Some volcaniclastics are often seen on the bedding plane of the trough cross-stratified unit.

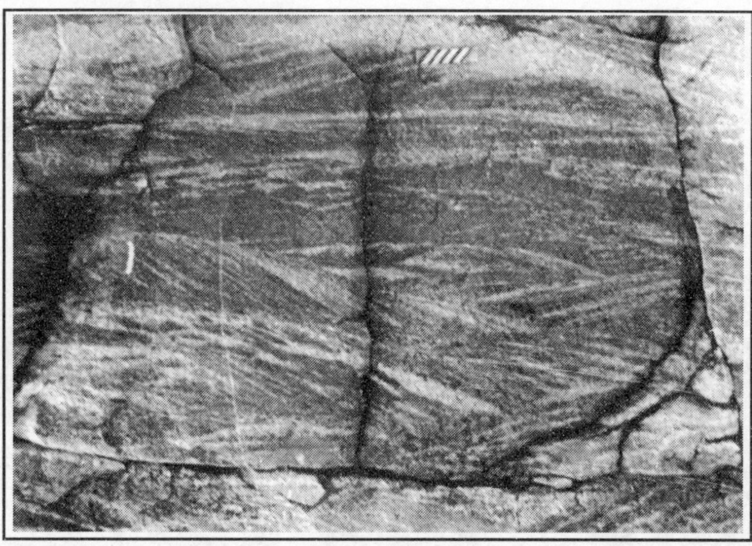

Fig. 7: Medium to coarse-grained ferruginous sandstone with torrential cross beds, Gulcheru Formation.

The frequency of erosion surfaces, trough cross-stratified channel-fills, scarcity of fine-grained sediments and the very rapid lithological and textural changes suggest sedimentation which was characterised by large discharge fluctuations, rapid channel-filling and abandonment, and transport through considerable surface topography. All of these features are typical of braided stream activity (Reading 1996).

The abundance of erosion surfaces suggests that the scarcity of the finest sediment may be due to erosion immediately after deposition. The abundant trough cross-beds form as dune bedforms migrate along the channel floor during intermittent high water (Allen 1965, Harms and Fahnestock 1965, Williams 1969,1971). The vertical sequence of fine-grained sandstone with climbing-ripple laminae overlying coarse-grained, flat-bedded and trough cross-stratified sandstone is comparable to that sequence which accumulated during heavy flowing of Bijou Creek (Mckee et al. 1967). Braided streams are a common feature of alluvial fan surfaces. The coarseness of these deposits and the common association and some dispersal directions with the debris flow and sheet flow deposits suggests that all these types of deposit accumulated on the alluvial fan. The anomalous angularity and maturity of the clasts in some of the braided stream conglomerate of the Gulcheru Formation can be accounted for by a weathering process known to operate on abandoned segments of alluvial fans. Desert pavement processes produce angular, sutured rock fragments by the disintegration of mechanically or chemically weak clasts. Some of the deposits of facies 4 may be the reworked product of ancient desert pavement. Good preservation of aeolian bimodality corroborates this contention.

FACIES 5 :MSL

This facies is characterised by a simple alternation of laterally extensive (tens of metres) very thinly bedded (centimetres) red mudstones and siltstones. They are associated closely with debris flow and sheet-flood deposits when the latter occur as fine-grained, thinly bedded units. i.e. in their distal part. The sediment in question, however, can be distinguished from the sandstone normally interstratified with the fan conglomerate by its finer grain size, common mud-cracks, curled mud flakes, ripple marks and small scale cross-stratifications etc. Bedded gypsum is occasionally associated with this facies.

The association and interfingering of facies 5 with fine-grained fan conglomerates suggests that the former developed near or beyond the toe of the alluvial fans, in areas which only occasionally received coarser fan sediment. Small bed thickness record low capacity currents while the fine-grain size suggests that these currents were relatively incompetent. This further suggests that the sediment accumulated from the fans themselves and not from another system merely infringing on the fan toes. The modern playa environment has analogous situation (Reading, 1966) and these sediments are probably playa deposits. The curled mud flakes and typical Proterozoic mud-cracks in the sequence and bedded gypsum at Parnapalle records infrequent sedimentation and the subaerial exposure characteristic of a playa.

FACIES 6 : Shale

This facies comprises laterally continuous, internally stratified sheets of sand. The units are usually 0.25cm to 0.50cm thick and bounded by planar, non erosive surfaces. Associated with the sheets are occasional stringers of pebbly gravel. The facies also occurs as sand lenses within the GMS facies.

This facies results from deposition of longitudinal bars under upper flow regime energy condition (Williams 1971).

Facies Association

Occurence of each facies varies spatio-vertically. However, the GMS facies dominates the alluvial architecture. Vertically stacked mudstones alternating with siltstones of MSL facies is located towards the western side of the Chitravati river adjoining the recently constructed dam, near Parnapalli. Bedded gypsum is occasionally associated with the facies.

Fivehundred meters SE of the dam in Parnapalli, is a well-developed, section on the western side of the Chitravati river which is dominated by St facies. Located 2 Km. east of the above, a cliff-section on the eastern side of the Chitravati river (Dadithota section) is a 38m thick multistory amalgamation of various sedimentary facies (Fig. 8). A

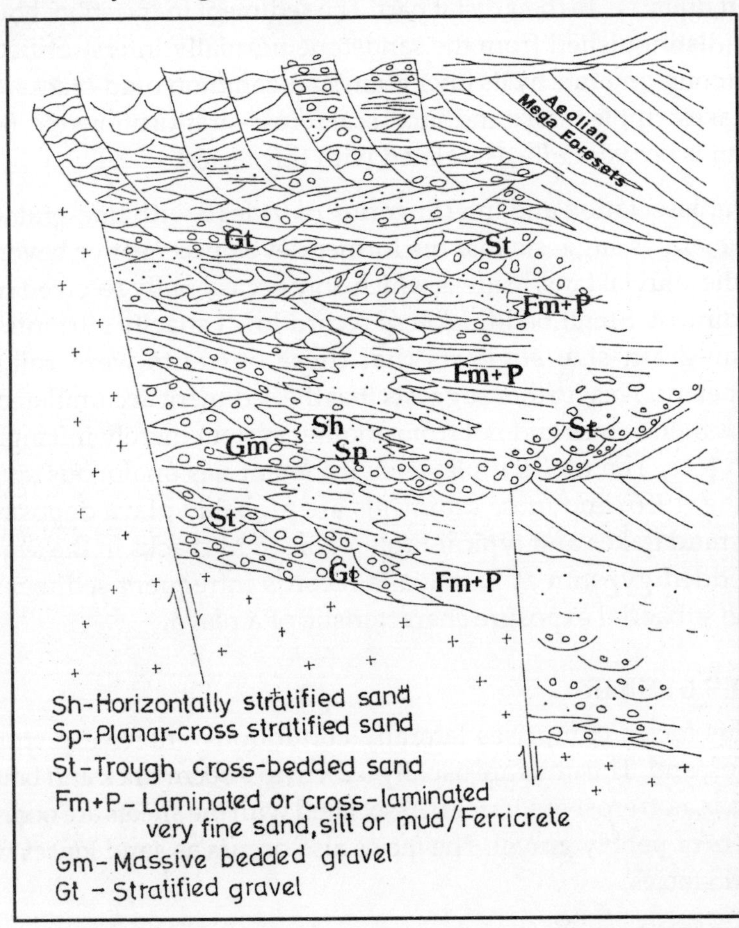

Fig. 8: Schematic diagram showing interrelations among different facies, Gulcheru Formation.

generalized section near Dadithota reveals juxtaposition of Sh-, Sp, St-, Fm+p-, Gm and Gt facies (Fig.8). Another Dadithota section further north near the basement illustrates various characteristics of facies developments, both spatially and vertically in a 42m thick section, exposed in a road cutting (on Parnapalli-Anantapur high way). It shows preservation of a succession of a ubiquitous packet of GMS facies along with BGS, GSR, St facies. The section starts with BGS facies and ends with St facies. The GMS facies along with GSR facies occur intercalated between BGS and St facies. The alluvial architecture, when traced over long exposures using panoramic photographs, reveals the major contribution of the GMS facies.

Spatio-vertical arrangements of different Gulcheru facies are discernible in and around the Chitravati gorge, Parnapalli and Dadithota. Planar cross-stratified sandstones, trough cross stratified sandstones, rippled sandstones, couplet beds, granule- and pebbly sandstones, orthoconglomerates as channel-fills as well as sheet flows along with sparsely developed laminated mudstones-sandstones and shales characterise the Gulcheru facies mosaic (Figs. 9, 10, 11 and 12).

Palaeocurrent Analysis

Detailed palaeocurrent analysis is essential in order to establish the major flow direction and basinal evolution of the Papaghni proto-basin over the Precambrian basement.

A total of 210 palaeocurrent data of regions (i.e. Chitravati-Parnapalli-Dadithota, Gandi Anjaneya, Roychoti-Cuddapah-Palkonda Ridge, Pellimeri-Narayanaswami sector) (Fig.13) from the cross-bed dip azimuths of the cross beddings were measured and subsequently these were corrected for the tectonic tilt. The corrected data were grouped together into 10°, 20° and 30° intervals and plotted in circular histograms.

Higher variance values of the cross bedding fore-set dip azimuths such as in Palkonda Ridge, near Cuddapah, clearly indicate a fluvial origin for the Gulcheru clastics. The higher values of the variance are generally associated with insignificant preferred orientation of palaeocurrent directions which are very much expected in case of a fluvial origin with converging tributaries from larger areas. The spread of palaeocurrent data around the vector mean of 277.1° very much support a fluvial origin of the Gulcheru sediments in an around Cuddapah. Palaeocurrent azimuths derived from large-scale cross-stratifications are unidirectional, and commonly have a range through the vertical extent of a given storey of about 28°. However, there can be

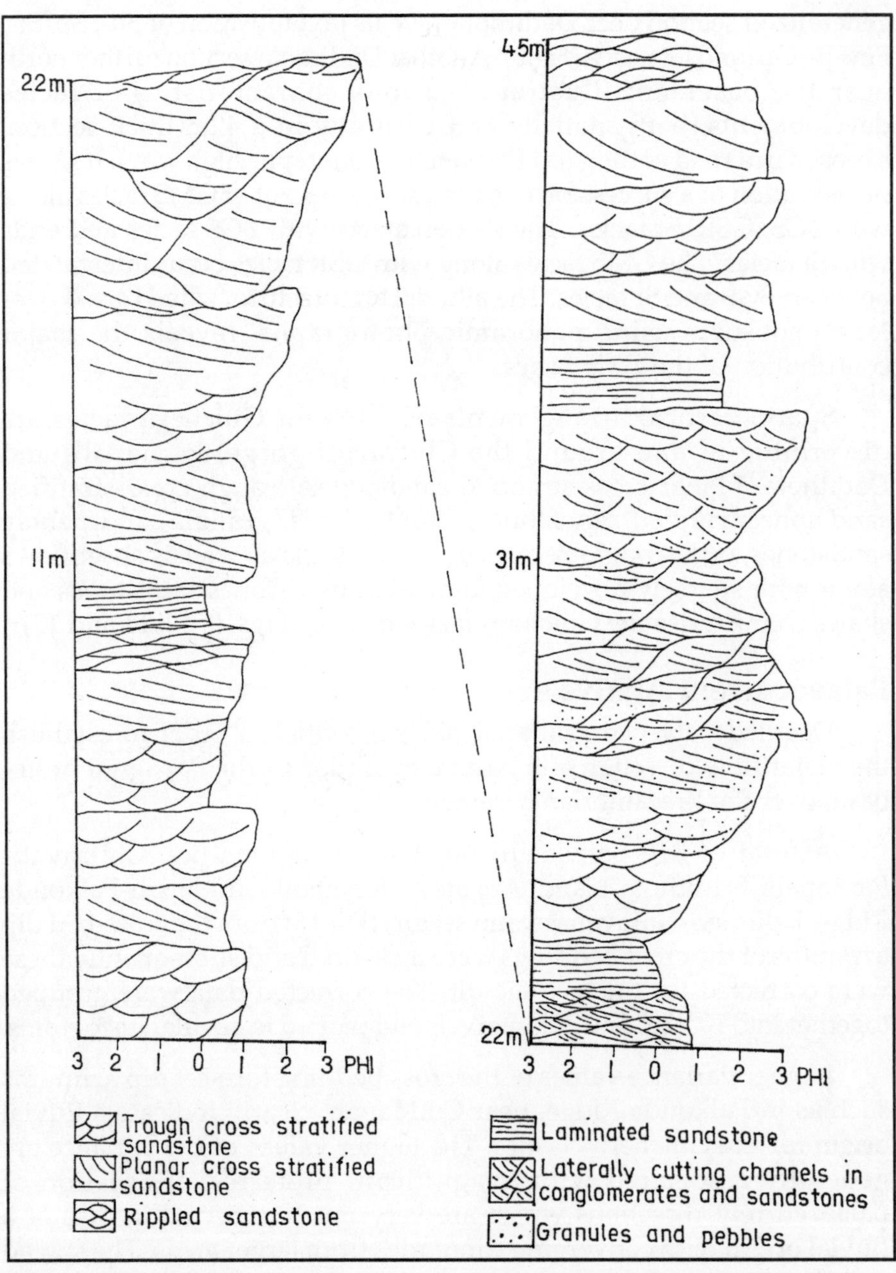

Fig. 9: Vertical facies associations in and around Chitravati gorge, Parnapallae, Dadithota and Palkonda ridge respectively, Gulcheru Formation.

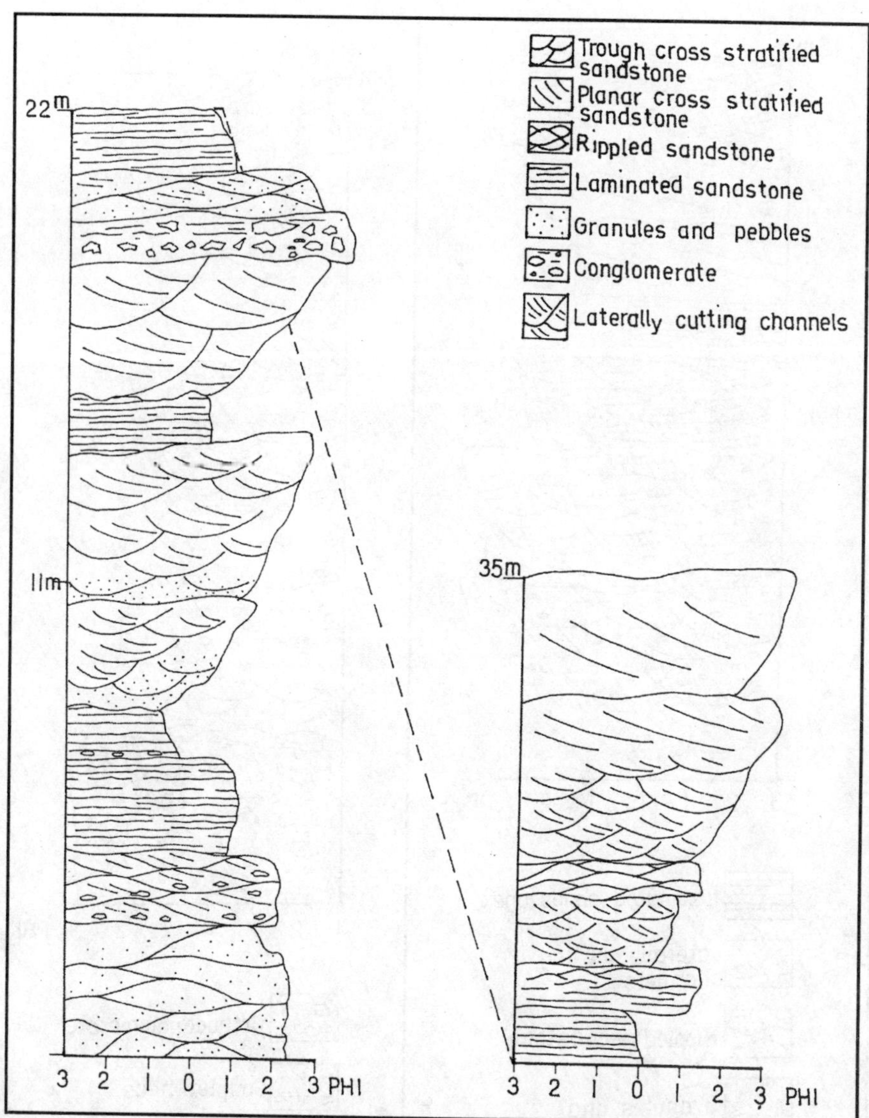

Fig. 10: Vertical facies associations in and around Chitravati gorge, Parnapallae, Dadithota and Palkonda ridge respectively, Gulcheru Formation.

local marked divergence from the mean of this range near the topmost part. The vector means of the current directions in the Chitravati Parnapalli-Dadithota area vary between 305° and 340° indicating a north-northwesterly flow.

A broad converging stream system during the Gulcheru time flowed from ESE to WNW (vector mean 277.1°) around Cuddapah within the

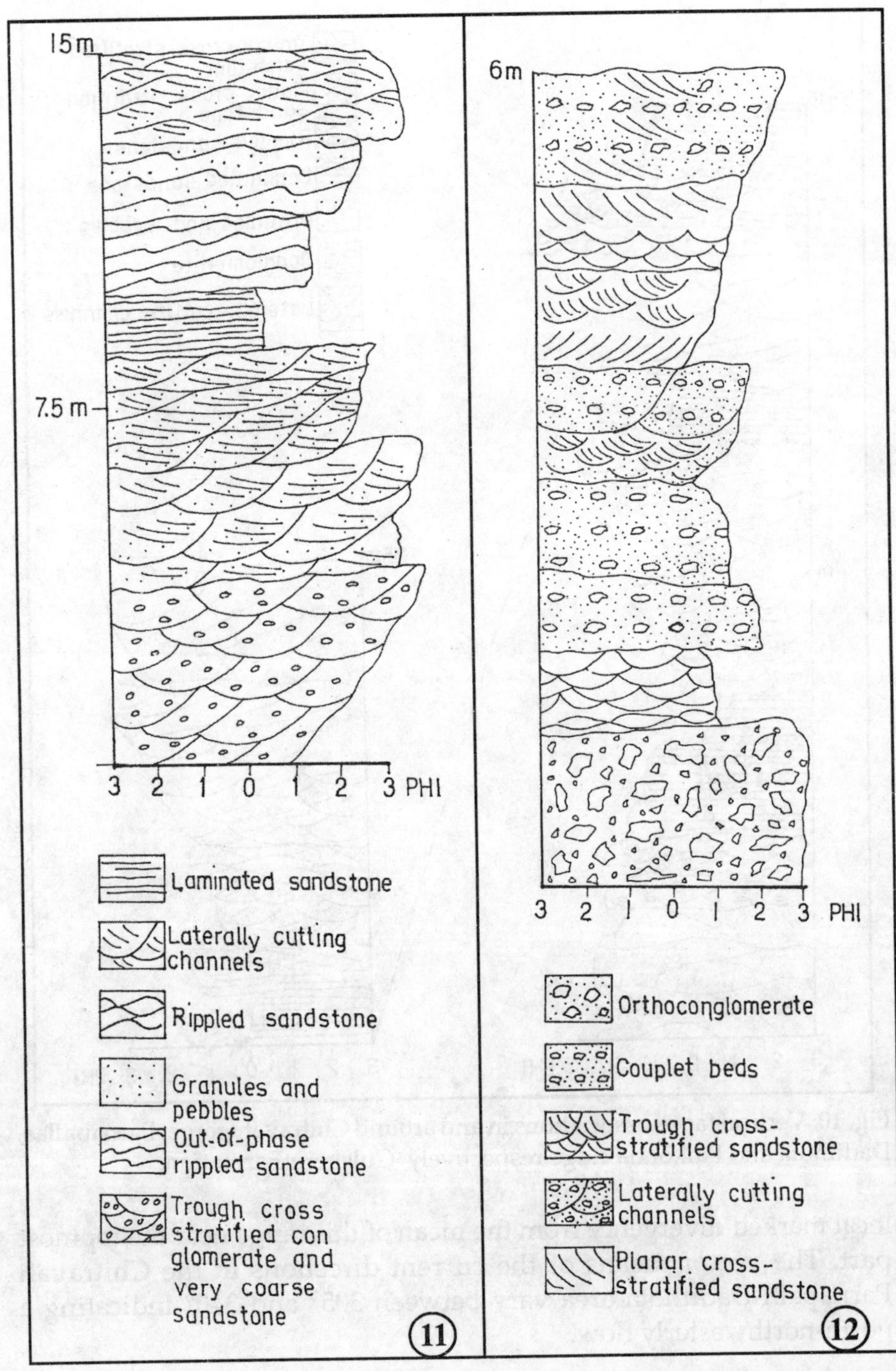

Fig. 11-12: Vertical facies associations in and around Chitravati gorge, Parnapallae, Dadithota and Palkonda ridge respectively, Gulcheru Formation.

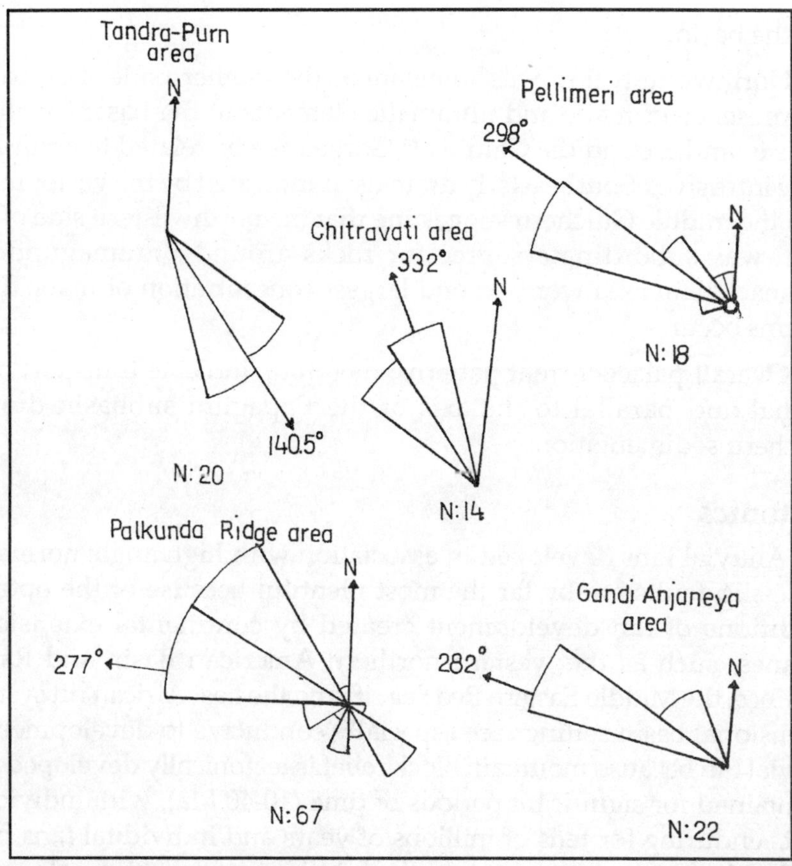

Fig. 13: Vector means from rose diagrams for paleocurrent directions, Gulcheru Formation.

Cuddapah-Roychoti-Palkonda Ridge region, the trunk flow of the main stream system was passing near Pellimeri-Narashinhaswami region and was flowing from SE to NW (vector mean of flow 298.4°) with low variance of 252.9°. Entire strike length of the Gulcheru rocks between Roychoti-Palkonda and Pellimeri-Narashinhaswami region is about 55 km. The basal deposits of Chitravati-Parnapalli were deposited in an alluvial fan (shed from the SSE to WNW).

Slightly bipolar distribution of the current direction means in and around Cuddapah-Palkonda Ridge probably indicates a swinging system underall an overall oscillatory condition of the basin.

An accentuated influence of swinging of flows controlled by local topography were observed near Tandra Prand; in this region vector means (130°-140.5°) probably indicates subordinate southeasterly flow into the basin.

Northwesterly flow was dominant on the southern side of the basin. The presence of mafic and ultramafic elements in the basal Gulcheru sands in and around the Chitravati-Gorge may be related to abundant mafic intrusives. Southeasterly drainage is indicated by the vector mean from the middle Gulcheru suggesting that the northwestern side of the basin was subordinate source for rocks around Bhramagundem-Dhronachalam area were, second largest concentration of mafic dyke swarms occur.

Overall palaeocurrent patterns, therefore, indicate transport both normal and parallel to the axis of the Papaghni subbasin during Gulcheru sedimentation.

Tectonics

Alluvial fans developed in association with high angle normal or strike-slip faults are by far the most plentiful because of the optimal conditions of fan development created by continental extensional terranes, such as the western northern American Basin and Range province, the Middle Eastern Red Sea rift and the East African rift system. Extensional basin settings are especially conducive to development of alluvial fan because mountain block relief is tectonically developed and maintained for significant periods of time (10-40 Ma), with individual faults enduring for tens of millions of years and individual fans from 0.2-7.0 Ma. The maintenance of relief is important for fan development because of the significant promotion of weathering created by the effects of gravity on slope erosion, as well as for providing ample time for drainage basins to evolve. The presence of relief, along with the typical drainage network that develops through time also is important in promoting catastrophic flash floods capable of moving poorly sorted and commonly coarse-grained sediment to the fan site. Seismic shocks associated with active tectonism also aid in fan construction by destabilizing the alluvium and bedrock slopes of the drainage basin. Moreover, tectonic fracturing related to faulting significantly promotes bedrock weathering both through its physical disintegration and its incitement of chemical weathering related to the increased rock surface area susceptible to wetting and oxidation in the fracture network (Blair and Mcpherson 1994).

Extensional basins are commonly bounded by steep normal faults that penetrate to atleast mid-crustal levels. Graben basins are bounded by such faults on both sides and half-graben basins are underlain by tilt blocks and a bounding fault on the other side. Tectonic slopes are produced by a combination of footwall uplift and hanging wall subsidence. The footwall slope is blanketed by coarse debris in form of alluvial fans (Miall, 1996). Alluvial fans are generally promoted by extensional faulting (Bull 1977, Blair and Mcpherson 1994). Such a physiographic setting was provided by large-scale fault system during Gulcheru time.

Two sequences on the scale of few metres thick exist in the Gulcheru alluvial-fan sediment near Parnapalli. Each sequence consists of a coarsening-upward cycle followed by a fining-upward cycle. The first coarsening upward sequence begins with facies 2 (granulestone). Clasts are extremely angular and have a maximum size of 10cm. This sediment probably represents grains that were transported only a short distance or not at all during the incipient stages of uplift. Significant increase in grain-size in the first sequence corresponds to the appearance of debris-flow conglomerate suggesting sedimentation in a proximal fan setting. The fining-upward part of the first sequence corresponds to a decrease in maximum grain size to only a few centimeter in diameter. Facies 2 is replaced by the laterally continuous braided fluvial facies (Facies 4), which indicates mid-fan deposition.

The second coarsening upward sequence, also exhibits a significant increase in grain size above 32mm in diameter. The grain-size increase is accompanied by the appearance of laterally continuous sheet flow facies reflecting deposition near the intersection point and proximal fan debris-flow facies. The second fining-upward sequence exhibits a decrease in grain size and is dominated by braided stream and sheet flow facies.

The two sequences in the Gulcheru Formation are interpreted to be result of two tectonic pulses. As the source area was uplifted along the basin-margin fault, coarse proximal alluvial fan facies prograded over the distal facies. The fining-upward part of each sequence resulted from source-area retreat during a tectonic quiescence. Fan regression ultimately superimposed distal facies over the proximal facies.

Petrography

The basal Gulcheru conglomerate and granulestone consist of myriad types of subrounded clasts such as pebbles of quartzite, quartz, BIF, quartz-monzonite and vein quartz. Matrix essentially comprises of quartz and feldspars with labile chemical constituents such as olivine, pyroxene, leucite etc.

The basal Gulcheru sands in and around Parnapalli have significant amounts of olivine, pyroxene, chromite, and rutile in the coarse population coarser modes of the bimodal granule sandstone (BGS facies) fall within 0.87mm to 2.00mm range whereas finer modes, lie between 0.12mm to 0.3mm.

Presence of such chemically unstable mafic-ultramafic elements in the framework is highly suggestive of transport through a relatively short distance. This might also indicate arid condition in the source area (Folk 1968).

Petrologically, the middle to upper Gulcheru sands show the dominance of clastic quartz framework with marks of wind biting (seen under SEM) with some geopetal siliceous ooides and siliceous oolites. These are replaced by calcareous oolites and all stages of transformation can be seen under the microscope. Textural and curve-shape analysis of Gulcheru clastics (Fig.14, 15 and 16) show the ranges of different textural parameters and overall energy conditions.

Provenance

From a thorough analysis of the various rock fragments of the framework constituents of the Gulcheru clastics, it is conjectured that the source area must be dominated by mafic and ultramafic rocks. Here, it is very pertinent to think about the ubiquitous presence of mafic and ultramafic dykes and dyke-swarms adjoining the western part of the basin. Perhaps, these dykes and dyke-swarms might be the source area of the Gulcheru clastics. Palaeocurrent analysis also corroborates this contention.

Conclusions

The Gulcheru sedimentation starts with deposition of granulestones and ends with deposition of evaporites. Around Parnapalli, in Chitravati Gorge, there is a linear terrestrial fan-complex (Figs.17,17a & 17b). This fan-complex is characterised by subrounded to rounded clasts. This rounding is due to the elongated catchment area upstream of the fan apex, as clasts are rounded during prolonged bed-load transport and are temporarily arrested upstream of the fan apex as channel bars. These clasts are remobilised and entrained in debris flows on the fan during events of anomalous discharge. A schematic diagram shows the basin evolution of Gulcheru sediments in relation to its source area in response

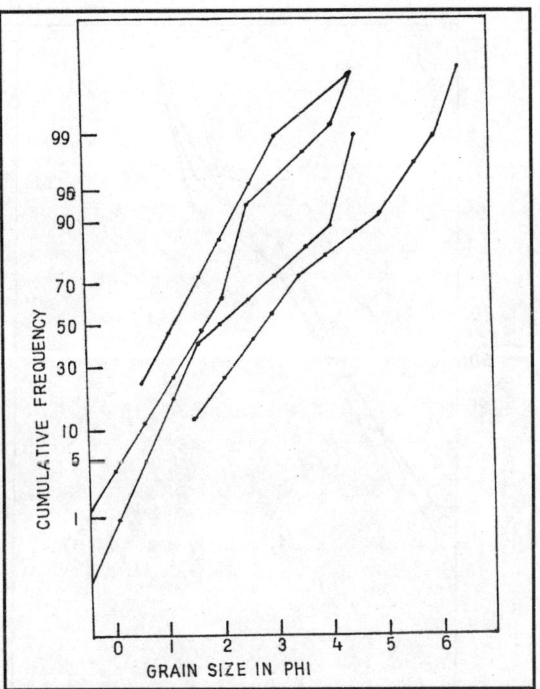

Fig. 14: Cumulative frequency curves of Gulcheru sands.

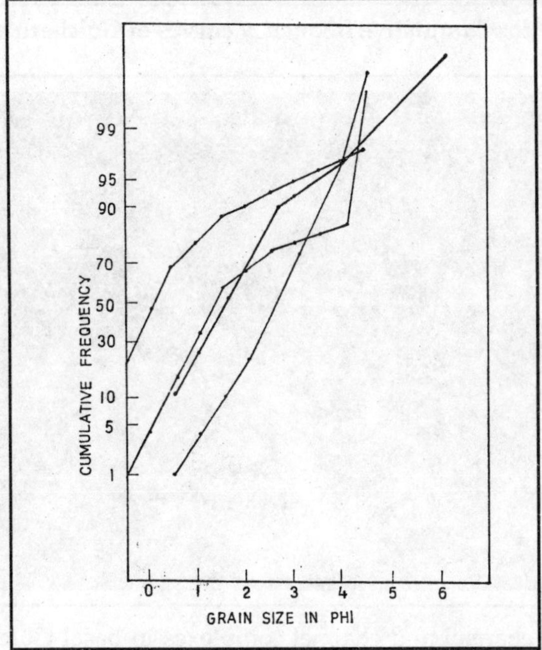

Fig. 15: Cumulative frequency curves of Gulcheru sands.

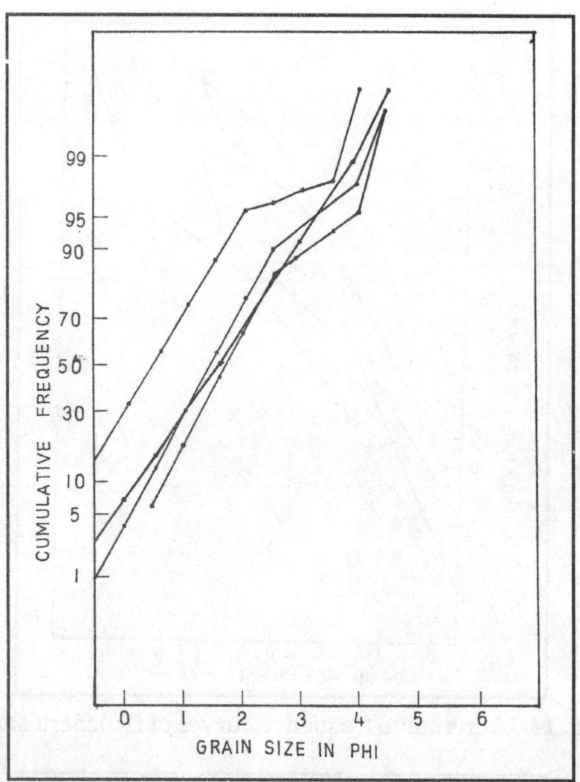

Fig. 16: Cumulative frequency curves of Gulcheru sands.

Fig. 17: Large channel and channel complexes in basal Gulcheru succession over the Archaean basement.

Fig. 17a: Large channel and channel complexes in basal Gulcheru succession over the Archaean basement.

Fig. 17b: Large channel and channel complexes in basal Gulcheru succession over the Archaean basement.

to basin tectonics (Fig. 18). Tectonism, no doubt, has influenced the location of depositional site, has provided the necessary physiographic contrast and has played an important role in the erosion of the fan. But the crux of the problem lies in deciphering the nature of tectonism. Seismicity associated with volcanism may play an overwhelming effect on the nature of tectonism that dictates the fan aggradation. The facies association, found in Gulcheru Fan-complex is unique in Proterozoic scenario all over the world, and so it demands unique type of tectonism to explain its origin. The convulsive state of the Cuddapah basin during Gulcheru time, being the manifestation of so-called passive rifting, may be deduced from the detailed study of the sedimentary structures

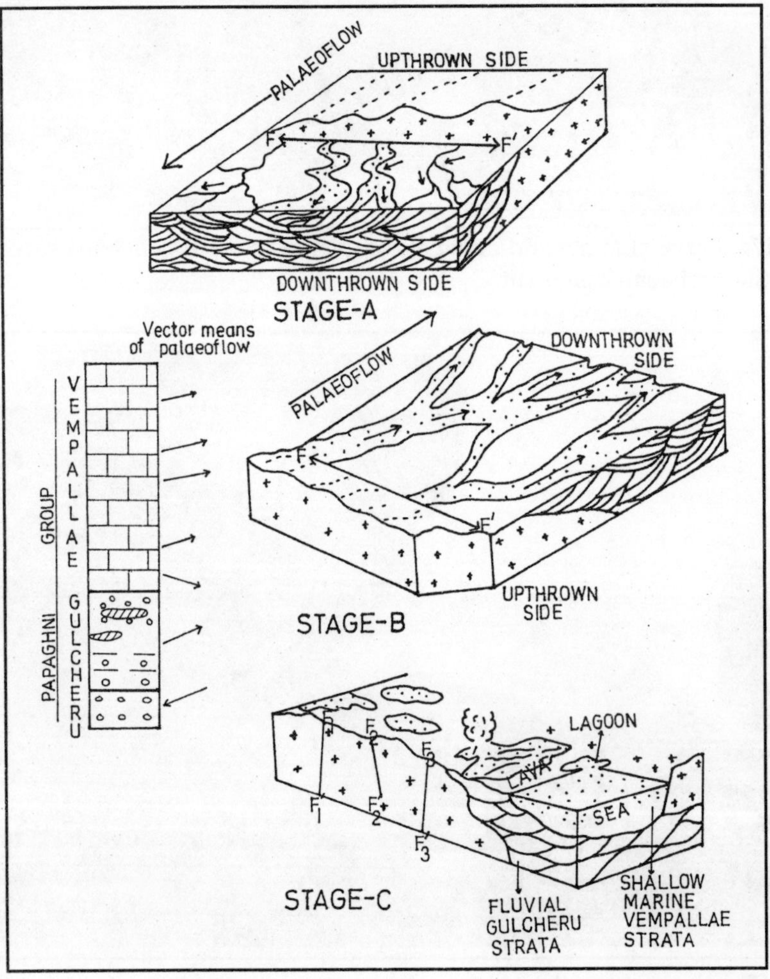

Fig. 18: Schematic diagram showing paleogeographic shifts during basinal evolution of the Gulcheru subbasin.

representing the locked-up vigorous energy coming from the deep interior of the mantle during eparchaean interval. The subcrustal arrest of the mantle plume may not be the sole answer to the enormous amount of energy that generates such jig-saw puzzle type sedimentary structures.

The Vempallae Formation

Outcrops of Vempallae carbonates extend strikewise roughly north-south (Fig.19). It covers a distance of hundreds of kilometers with a moderate vertical thickness of ca. 2300m onlapping the Gulcheru siliciclastics. Best of the outcrops are found in and around Vempallae, Pulivendla and way to Parnapallae. Facies analysis of the Vempallae

Fig. 19: Outcrops of N-S trending Vempallae stromatolitic carbonates, Kadiri-Vesthaveripallae road section, Vempallae Formation.

rocks reveals twenty one facies (Figs. 20).

Facies 1 Intraclastic Dolomitic Packstone

These intraclastic dolomitic horizons are seen to overlie erosional surfaces with units 1-2m thick. The facies shows variation in grain size and are mainly composed of angular to rounded grains. The rounded intraclasts indicate transportation and are often associated with detrital quartz silt (Fig.21). The micritic intraclasts are 0.5 to 1.5mm in length and thickness between varies between 0.1 to 0.3mm which suggest their

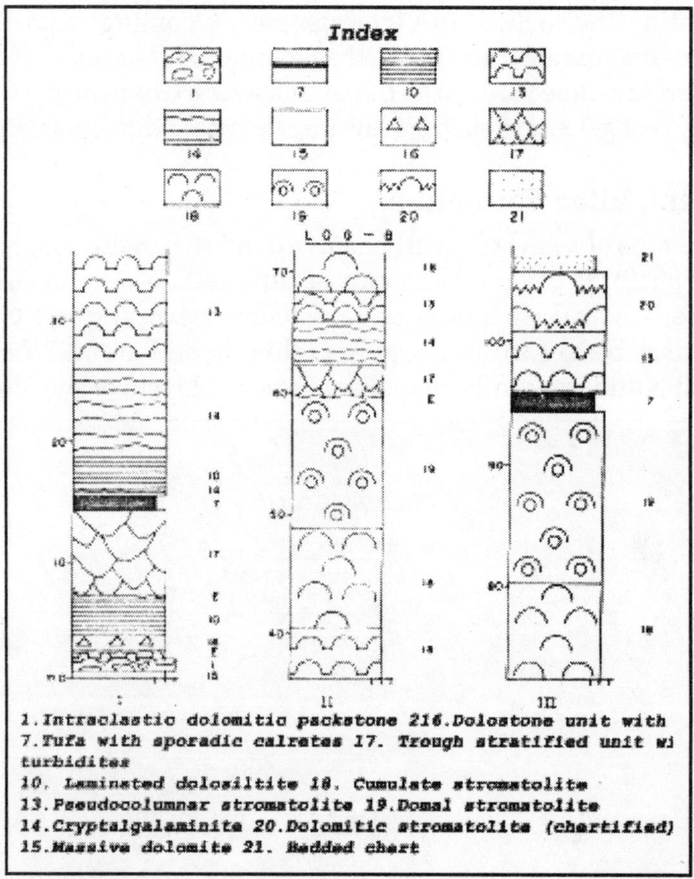

Fig. 20: Vertical facies association near Tumalapallae, Cuddapah district, Vempallae Formation.

breaking down and almost immediate entombment with little or no reworking. Carbonate mud sediment is present as thin layers alternating with intraclastic horizons are also associated with the intraclasts. The facies also contains dolomite spar cement (mean size=7.6 microns; S.D. 1 micron) (Folk, 1959).

Facies 2 Intraclastic Dolomitic Packstone with sparsely develped stromatolitic laminae

This facies, according to the established schemes of classifications (Folk, 1959; Dunham, 1962), belong to the same unit as Facies 1. But there are distinct evidences of intraclasts being linked to algal growth mats. Wavy, discontinuous short algal laminae have been broken down

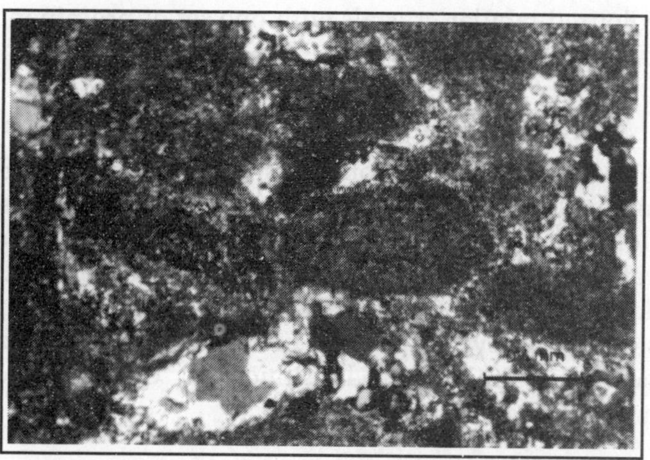

Fig. 21: Rounded intraclasts of micritic mud indicative of subaqueous transport, associated with detrital quartz silt, Vempallae Formation.

and instantly buried with almost negligible transportation. Palaeobasinal controls in generating intraclasts of laminites, algalmats and carbonate muds, forming packstones are discernible near Tumalapalli (Fig.22). The intraclasts show typical cryptalgal wavy, grain size differentiated horizons of micrite and layers of intermixed sparite and micrite. Though it is a packstone (Dunham, 1962) or a intrasparite (Folk, 1959), it is more suitably named as intraclastic dolomitic packstone with sparsely developed stromatolitic laminae. This facies is associated with some ash fall layers along with some emergent features. Minor chert often recrystallises to quartz and is also associated with some evaporitic beds (anhydrites) showing blue

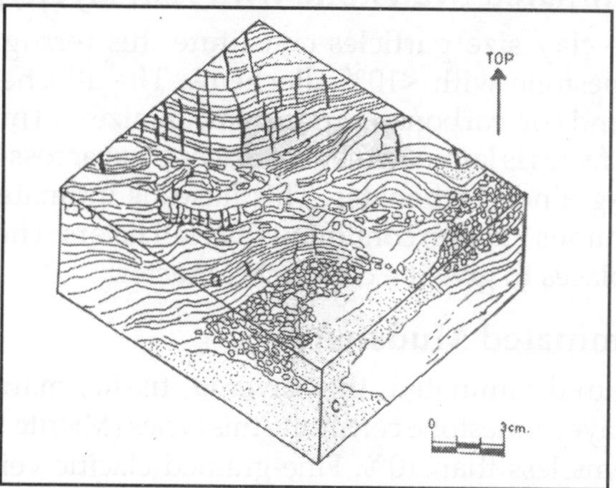

Fig. 22: Palaeobasinal controls in generating intraclasts of laminites, algalmats and carbonate muds, forming packstones, discernible near Tumalapalli.

to violet coloured tabular crystals with one set of prominent cleavage plane.

Facies 3 Laminated Dolosiltite/Dololutite with Ash pockets

The facies (interlayered dolomicro-sparite/ dolomicrite, Folk, 1959) shows alternate layers of thin dololutite and adjacent relatively thicker layers of dolosiltite mixed up with some micritic mud. Laminations are geometrically precise and often cut across by thin veins filled up by sparry calcite. This facies often contains small fenestrae which are on rare occasions, bird's eye shaped and filled up by later emplaced sparry calcite (mean size=9microns; S.D.=2microns). Occasional evaporitic crystals (anhydrites mainly) are present.

Facies 4 Ash Beds

Thin layers (1-1.5cm) of ashy material, laminated and rippled, occur as distinct beds, either separately or associated with mudstone facies. Massive pyroclastic flow deposits over- and underlying the cross-bedded rippled surge deposits are discernible. Incomplete Bouma sequences Tbc, Tcd, Tc/c are discernible within the low-energy volcaniclastic turbidites, exposed in outcrops along Vesthaveripallae-Kadiri road which is intimately associated with laminated mudstone facies (Figs. 23 & 23a). These fine-grained volcaniclastics have often been weathered to ochre yellow and are often chertified and hence welded together. The facies association has been interpreted as low-energy lake turbidites.

Facies 5 Laminated Mudstone with Ash layers/pockets

Very fine clay size particles constitute this ferruginous matrix supported limestone with <10% allochems. The allochems are small siliciclastics and/or carbonate grains (mean size=5.1microns; S.D.= 1micron). The facies is laminated but at places is cut across by thin grains filled up fine-grained carbonates. This facies is intimately associated with discontinuous to often continuous ash fall layers. The ash has been chertified at places to give it a compact character.

Facies 6 Laminated Mudstone

Red coloured laminated, thin layered, fissile, matrix supported ferruginous clayey limestone constitute this facies (Micrite by Folk, 1959). It has allochems less than 10%. Fine-grained clacitic veins criss-cross with atleast three generations of vein emplacements.

Fig. 23: Massive pyroclastic flow deposits over- and underlying cross-bedded rippled surge deposits, (base not seen). Incomplete Bouma sequences Tbc, Tcd, Tc/c within volcaniclastic turbidites, exposed in outcrops along Vesthaveripallae-Kadiri road, Vempallae Formation.

Fig. 23a: Laminated mudstones, along Vesthaveripallae-Kadiri road, Vempallae Formation.

Facies 7 Tufa

This is a purely carbonate facies consisting of micritic carbonate grains. Thin sections show pink red coloration on staining with Alizarin Red-S. Samples of tufa effervesce vigoursly with dilute HCL acid and when digested in the acid, these leave very little residue. It is extremely pure calcium carbonate associated stratigraphically with tepees and other emergent features of shallow subaqueous origin (Grotzinger, 1986a). Allochems constitute less than 10% of the rock and consists mainly of mudcracked shale clasts (Fig.24), evaporites (anhydrites mainly), volcaniclastics associated with rounded lapillae and fine-grained quartz crystals (mean size= 10microns; S.D.=4.5microns). Quartz crystals show development of microsparitic cement as syntaxial rim or outgrowth. Volcaniclastics and evaporites of the facies generally occur in small vugs within the rock filled up later by sparry calcite cement. Shale clasts are the major allochems with prominent mudcracks filled up by carbonate mud.

Presence of colloform texture in calcretes within tufa horizons point towards a possible soil formation during this period of tufa precipitation and subsequent subaerial exposure. Microstructures in rocks that are indicative of gel origin are colloform structures expressed by curved, reniform or spheroidal forms.

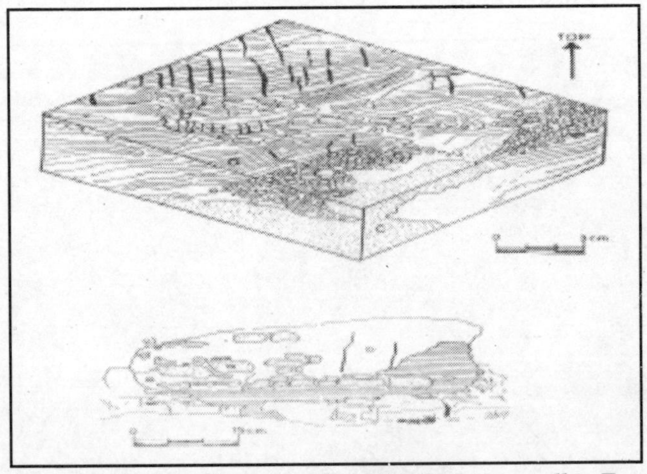

Fig. 24: Tufa with shale clasts near Tumlapllae, Vempallae Formation.

Stromatolitic Dolomite (Facies 8,9,13,14,18,19&20)

Laminations found in recent stromatolites, an intrinsic feature of many of them, comprise a number of sediment rich and organic rich couplets (Park, 1976). In contrast to this, Precambrian stromatolites of the Vempallae Formation have laminae which are usually alternations of micritic mud and sparry sediment. The laminae, which are megascopic features, at times, are of few millimeters thick. Microfabric shows thin laminations comprising alternate micritic mud and dolomite neospars respectively. They are often cut across by sparry calcite veins which also fills up the fenestrae in the sections (Mean size=9.6microns; S.D.=2 microns). Nature of the layering and their external morphology have grouped them into a number of facies (Facies 8,9,13,14,18,19&20) and this makes them discrete from other facies.

Facies 8 Pseudocolumnar stromatolites associated with columnar varieties

Colony of slender pseudo columns, slightly elongate to often circular in cross-section with associated patches of colonial concave-up laminated columnar stromatolites (Fig.25). Stratigraphically, higher levels show preservation of convex-up laminated, branching columnar stromatolites with uneven margin and high synoptic relief, identified as the genus Kussiella (Cloud and Semikatov, 1969) occur associated with more elongated pseudocolumnar stromatolites.

Fig. 25: Pseudocolumnar stromatolites associated with columnar varieties, near Pulivendla, Vesthaveripallae-Kadiri road, Vempallae Formation.

Facies 9 Columnar Stromatolites

The facies consists of discrete, thin horizons of discontinuous, club shaped, concave-up laminated, rough margined stout, short columnar stromatolites.

Facies 13 Pseudocolumnar Stromatolites

Interlinked laminae define pseudocolumns with slender to moderate thickness and large synoptic relief. These are generally elongated and often closely linked with columnar-layered varieties (Fig.26).

Facies 14 Cryptalgalaminites

Planar, wavy, discontuous laminae with minor unconformities occurring as encrusts; presence of irregular cracks and filled up by large

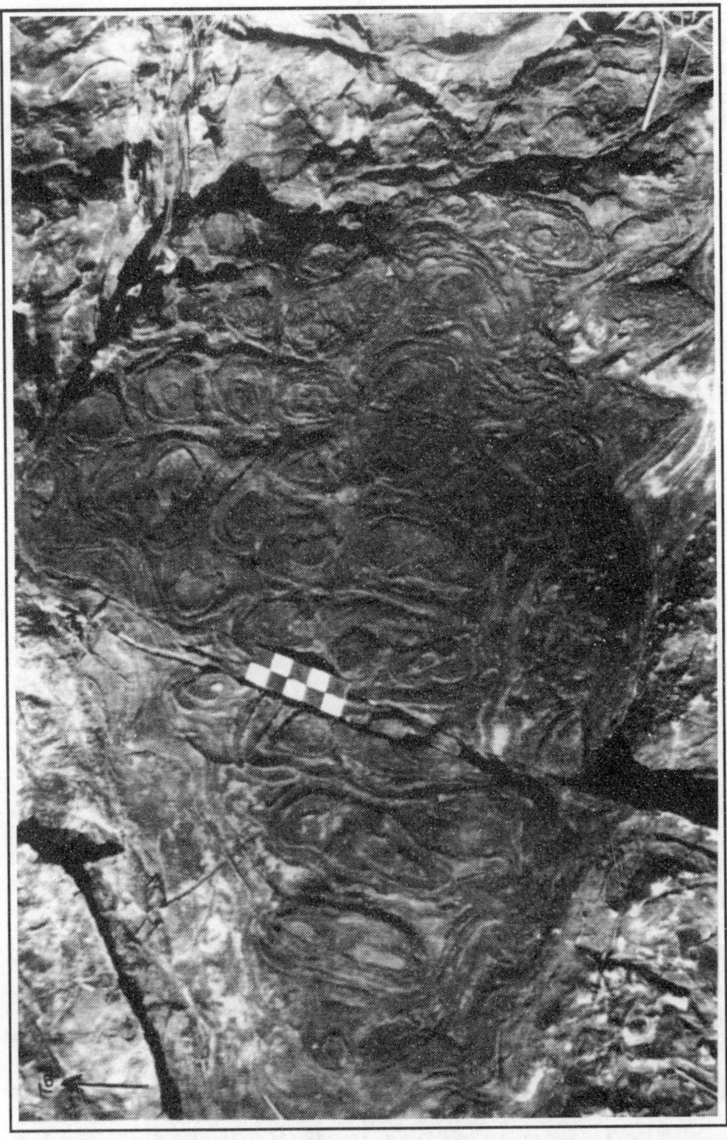

Fig. 26: Pseudocolumnar stromatolites, Vempallae Formation.

sparry calcites (mean size=19.5microns, S.D.=4.5microns) and rare evaporite minerals (mainly anhydrite).

Facies 18 Cumulate Stromatolite

This is a special type of noncolumnar stromatolites categorized as cumulate stromatolites by its shape, size and bulbous-up morphology between the pseudocolumns and the large domals.

Facies 19 Domal Stromatolites

Large bulbous mounds of interlinked stromatolites; highly elongated (1.6m) with low synoptic relief and associated with veins filled up by sparry calcites are peculiar to and characteristic of the facies.

Facies 20 Partially Chertified Dolomitic Pseudocolumnar Stromatolites

Smaller pseudocolumnar stromatolites, dolomitic in composition and partially replaced by chert horizons characterise the facies. These are possibly volcanogenic in origin.

Facies 10 Laminated Dolosiltite-Dololutite

The facies consists of more or less continuous alternating layers of dolosiltite and dololutite. Algal laminae show a much regular arrangement compared to those of other facies. Often micro-laminae delineate small bodies of micritic mud. Layers show dessication cracks filled up by fibrous gypsum, designating signature of evaporite facies. Originally clasts with hydroxidic iron were initially deposited in a relatively strong pulse of current. Subsequent diagenesis crystallized iron oxides and therefore, transverse sections of cubes in form of squares, triangles and rectangular cross sections are discernible.

Facies 11 Oolitic Chert

Oolitic chert beds are punctuated by thin (10cm approximately) bands of oolitic chert. Ooids have well developed radial structures, often obscured due to intense micritization. Nuclei are generally small and micritic with a thick cortex. Ooids are on an average 0.6mm in diameter with the intervening spaces filled up by clear and cloudy zones of radial fibrous chalcedonic quartz. This occurs as a possible pore fill-up rather than replacement as there are straight boundaries between adjacent growth of chalcedonic quartz. Thin cortexed superficial ooids are rare with some peloids which again shows partial chertifiation (Fig.27).

Facies 12 Oolitic Dolomite

The nature of the oolites is the same as in facies 11, with peloids cemented together by dolomite, spar cement (oo-pel-sparite, Folk, 1959). They also occur as rare, very thin laterally continuous units or associated with oolitic chert separated by thin cherty layers.

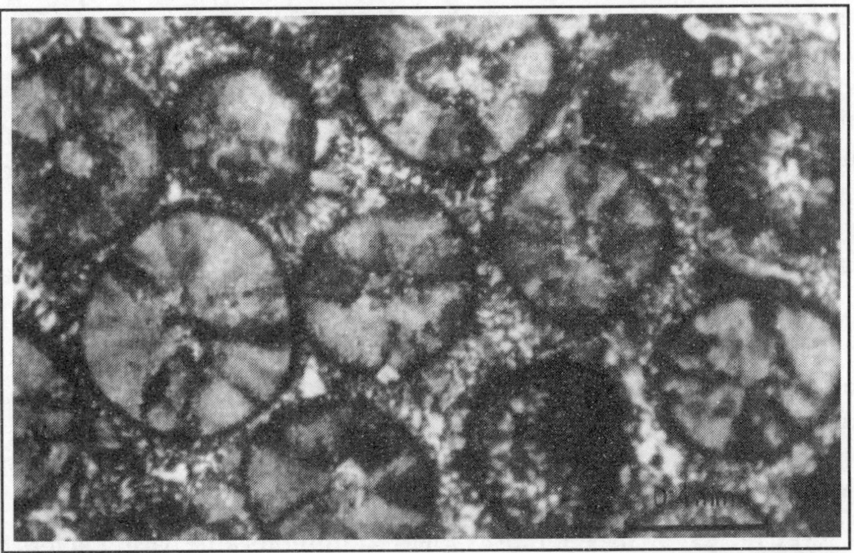

Fig. 27: Partially chertified radiating oolites, from oolitic dolostone, Vempallae Formation.

Facies 15 Massive Dolomite

Microsparitic dolomite constitutes this unit of massive structureless dolomite. It is mostly an assemblage of anhedral to rhombic dolomite grains with complete replacement evident by absence of calcite grains except some secondary saccharoidal growth. This has been confirmed by almost no colouration of the thin section on staining by Alizarin Red-S. Allochems are <1% and no structures are evident even in microscopic level.

Facies 16 Dolostone with Ripple phase

The facies is laminated with microsparitic dolomite alternated with micrite and show development of a climbing ripple phase generating a ripple-drift lamination. Variability of the climb is marked in vertical sections and angle of climb of the ripples also differs. The facies overlies a scoured base and are often cut through by typical Vempallae channel sequences with another erosive contact.

Facies 17 Trough-cross-stratified dolosiltite

Numerous onlapping channels of average length : depth ratio of 3.1 to 4.1 (observed values are 4m : 1.6m; 25cm:9cm; 4.2:1.1m) characterise the facies. Channels fills are almost entirely dolosiltite mixed with subordinate doloarenite and carbonate mud. Planar- and trough-cross-stratifications and ripples with positive and negative climb constitutes the channels mainly, indicative of oblique bars as channel fills. These again contain packages of incomplete Bouma sequence such as graded unit (Ta), overlain by laminated units (Tb) and rippled units (Tc). This point to the nature of complexity in the growth of the Vempallae channel bars.

Facies 21 Chert

Two types of chert have been identified from the Vempallae strata. While one type is syndepositional chert with dolomites, the other type is locally found chert, diagenetically altered from volcaniclastics. Microscopic studies reveal the different stages of gradational changes from volcaniclastics to chert. Chert of inorganic origin has been reported from many ancient analogues (Pettijohn, 1975).

Interfingering vertical gradation, laminated and bedded nature having sharp contacts with dolostones and as well as discrete nodules having irregular shape and size floating within the dolostone probably suggests sydepositional origin with dolostone. Lithosomes of chert, when traced also offers scope to suggest independent chert sedimentation controlled by local basinal depressions and paleoslopes. Krumbein and Sloss (1963) have dealt exhaustively on such chert sedimentation. Primary bedded chert of the Precambrian is also a topic in itself. Paleoanalouges indicate a strong possibility for the existence of syndepostional chert with dolostone in the Vempallae set-up.

Salient petrological attributes of the Vempallae rocks

Clastic quartz with extreme broken roundness is mixed up with unstable volcaniclastics and intrabasinal micritic mud. Lithification and load pressure generates pseudomatrix from volcaniclastics. Medium to fine silt-sized quartz defines approximately clear contact with micritic mud and subordinate very fine-grained clastics.

In response to intrabasinal erosion and flowage, shaly layers develop whirls of intermixed shale laminae with micritic mud. Detached ferruginous shaly bands terminate at pointed or pointed penecontemporaneously deformed ends. Slumping generates intraclasts of already deposited crudely laminated micritic mud with subordinate quartz silt.

There are different cases of sudden changes in intrabasinal depositional condition of energy. Low energy crudely laminated ferruginous shales are overlain by graded layers. Grading indicates sudden drop in the energy level. Extreme low-energy situation is defined by succeeding crudely laminated ferruginous shales. Ferrugenous cements probably obliterate signatures of fine laminae. Intrabasinals overstep the land supply. Persistent micritic top is the response.

Turbulence in the intrabasinal condition is well documented through different stages of intraclasts and pellets. Textural maturity is revealed from the shape, size and roundness of micrite intraclasts. This is also the same in case of pellets. Derivation of intraclasts through turbulence is evident from intraclasts, intramicrite with textural signature of turbulence.

Textural immaturity of pellets exists side by side with submature-subrounded micrite intraclasts (Fig. 28). Well rounded micrite clasts

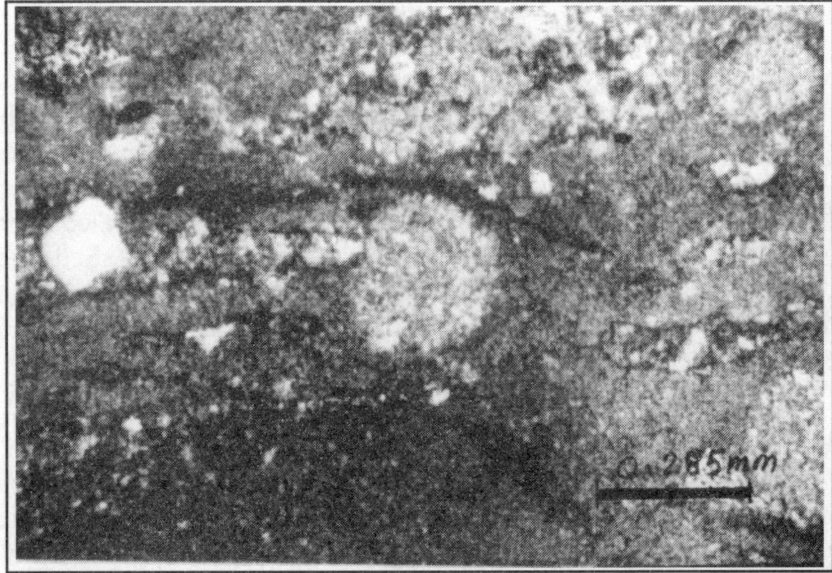

Fig. 28: Textural immaturity of pellets exists side by side with submature-subrounded micrite intraclasts, Vempallae Formation.

are deposited possibly after considerable transport from sites of its origin (Fig. 29). Generation of intraclasts from ferruginous mudrocks (Fig.30) is promoted. Large well-rounded intraclasts of micritic mud indicates considerable transport and erosion (Fig.31).

Micro-ripple-drifts in graded-laminites with admixtures of very

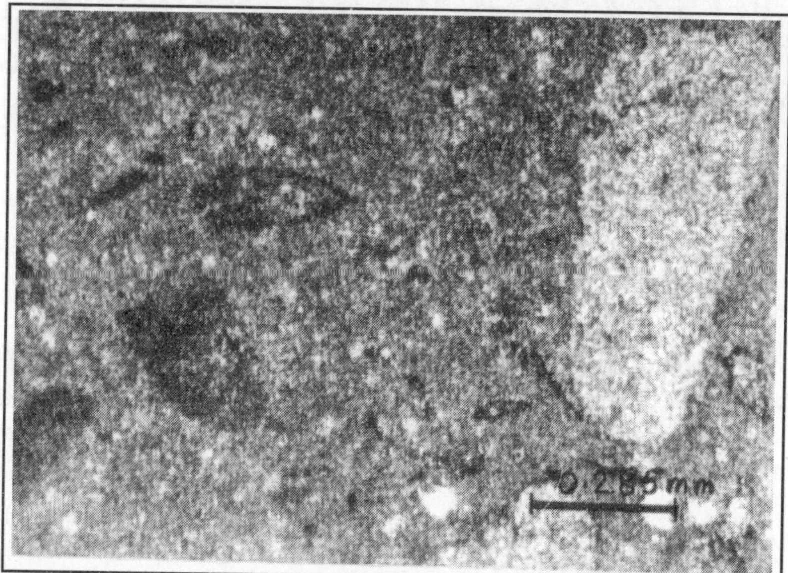

Fig. 29: Well rounded micrite clasts are deposited possibly after considerable transport from sites of its origin, Vempallae Formation.

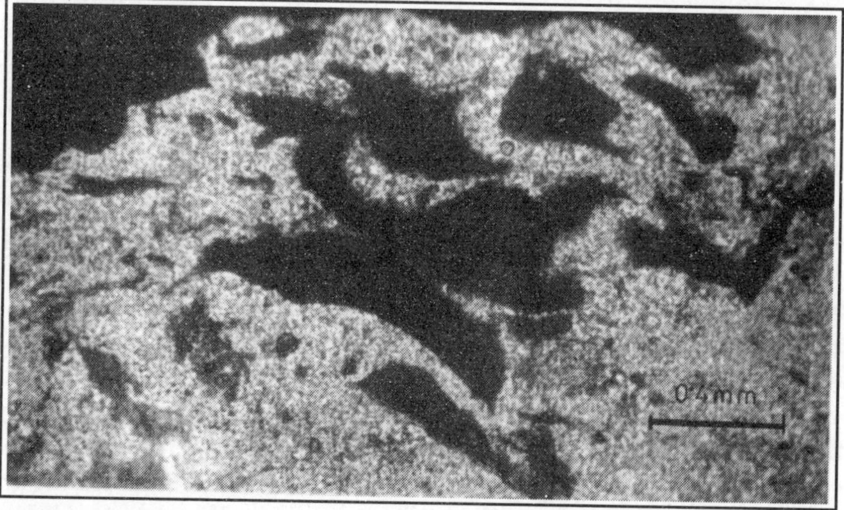

Fig. 30: Generation of intraclasts from ferruginous mudrocks, Vempallae Formation.

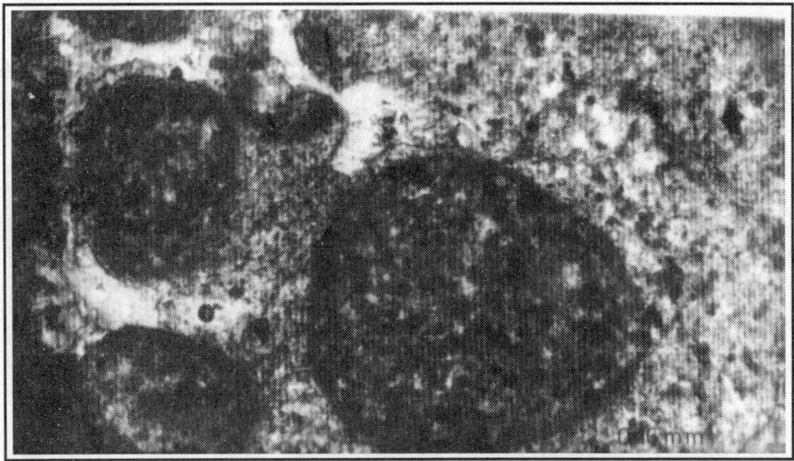

Fig. 31: Large well-rounded intraclasts of micritic mud, Dolostones, Vempallae Formation.

fine clastics, carbonates and volcaniclastics are also identified. Fine-grained clastics, in apparant chaos highlighted by slump structures, take ground of carbonate microfacies. Chaotic slump facies is however, common to both clastics and nonclastics.

Conclusions

With the exception of the Vindhyans, very little is known on facies attributes of Proterozoic carbonates from Indian intracratonic basins. The Vempallae carbonates of the Papaghni subbasin holds a significant clue in its intrabasinal sedimentation. Facies analysis along chosen sections of the Vempallae Formation revealed twenty one lithofacies with strong volcaniclastic elements and points to an overall pulsating basin dynamics. In general, the depositional environment appears to be that of a low-energy platformal (shelfal) set-up punctuated at intervals by periods of relatively high energy. Mudstone dominated calcareous sequences with apparent sheet geometry characterise low-energy shallow-marine facies. Presence of laminated dolostones, evaporites and associated stromatolitic growth also corroborate a shallow-marine to lagoonal environment. The low-energy condition was occasionally punctuated by high-energy phases as indicated by the presence of oolitic chert and/or dolomites apart from the columnar varieties of stromatolites. The most intriguing feature of this subbasinal development is the associated volcanism, impress of which is evident in majority of the Vempallae facies. Basinal condition in response to sedimentation during volcanism did not appear to be congenial to the stromatolitic growth.

The Trappean-Intertrappean Sequence and its stratigraphic significance

The Vempallae Formation is unconformably overlain by a trappean-intertrappean sequence comprising basic- grading to felsic volcanics, volcaniclastics, terrigenous clastics, argillites and carbonates. Earliest phase of volcanism in Vempallae was however, basaltic, which got emplaced as sills and dykes in the Vempallae sediments. Their contact with aluminous shaly materials produced asbestos, which is commercially viable. The basalts were also subaerially emplaced attested by columnar joints, pumiceous layers and lapilli. Felsic trappean-intertrappean sequence is well exposed in the Brahamanapalli 70m representative section (Fig.32). This is unconformably overlain by the Pulivendla clastics of the Chitravati Group.

Vertical and lateral facies associations of the trappean-intertrappean sequence, designated as the Brahamanapalli trappean-intertrappean sequence is discernible in and around the Brahamanapalli mines near the Pulivendla town in Cuddapah. Scattered exposures of epidotised and/or serpentinised basalts and acidic volcaniclastics are identifiable over an area of about 20sq.km. Main trappean-intertrappean sections

Fig. 32: Felsic trappean-intertrappean sequence, well exposed in the Brahamanapalli section, Trappean-Intertrappean Formation on top of Vempallae Formation.

are exposed in small stream cuttings and mine excavation in mining pits and trenches. The sequence is exposed for a strike length of 8km. Field relations are well preserved and discernible on exposed surfaces where sedimentological details, depositional or erosional surfaces have been clearly observed.

Vertical facies analysis emerged over decades as a very powerful tool for understanding the facies delineations, their attributes, ultimately leading to establishment of environmental set-ups for respective lithofacies. Vertical facies associations can lead to development of three-dimensional models for certain set-ups which through further detailed analysis may bring forward the intricate interplays among the different environments.

The trappean-intertrappean sequence is unique in the sense that no such facies has been observed so far from other Cuddapah strata.

Facies analysis of selected sections in the eastern part of the Saraswati mines have delineated five distinct lithofacies which are as follows :

a) Carbonate lithofacies including dolostone, impure limestone.
b) Mudstone having intermixture of siltstone and shale in different proportions with intermittent emergent features such as mudcracked surface, shale-pebble conglomerates and salt pseudomorph shales.
c) Basaltic flows, mainly layered ones, belched out in succession with intervening quiescence. The top surfaces of flows contain vesicles at times as well as lapilli tuffs, welded tuffs and lithic tuffs.
d) Ash fall tuffs with tear drop lapilli.
e) Tufa in association with dolostone, chert, limestone and a facies itself shows a carbonate backdrop lithology. Presence of translatent strata with low-angle cross-stratification having subcritically climbing wind ripples with negative as well as positive climb bears the signature of aeolian set-up. These aeolian strata are made up of impure muddy carbonates. The facies attributes attest to an aeolian clay dune origin.

All the five lithofacies have sharp contacts among them.

For carbonate facies analysis, it is advisable to study in detail a single vertical section and as such a single trappean-intertrappean mixed sequence would reveal the history of development of the facies associations with different environmental signature. Therefore, a Brahamanapalli section was chosen for facies analysis and studied upto subfacies level.

Detailed log for the section has been arranged in a stratigraphic succession (Fig.33) which runs as follows :

Unit 19 (10.4m thick, the topmost unit) : Chertified rock with some greenish to reddish minerals.

Unit 18 (5.6m thick) : Volcaniclastic facies intensely block fractured containing lapilli.

Unit 17 (8.4m thick) : Rippled ash fall tuff beds, epidotised units with distinct traces of rhyolitic flow texture; profuse carbonate replacement.

Unit 16 (0.2m thick) : Red shale facies.

Unit 15 (1.3m thick) : Beds comprising lapilli tuff.

Unit 14 (0.6m thick) : Volcaniclastic with profuse chertification; development of greenish and some reddish minerals.

Unit 13 (0.7m thick) : Welded ash fall tuff sequence with all characteristics same as those of previous ash fall tuff unit.

Unit 12 (1.1m thick) : Red shale-mudstone facies with intercalations of dolomitic shale and mud.

Unit 11 (6.6m thick) : Shale-pebble facies, patchy in development.

Unit 10 (0.9m thick) : Volcaniclastic facies with presence of lapilli, round and tear drop in shape; crudely graded lapilli, highly block-fractured and epidotised.

Unit 9 (1.7m thick) : Rippled dolomite facies, large crested current ripples; in-phase ripples indicative of rapid sedimentation due to probable reworking by rapid wash outs at peripheral part of the basin; the structural package associated with low-angled translatent strata.

Unit 8 (0.2mthick) : Tufa dolomite interlayered facies

Unit 7 (2.1mthick) : Partially chertified ash fall tuff facies with medium to small lapilli layers in patches. The chertified patches elliptical in shape with major axis = 16cm, minor axis = 15cm.

Unit 6 (2.1m thick) : Lapilli bearing calcareous red mudstone facies.

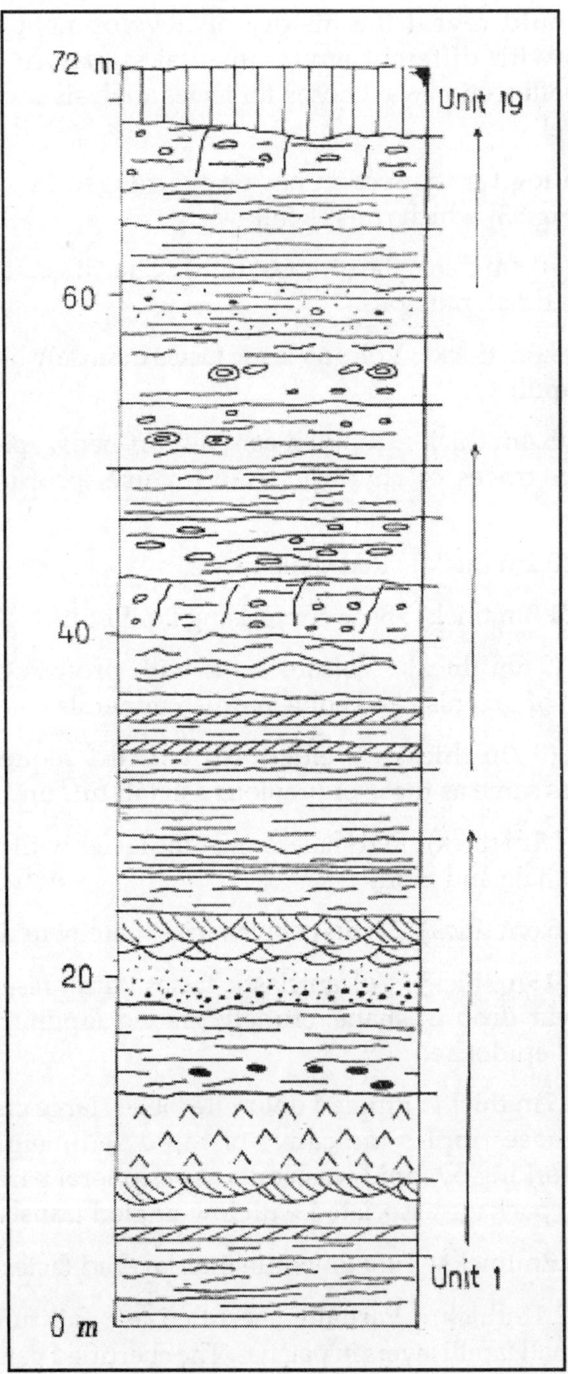

Fig. 33: Vertical facies association of Trappean-Intertrappean Formation on top of Vempallae Formation, near Brahamanapalli.

Unit 5 (5.3m thick) : Ash fall tuff facies with a 4cm thick well graded lapilli layer. Welded tuff facies with normally graded lapilli and devitrified glass with facies attributes such as current ripples, primary lineations and trough cross-stratifications are indicative of flowage.

Unit 4 (1.1m thick) : Calcareous mudstone facies with thinly laminated shale-pebble layers.

Unit 3 (13.6m thick) : Subaerially emplaced epidotised basalt with vesicles, flow texture, patches of lapilli and ripples on top surfaces; discontinuous layers of ash fall tuffs and lithic tuffs mixed up in various proportions.

Unit 2 (1.7m thick) : Fine-grained variegated calcareous tufa rich facies interlayered with dolostone facies (0.1-0.3m) and thin limestone channel facies having infillings of oblique bars comprising different ripple phases.

Unit 1 (8.4m thick) : Crudely laminated fine-grained dolostone facies interlayered with granular thin layers; occasional ripple-drifts, starved ripples and shale-pebbles as markers.

Spatio-vertically, the trappean-intertrappean facies associations in and around Saraswati mines and Brahamanapalli are summarized:

Tufa facies with floating dolostone.

Basaltic flows showing development of palaeosols in thin layers.

Red mudstone facies with mud-cracked surfaces, shale-pebble conglomerates, salt pseudomorph shales.

Ash fall tuff with distinct Lapilli layers.

Red shale facies

Ash fall tuff facies serpentinised with pockets and patches of chertified fine-grained ash.

Tufa facies showing intercalations of dolomite layers.

Carbonate facies with dolostones, teepes structures, laminated dolomites, ripple-drifts laminations, occasional mudcracked surfaces, interfering current ripples, carbonate flats and thin tufa layers with variegated colours.

Volcaniclastic facies with lapilli tuff in profusion.

Mudstone facies intermixed with volcaniclastics, mudcracked surfaces frequented by interclasts of shale-pebble conglomerates.

Red mudstone facies with intercalations of dolomite and red siltstone.

Ash fall tuff with subordinate amount of lapilli tuff and lithic tuff, fine-grained volcaniclastic facies showing profuse chertification with occasional tear-drop and lithic tuff.

Lapilli tuff facies dominated by tear drop lapilli frequently forming coalescence patches; red mudstone beds, fine sandstones and siltstone intercalations.

Ash fall tuff facies with lapilli; ash fall and lithic tuff form fine volcaniclastic facies which is profusely chertified. This facies is often serpentinised or epidotised.

Hyalopilitic lavas with felted microlites, glass and fine-grained lavas in part devitrified showing dark, globular lapilli, lava with spherulitic growth showing conjugate spherulite, partially devitrified glass mixed with chertified fine ash; axiolitic texture comprising a fine linear intergrowth of quartz and feldspar, perlitic cracks and tear-drop lapilli with irregular corrugated boundary and epidotised and/or serpentinised tear-drop lapilli are some of the features which highlight subaerial emplacement of the volcaniclastics (Fig.34). The volcaniclastics are subaerially emplaced welded tuffs. The welded tuffs show dominance of tear-drop lapilli, indicative of free fall through atmosphere during volcanic explosion.

Salient aspects of the sequence

Subaerially emplaced volcaniclastic rocks and basaltic flows are visible around Brahamanapalli from the trappean-intertrappeans with carbonates and siliciclastics. Megascopically volcanic rocks are in most cases, epidotised or serpentinised. The stratigraphic section of the trap rocks in trappean-intertrappeans, unconformably overlying the Vempallae Formation of the Papaghni subbasin records mafic and acidic volcanism. Subaerially basaltic lavas were emplaced successively in intermittent phases. During the quiescence, there was formation of red soil or red bole due to weathering and erosion of basic lava. Thin layers of red bole ranging in thickness from few millimeters to 6 cm indicate time intervals among different basaltic flows.

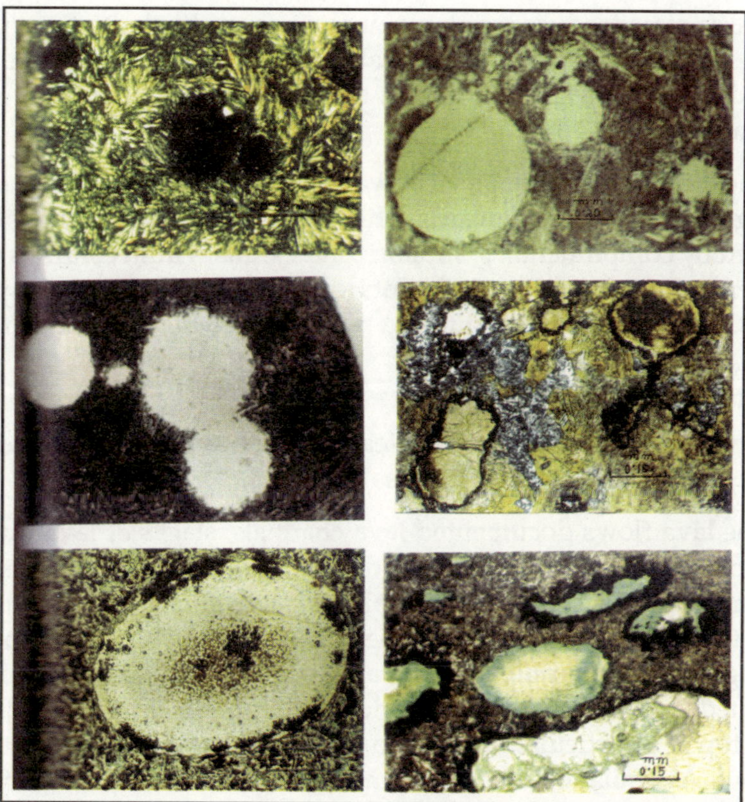

Fig. 34: Hyalopilitic lavas with felted microlites, glass and fine-grained lavas in part devitrified showing dark, globular lapilli, lava with spherulitic growth showing conjugate spherulite, partially devitrified glass mixed with chertified fine ash; axiolitic texture comprising a fine linear intergrowth of quartz and feldspar, perlitic cracks and tear-drop lapilli with irregular corrugated boundary and epidotised and/or serpentinised tear-drop lapilli, Trappean-Intertrappean Formation, Cuddapah Supergroup.

The basic lava flows together constitute 13.6 m thick sequence and do not show any significant subaqueous deposits except few patches of basic lapilli tuff intermixed with fine-grained red clastics and subordinate impure carbonates. Such aqueous deposits are also found in the upper part of the basaltic flows.

Columnar joints

The basal basaltic traps show evidence of cooling which was accompanied by contraction and took place from the top and bottom surfaces inwards. The tensional stresses set up during contraction in such cases may produce regular joint sets perpendicular to the cooling surfaces and usually these are vertical to subvertical in orientation. Well defined intercepting joint sets in the basalts produced polygonal columns.

The joint faces propagated inwards from cooling surfaces as the cooling front advanced inwards. This propagation probably has been reflected by complimentary, subhorizontal joints within columns or by sedimentary pattern on vertical joints reflecting successive propagation joints. Columnar jointing exhibits a two- or three-tiered arrangement.

Structures in Basaltic Lava

Lava channels and other flow features such as lava trough and lava ripples are generally lacking in basalts.

The lava flows documented lava channels, stacks of lava troughs and ripples. Basalts mixed up with basic volcaniclastics generated climbing ripple trails and ripple-drift laminations. Sharp to round crested ripple trails have also been documented. Troughs show some partial infillings by basic lapilli tuff and ashes. There are thin-bedded and laminated layers of mixed ash fall and lapilli tuff which are either found on the top of a particular basaltic flow or occur in patches on the top of the flow. Assimilation of fine- to medium-sized ash by basaltic flow made it prone to chertification. The basaltic layers with assimilated ash appear as brownish grey to pink brown on weathering. Coalescence of lapilli forms clusters.

Subaerial Basaltic lava

There are exposures of flat-lying basaltic flows below which hybrid rocks comprising carbonates, epiclastics and pyroclastics occur. There are hyaloclastic breccias interfingering and alternating with massive lava bodies and are also found covered by massive subaerial basaltic lava.

If lava has high viscosity and high yield strength, it may interact differently with water and may flow under water maintaining the continuity. Some of the important constituents of hybrid rocks are lapilli, pumice, broken or clusters of crystals, impure carbonate and terrigenous materials. Spatially continuous parallel thin basaltic layers are found in association with hybrid sediments and pumiceous layers covered by basaltic flows which might have been subsequently eroded out.

Facies associations supported by ancient and recent analogues strongly suggest that some of the basaltic flows were of subaqueous origin.

Emplacement of subaerial welded tuff

Asf fall tuffs, lapilli tuffs or lithic tuffs are found as laterally continuous beds which are in most of the cases epidotised or serpentinised along certain preferred surfaces which define either successive curvilinear layers or climbing curvilinear bedforms. These are current ripples over the top of subaerially emplaced rhyolitic flows. Ash, lapilli and/ or lithic fragments might have formed thin veneers on the top of rippled surfaces.

Chertification is promoted by fine-grained ash layers or their patches. Types of chert are light grey to greyish brown in colour which are typically associated with chertified ashes. This feature serves as a tool to identify thin layers of fine-grained ash and/ or lapilli tuff over the top of the rhyolitic flows and basaltic flows. This layering is frequently folded. Microscopically spherulitic growth has been identified. Spherulites are radiating aggregates of alkali feldspars with or without high temperature quartz like trydimite or crystabolite.

Vesiculated pumiceous layers may occur as bands with obsidian and spherulitic layers. The rhyolite lavas of the formation are often capped by blocks of pumice or pumiceous layers.

Different types of rhyolitic flow textures are also seen. The lavas are steep at the flow fronts and typically terminated by tallas apron. Aprons of brecciated lavas and pumice blocks are built by the auto-brecciation of the lava. These advance with great speed and have a great role in building up such aprons. Auto-brecciation is made possible due to differential stress field which is set up in a simple lava flow. The lowermost part of the lava being concealed from the atmosphere, remains hot and fluid unlike the upper part which gets solidified and or semisolidified due to radiation under subaerial condition. Therefore, relative motion is set in a single flow under differential stress field. This could cause shattering of the upper part of the solidified flow into blocks by auto-brecciation. Other features of the rhyolitic lavas include great prevalence of pumice.

Association of excessive volumes of pumiceous materials with rhyolitic lava is an evidence for great explosiveness of the eruption.

Dynamic processes operating within the system itself did modify proportion of pumiceous materials with the rhyolitic lava and associated sedimentary piles.

Stratigraphically basaltic eruption was followed in the latter phase by rhyolitic volcanism in phases, punctuated by quiet condition and witnessing sedimentation. Intrabasinals and terrigenous clastics and epiclastics were deposited in pools and ponds during such quiescence. Aeolian action might have been operative sometime as evidenced by the presence of subcritically climbing translatent strata.

Volcanic textures

Axiolitic structure : Fine linear intergrowth of fibrous quartz and feldspar arise out of devitrification of glass shards. Long axis of crystals is perpendicular to shard boundary. Axiolitic structures in the ash flow are rare. Ancient analogues also support the observation (Cas and Wright, 1987; Furnes et al., 1980).

Hyalopilitic texture : It is a mesh work of randomly oriented microlites with interstitial glass. Such texture has been frequently found in rhyolitic welded tuff which occurs in the upper part of the stratigraphic section within the welded acid tuffs.

Perlitic structures : These are concentric onion-like partings in glass produced by hydration and exfoliation of obsidian. Obsidian is occasionally met with the flows in the Brahamanapalli area and reveals these structures megascopically as well as microscopically.

Pylotaxitic texture : This is observed in the welded tuff of Brahamanapalli and comprises mesh work of randomly oriented microlites. Glass might be present along the crystal meshwork.

Vitroclastic texture : These are typical fragmental structure of glass-rich volcanic rocks and glass shards. Such layers are found in association with obsidian in Brahamanapalli section.

Welded tuff with various types of flow structures, consolidated volcanic ash comprising various types and fine-grained glassy light coloured assemblage of quartz and alkali phenocrysts in groundmass of glass form the typical rhyolitic flows.

Relicts of ash structure in acid flows, fragments of fluidal glass in ignimbrites are vitroclastic. Character of groundmass of ignimbrites, typical ash texture and flow structure, spherulitic structure in tuff lavas, lava breccia, circular and elliptical pepperites (globular and elliptical glass) with internal infillings of serpentinised pyroxene, tear-drop lapilli are some of the leading facies attributes of the Brahamanapalli volcaniclastic facies.

Intertrappean Facies

The facies comprises carbonates, terrigenous clastics, chert, hybridised rocks, impure doloarenite, dolosiltite, neomorphosed intrasparite and intramicrite. Hybridisation is due to intermixing of volcaniclastics in different proportions with carbonates.

Tufa with chert and mudcracked surfaces is a distinct facies. This association indicates oscillatory emergence of shallow-marine and/or lagoonal set-ups. Terrigenous clastic facies includes variegated laminated, microrippled shales, red coloured rippled and laminated siltstones, fine-grained laminated-, rippled- and cross-stratified sandstones with mudcracked surfaces. Subaerially formed subcritically climbing aeolian translatent strata within impure and fine-grained clastics have been occasionally found in association with subaqueous clastic and carbonate facies. Other prominent facies attributes in the sandstones and siltstones are sharp-crested current ripples, sharp-crested assymetrical current ripples, ill-sorted shale pebble clasts, small tabular cross-stratifications and mudcracked surfaces.

Petrological facies attributes of the Brahamanapalli volcaniclastics

Single lapillus shows reaction with hot glassy materials at its margin. The glass is epidotised there and partially devitrified. The groundmass shows development of pilotaxitic texture. Texturally the groundmass comprises volcanic rocks with a meshwork of weakly oriented microlites. Partially devitrified glass and patches of hydroxylic iron are added during subsequent alteration.

Partially devitrified glass shows a groundmass which got mixed up with fine ashy materials and subsequently chertified. Light brownish grey patches depict such chertified pockets. Axiolitic structures comprising a fine linear intergrowth of quartz and feldspar form arise out of devitrification of glass shards. Ross and Smith (1961) reported growth of crystabalite and feldspars in response to such devitrification. Such axiolitic growth is very rare in volcaniclastic facies. Globular pepperites float in the groundmass which is in the size range of 0.2 to 0.4mm.

Coaslescence of bulbous lapilli during hot stage is a confirmatory test for the presence of lapilli. Welded tuff showing flow structures is well preserved. This is indicative of free-fall of glass through air. Free fall of glass forming lapilli has been profusely reported from other analogues (Ross and Smith, 1961; Fisher and Schminke, 1984).

Felted microlites, spherulites with dark globular lapilli, eroded lapilli, fused lapilli, fibrous quartz are visible in the hyalopilitic lavas. One of the eroded lapilli is found embedded in a groundmass of lithic tuff comprising broken crystals of quartz and feldspar.

Chertified thin layered perlitic glass with lithic fragments is a characteristic volcaniclastic facies. Minute grains of quartz and feldspar crystals float within large lithic fragments of basaltic materials, which is also much devitrified.

Debris flow has been depicted by a high density, sediment water mixture containing large proportion of coarse fragments. Such debris flow might have been strongly interfered by subaqueous concomitant thin basaltic flows which ultimately hybridized the debris flow. Vescisular cavities, green epidotised and/or serpentinised cavities with lining of hydroxidic irons, tear-drop lapilli with irregular, corrugated boundary. Hyalopilitic mesh work of randomly oriented microlites with instertitial glass, some basaltic fragments with some subophitic textural attributes present some complex assemblage of facies attributes to such hybridized horizons in the Brahamanapalli stratigraphic sequence. Presence of calcrete and colloform texture increases the complexities of microstructure assemblage which indicates submergence and complex colloidal action.

Lithic tuff often forms important stratigraphic level in the sequence. One lithic tuff horizon comprises basaltic lithic tuff within rhyolitic flows. Basaltic clast show subophitic texture with partial enclosure of plagioclase laths by anhedral pyroxene. Vitrophyric texture is discernible within such lithic clasts. Partial devitrification is very common.

Pumice beds show highly vesicular silica-rich volcanic glass with porosity as high as 40% (volumetrically). Patches of fine ashes are chertified and have been identified from their grey brown body colour in plane polarised light. Glassy groundmass shows profuse serpentinisation. Pumiceous lapilli show internal lining of chertified fine ash and/or hydroxidic iron.

Shards with crude spherulitic growth and devitrified brownish and cherty patches, minute lithic fragments and crystalline tuff make thin layers of hydroclastite. This deposit consists of shattered small angular cherts when lava is extruded under water. Volcanic shards are incorporated in the subaqeous deposits.

Minute lithic fragments comprising quartz and feldspar, small shards got intermixed with fragments of fluidal glass which are much devitrified. These types of rocks differ markedly in structural and textural characteristics as a consequence of different degree of welding. The end members of this series are on hand, strongly welded varieties with the physical properties of liparites and on the other end, weakly welded to non-welded tuffs. The form of the lithoclastic fragment is found to depend on the degree of welding. In strongly welded ignimbrites, volcanic glass fragments have a flattened even elongated form having indistinct border.

Another variety of ignimbrite shows recrystallisation in varying degrees. Spherulitic and microfelsitic structures are very common. Some of these varieties might show light green groundmass and red brown fragment forming skeletal structures. Acid volcanic glass forms the major constituents of ignimbrites.

Thickest portion of the basaltic lava flow, growth of gabbroic texture, ophitic textural growth, olivine phenocrysts at places form porphyritic texture. In some of the olivine basalts, glomeroporphyritic texture is discernible. Carbonate lithofacies, tufa associated with chert and thin layered oolite define some characteristics of the shallow subaqueous carbonate facies in the intertrappeans.

Neomorphosed carbonates with distinct cleavage growths and poikilotopic assemblage of newly grown carbonate crystals clearly mark certain levels within micritic mud where differential strong neomorphism took place.

Discussion

Lava channels and other features such as troughs and ripples are generally lacking in flood basalts (cf. Cas and Wright, 1987). Presence of unidirectional structures such as channels, current ripples and troughs within the lava flows probably indicate a relatively small stage of eruption stretched over few kilometers only. Lack of volcanic bombs and large pumices are suggestive of some rift-controlled volcanism which belched out episodic lava flows. The trappean-intertrappean sequence of Brahamanapalli was successively emplaced with intermittent time gaps. The Precambrian scenario in an intracratonic set-up might have been an ideal stage for such fissure formation in an extensional realm. Role of extensional tectonics and associated basaltic emplacement from relatively shallow depths as documented in this case have often been reported from other Precambrian set-ups (Fisher and Schmincke, 1984).

A feeble touch of sedimentation was felt by the uppermost part of the basaltic flow. At successive higher stratigraphic levels, acidic tuff, lavas and related rocks were emplaced. The Brahamanapalli volcanism suggests a strong bimodal affinity.

Different major subfacies of the Brahamanapalli volcaniclastics were studied and cogent interpretations of the subfacies were attempted based on their spatial and vertical relations. Analysing different facies attributes of the Brahamanapalli acid effusives, it is clearly seen that volcaniclastics stratigraphically represent some types of surge-like flows, comparable to those of ignimbrites.

Eruptive phases are recorded in formation of lapilli tuff, lithic tuff, ash fall tuffs and/or their intermixtures at different levels of the section. There was significant amount of scours and assimilation of basic volcanic rocks, with the result, acidic tuffs compositionally became hybridized.

These volcaniclastics are often punctuated by subaqueous lithofacies, which at certain levels are volumetrically subequal. Most of the lithofacies shows intermixing with basaltic rock fragments and other acidic volcaniclastics.

Intermittent emplacement of basaltic lava was replaced by acidic tuffaceous flows during the closing phase of eruption. Volumetrically, the emplaced volcanic rocks are of limited extent.

These rocks were emplaced in an intracratonic set-up. Frequencies of ash fall tuff, lapilli tuff and thin flows of basalt, clearly demonstrate their derivation from a moderate to shallow depth in an intracratonic set-up.

Intracratonic bimodal volcanism of limited extent has been often reported from extensional regimes of the Precambrian (Windley, 1984; Condie, 1992; Goodwin, 1981). Age of the Vempallae basaltic flows have been found to be of 1887±10Ma (Gopalan personal communication).

Brahamanapalli trappean-intertrappean facies (1.9 billion year old) in all probability appears to have originated in an extensional regime, which promoted intermittent volcanic eruptions. Subaerially emplaced flows were basaltic in the beginning. Intermittent quiescence heralded subaqueous sedimentation within limited basinal extent. Sedimentation was either intrabasinal or land-derived. There was also touch of aeolian modification to the deposited sediments which were subaerially exposed intermittently. Final phase of volcanism also came in frequently but was marked by emplacement of welded tuffaceous flows of acidic nature. The Brahamanapalli trappean-intertrappean facies is a case in hand. It stands for a Proterozoic bimodal volcanism. In response to tectonic controls of sedimentation, cyclicities in sedimentation developed intrinsically.

Conclusions

1. A trappean-intertrappean sequence capping the Proterozoic allocyclic Vempalle carbonates in eastern part of Saraswati mines in Brahamanapalli, southwest of Pulivendla in the Cuddapah district of Andhra Pradesh has been covered by facies modeling.
2. Five main lithofacies identified are : impure carbonates, impure mudstones with frequent emergence features, periodically emplaced subaerial basaltic flows, acid volcaniclastics and tufa associations.
3. Facies associations in its 70m type section show repetitive relations among different facies.
4. Periodically emplaced subaerial basic flows forms the main basal phase of volcanic eruption.
5. Seven low energy subaqueous facies with intermixture of elements such as impure carbonates and/or fine siliciclastics with basic or acid volcaniclastics in variable proportions and periodically emplaced nine phases of subaerial acid vocaniclastics.
6. Stratigraphically basaltic eruption was followed in the later stage by rhyolitic volcanism in phases, punctuated by quiet conditions often witnessing sedimentation. Intrabasinal and terrigenous clastics and epiclastics were deposited in pools and ponds during quiescence. Aeolian action might have been operative sometime as evidenced from the presence of subcritically climbing translatent strata.

The Chitravati Group

The Chitravati Group comprises the Pulivendla Formation in the lower part, the Tadpatri Formation in the middle and the Gandikota Formation in its upper part. The Pulivendla unconformably overlies the Vempallae rocks of the Papaghni Group and is again unconformably overlain by the Tadpatri volcaniclastics and volcanics. The Gandikota Formation overlies the Tadpatris and comprises sandstone-mudstone sequences.

The exposures of Pulivendla Quartzite near Talapallae and Tumalapallae village, Andhra Pradesh reveals few distinct facies assemblages (Fig.35).

Fig. 35: Exposures of Pulivendla clastics near Talapallae and Tumalapallae village, Andhra Pradesh. Fining-upward braided-fluvial deposits showing laterally cutting conglomeratic channels in the lower part, grading up into sandy channels, Pulivendla Formation.

Facies analysis of a well exposed section reveals the following facies:

8. Channels with couplet beds in 50cm thick crudely bedded orthoconglomerate (Topmost unit)
7. Sandstone unit (54cm thick); lower part is trough-cross-stratified and upper part comprises planar cross-stratifications.
6. Sandstone unit (45cm thick); lower part is planar cross-stratified and the upper part is trough cross-stratified.

5. Couplet beds, 85 cm thick; lower part is trough-cross-stratified and upper part comprises flat-bottom channels with couplet beds of medium- to fine-grained sandstones alternating with conglomerates.
4. Couplet beds, 55 cm thick
3. Alternating coarse and fine layers of couplet facies, 28cm thick.
2. Sandstone unit; 68 cm thick; lower part lower part is rippled and the upper part is trough cross-stratified.
1. Orthoconglomerates, 57cm thick; forms the bottommost unit of the stratigraphic section.

Facies Attributes

Couplet Beds : vertically alternating subhorizontal 45cm thick planar bedded couplets with a discontinuous sheet like geometry (Figs.36 & 37). Boulders, pebbles, gravels and cobbles are interstratified with coarse sand. The finer grained beds are more extensive than their coarse-grained counterparts. Bedding is irregular in this facies as there are obstacles like gravel clasts along the flow paths. Nesting of clasts is also visible within the finer grained counterpart. This particular facies attribute can be regarded as an effect of unconfined water flows like sheet flooding (Blair, 1985, 1987).

Fig. 36: Planar bedded couplets with a discontinuous sheet like geometry. Boulders, pebbles, gravels and cobbles are interstratified with coarse sand, Pulivendla clastics, Cuddapah Supergroup.

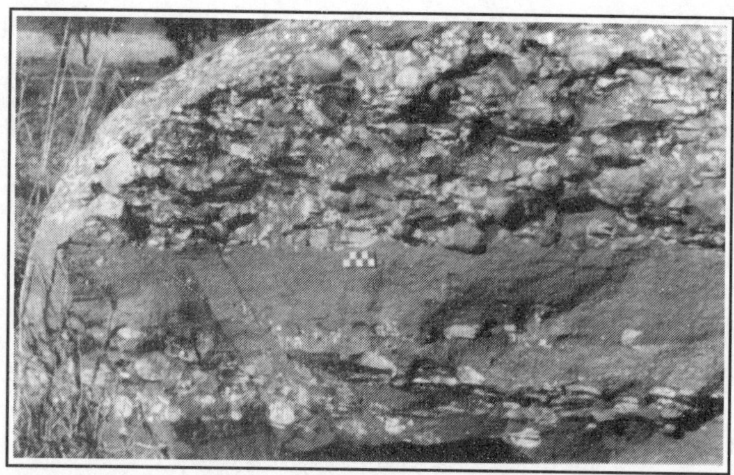

Fig. 37: Planar bedded couplets with a discontinuous sheet like geometry. Boulders, pebbles, gravels and cobbles are interstratified with coarse sand, Pulivendla clastics, Cuddapah Supergroup.

Planar cross-stratified sandstones : this is a 30-35 cm thick reddish brown medium-grained sandstone (Fig.38), sandwitched between trough cross-stratified sandstones and couplet beds. Inclined sets of strata at high angle (25°-30°) abut the trough cross-stratified layers.

Trough cross-stratified medium- to fine-grained sandstone : low angle trough cross-stratified sandstone is found interstratified with planar cross-stratified beds in the upper part and rippled sandstone

Fig. 38: Planar cross-stratified channel with alignment of clasts in the upper part, Pulivendla Formation.

and couplet beds in the lower part. The sandstones are reddish brown in colour indicating its high ferruginous nature. Deformed- and convex-upwards foresets are occasionally present in the beds. Troughs are at places 76cm in length. Infillings of normally graded medium- to fine sand within the troughs have been noticed. Inverse grading in the infillings of the troughs are also present.

Rippled medium- to coarse-grained sandstone : the facies is 30-35 cm thick and occurs at the lower portion of the sequence. It is overlain by trough cross-stratified sandstones and underlain by orthoconglomerate horizon. The ripples are large, cross-laminated in some places with asymptotic termination at the bottom or doubly crested, 18-20 cm in length (occurs at the bottom part of the facies) or having a deeply scoured base, flat crested, 19-20cm wavelength and 3-5cm in amplitude. The ripples are climbing in nature. Straight crested, symmetrical wave ripples are also seen.

Orthoconglomeratic facies : the facies is 55 to 58cm thick and composed of triangular, subangular to subrounded clasts of elongated cherty pebbles 3.2 cm to 9 cm in length. This facies occurs here as flat-bottom subhorizontal channels at the base of trough cross-stratified rippled sandstones and has a sheet-like appearance. The pebbles of chert are virtually indestructible and may be reworked and redeposited through several cycles of sedimentation. Such a sheet of orthoconglomerate at the base of sandstone may indicate an extremely high energy fluvial sequence followed by comparatively low energy conditions.

Channelised sandstones : in addition to the mega channels of Pulivendla clastics of the Talapallae exposure, several smaller and larger channels are common attributes of the Pulivendla Formation. Channel fills include ripples, troughs, planar cross-beds, deformed laminations and convolute laminations (Fig.39). Sandstone and laminated mudstone within the channels have also been observed. Channels with couplet beds are conspicuous in the exposures. Incised channel facies with thick beds of bouldery deposits, resulting from debris flow have also been reported. Development of channel facies might develop in different environments such as terrestrial fans, meandering, braided-fluvial, fluvio-deltaic, shallow-marine and deep-water depositional set-up. The Pulivendla clastics with all the above facies attributes attest to a high-energy braided-fluvial environment.

Fig. 39: Channel fills comprising ripples, troughs, planar cross-beds, deformed laminations and convolute laminations, Pulivendla Formation.

Petrological Attributes of the Pulivendla clastics

The major rock types of the formation are orthoconglomerates and reddish brown sandstones. The Pulivendla orthoconglomerates consists of well rounded elliptical to semicircular chert pebbles, volcaniclastics rock fragments and mud cracked shale pebbles, 3-9cm on an average with matrix (less than 15%) containing quartz and subordinate feldspars mainly. The reddish brown colour of the matrix suggests its ferruginous nature. The sandstones are reddish brown in colour, medium- to coarse-grained with moderate specific gravity and contain quartz and subordinate amount of feldspar. Some of the sandstones are massive, while some are laminated. Deformed laminations and convolute laminations are also commonly seen. Circular structures with concentric rims of successive quartzofeldspathic or cherty- and ferruginous materials are conspicuous of these sandstones, specially those with deformed laminae. Microscopic features reveal well sorted grains of quartz, different kinds of rock fragments with subordinate presence of potash feldspar and occasional phyllosilicates. The rocks show an overall interlocking texture. Framework, matrix and cement are recognizable. Well rounded quartz and rock fragments mainly constitute the framework. Some of the quartz grains show signatures of pressure solution, authigenic growth. Wavy (undulatory) extinction suggestive of deformation is very common among the quartz grains. Microrippled cross-laminations are present in some of the framework. Some feldspars are deformed. Rock fragments mainly include chert i.e. cryptocrystalline quartz. Microcrystalline quartz and polycrystalline quartz are also common among the rock fragments. Volcaniclastic rock fragments with signatures of later chertification also constitute the framework. In some cases, radiating oolites have also been identified in the framework proportion. The matrix of the rocks is composed of quartz, occasional feldspar and phyllosilicates, which include mainly chlorite and muscovite. Three kinds of cementing materials have been identified. Silica cement, ferruginous cement and brown chlorite which is indicative of plastic flowage when occupies the microdomains within the rock. Distinctive mineralogical maturity characterizes the Pulivendla clastics. It consists of mature elements of quartz although interlocked polycrystalline quartz shows a contradiction. Textural maturity is also visible in some microdomains. This is represented by segregates of well sorted detrital quartz and/or similar shale fragments. From mineralogical and textural point of view, the rock can be named orthoquartzite or quartz arenite and sometimes sublitharenite.

Discussion

The Pulivendla clastics have suffered intense quartz diagenesis and secondary silicification which has obliterated the sedimentological details. The exposures of the formation, in panoramic view and at close proximities represent an apparent sheet geometry resembling shallow-marine bedforms which led to workers in the past to suggest shallow-marine origin for these rocks. Selected exposures of the Pulivendla Formation reveal that the clastic beds are actually lenticular; dominance of couplet beds and concave-up scoured surfaces showing stacked and interfingering arrangement precludes the sheet geometry and shallow-marine origin of the Pulivendla clastics.

Presence of orthoconglomerate (oligomictic) indicates that it has been derived from areas dominated by sandstones, quartz veins and other granitic rocks. Channel fills of such conglomerates in association with mud cracked surfaces develop within spatially related mudstone facies, vertically grading into trough cross-stratified sandstones with channel facies and indicates deposition under a high-energy braided-fluvial condition. The terrigenous infills comprising very coarse- to coarse-grained trough cross-stratified sandstones, couplet beds and laterally cutting channels, indicate a high-energy braided fluvial environment (cf. Reading, 1996; Blair and Mcpherson, 1994; Miall,

1996; Coleman, 1969). Under conditions of bed shear stress, only the coarser clasts are deposited leading to clast supported texture. The finer clasts and sand infiltrate the clast framework at lower flow velocities.

Presence of couplet beds indicates an extremely high energy condition followed by comparatively low energy condition during flash floods, which was common during the Precambrian. When faulting takes place and raised topography is created in an extensional basin, very coarse grained fraction of a couplet bed is deposited and is typical of fluvial fan complex.

Trough cross-bedded sandstones are produced by downstream migration of transverse to sinuous crested dunes and ripples respectively. Ubiquitous presence of trough cross-beds and laterally extensive erosional surfaces point to a fluvial origin for the facies. The planar cross beds were generated by two dimensional bedforms and laterally accreted bars (cf. Allen, 1983).

The small symmetrical trains of ripples are lower flow regime structures within the fluvial channels and reflect a low energy condition. The vertical transition of the orthoconglomerates and couplet beds to trough- and planar cross-bedded sandstones records waning from upper to lower flow regime condition in the study area.

Laterally cutting channels with couplet beds, profuse trough cross-stratifications, planar cross-stratifications, ripples and randomly oriented disorganized clasts, an overall medium-to coarse grain size all point to a high-energy braided fluvial condition, often interfered by flash floods. Channel fills indicate high degree of turbulence, sliding, slumping and autobrecciation, sudden and quick influx of sediments. Shale pebble clasts were suddenly dumped into the channels. Effectively their encasement within loose sediments and sudden dumping did not allow the unstable shale clasts even to disaggregate along the mudcracks in them. This is supported by recent and ancient analogues (Reading, 1996; Coleman, 1969; Miall, 1996; Allen,1983). Terrestrial fan complex of ancient and recent times does possess such disorganized, chaotic arrangement of textural and mineralogical elements (Blair and Mcpherson, 1994; Blair, 1985, 1987; Well and Harvey, 1987; Reading, 1996).

The Tadpatri Formation

The Tadpatri is dominated by thick ignimbrites, various types of pyroclastics flows and surges interbedded with volcaniclastic-mixed mudstones, shales with thin intercalated sandstones. Mud dominated stromatolitic dolomites, ash fall/flow tuffs with mud, limestone-dolomite-shale interbedded units, quartzite-shale interbedded layers are the leading lithofacies attributes of the formation. Fluvial-, lagoonal- and shallow-marine depositional milieus along with a dominant terrestrial felsic volcaniclastic environments are documented in the Tadpatri. The Tadpatri records a major felsic volcanic episode in lower Cuddapah. Volumetrically as well as spatially, Tadpatri record different phases of sedimentation juxtaposed with felsic volcanics and volcaniclastics. The Tadpatri wedge with its flows and surges, various types of welded tuff and breccia stand for a huge volcanic domain.

Partially welded, chertified, planar-bedded tuffaceous facies (F1)

The layers constituting facies F1 are partially chertified and are of relatively high density. It is yellowish- to reddish in colour on fresh surfaces and occasionally graded. It is associated with thin felsic layers showing ferruginisation in lower and upper contacts (Fig.40). Flattened pumice fragments, lapilli, ropy pumice, shards, fibrous pumice fragments and glass shards within some parts of the unit define eutaxitic texture. Ignimbrite surges with drag folds and shrinkage cracks are distinct facies attributes in many of the ignimbrite beds (Figs.41 & 42). Ellipsoidal voids are preserved in the ignimbrite beds owing to partial erosion of subvertically emplaced volcanic bombs (Fig.43). Clusters of chertified volcanic bombs set in the ignimbrites indicate veracity of volcanic eruption during some ignimbrite surges (Fig.44). Spatio-vertically in the middle part of the Tadpatri stratigraphic section, near Kondapuram village on the way to Tadpatri, couple of calderas is delineated in association with the ignimbrites (Fig.45).

Convexo-concave lenses of very low-angle cross-stratified ash within planar-bedded facies F1 of tuff show occurrence of micro-ripple forms. The small ripple forms show cross beds ranging from less than angle of repose to steeper than angle of repose. Non-welded equivalents show frequent presence of thinly laminated tuff layers with accretionary lapilli, tear-drop lapilli, non-deformed tephra and pumices. Tuffaceous facies in the non-chertified portions show densities of rock comparable to that of cork, which is an important facies attribute of light ash.

Welded tuffs (F1) with the associated facies attributes are common in geological record (Cas & Wright, 1987; Reading, 1996; Chapin & Elston, 1979). Welding is documented in huge tuffs where lava induces

Fig. 40: Partially welded, chertified, planar-bedded tuffaceous facies, Tadpatri Formation.

Fig. 41: Ignimbrite surges with drag folds, Tadpatri Formation.

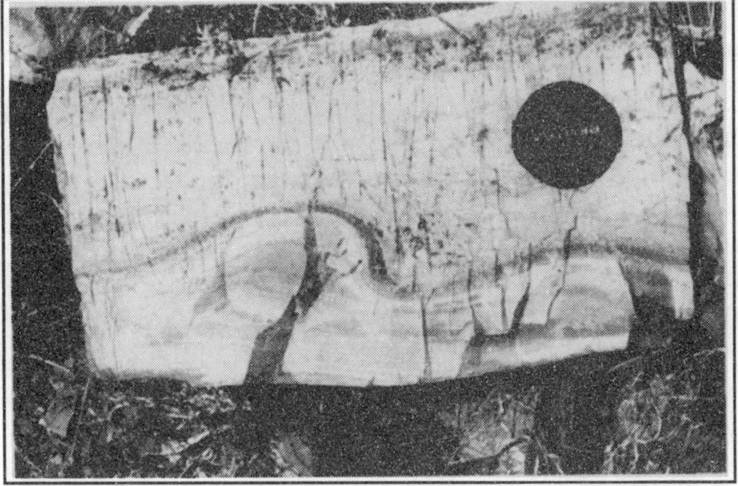

Fig. 42: Ignimbrite surges with drag folds and shrinkage cracks on top, Tadpatri Formation.

Fig. 43: Ellipsoidal voids, preserved in ignimbrite beds owing to partial erosion of subvertically emplaced volcanic bombs, Tadpatri Formation.

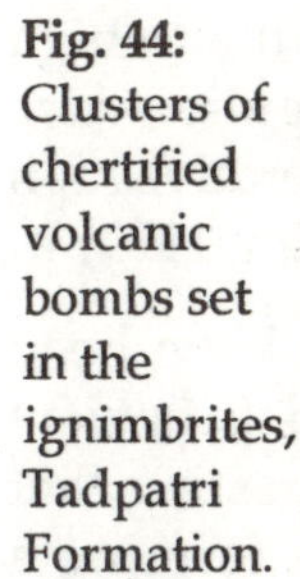

Fig. 44: Clusters of chertified volcanic bombs set in the ignimbrites, Tadpatri Formation.

Fig. 45: Calderas in association with Tadpatri ignimbrites, Tadpatri Formation.

welding of underlying air-fall layers (Schmincke, 1974). It is found from recent and ancient analogues that volcanic flows gradually thin out as it is formed further away from the source. Therefore, thinness of the felsic flows with contact baking effects are explained to be hot volcanic flow units emplaced much away from the vent, producing baking effects at its lower contact that are highly ferruginised. Presence of smectitic clay attests to the occurrence of volcanogenic clays (Fisher & Schmincke, 1984).

Shape of the tear-drop lapilli in the tuffaceous facies is indicative of free-fall through air. Such tear-drop lapilli from palaeo- and recent anlogues indicate subaerial emplacement (Fisher & Schmincke, 1984; Sparks & Wright, 1979; Self & Sparks, 1978). Differential cooling from bottom to top in subaerial pyroclastic surges results in formation of drag folds and shrinkage cracks of the beds. Preservations of elliptical voids in the ignimbrites are reminiscences of free fall of volcanic bombs though the atmosphere. Thin lenses of very low angle cross-stratified ash within the planar-bedded facies of tuff represent micro-ripple-forms which were formed due to migration of bed forms accentuated by ash fall. The ripple-form intervals cross-laminated reflect the cohesive nature of ash (cf. Cas & Wright, 1987). Low angle cross-stratification is due to high velocity driven grain layers being sheared over an irregular wave surface, which itself has a relatively low relief due to the operative high bed shear stress (cf. Cas & Wright, 1987; Wohletz & Sheridan, 1979).

Ignimbrites are often associated with calderas that have been recognized for quite a long time and were well documented. In ancient geological records, ignimbrite to caldera ratio is low, suggesting that many eruptions are monogenetic outpourings with each forming a caldera. Moderate to large Tadpatri calderas are exposed in middle part of the Tadpatri stratigraphic section and clearly lend support to the dimensions of magma chambers at depth.

Partially Non-welded layered chertified facies (F2)

Chertified lithic units are dense with high specific gravity. Again, associated non-chertified ash-portion depicts density of the lithounits as that of cork. Colours of the layers are brown, yellow, white or even red (Fig.46). The layered tuffaceous facies is composed of well-sorted, very fine- to medium (occasionally coarse) grained vitric- and lithic ash with chertified tear-drop-, conjugate- and accretionary lapilli. Angular pumice, laminae interspersed with lithic fragments occur in discrete layers or with gradational contacts. Bedding is defined by changes in grain size and lithic variations. A well-sorted matrix of ash-size fragments, pumice, shards and microscopically identifiable globules generally form a minor part. Overall layers with much finer ash are discernible. Such facies show lateral continuity in hundreds of metres. Thickness of the deposit is 0.6 to 2.5m. Individual layers have usually non-erosional basal surfaces, internally massive or have diffused thin parallel laminations and are normally or reversely graded. Lamination thickness varies from 0.1-0.8 cm.

Such layered tuffaceous facies of F2 stands for the product of explosive volcanism with high viscosity, spread over limited areas (cf. Smith, 1970; Smith & Roobol, 1982). Normal and reverse grading, occasional presence of tear-drop lapilli, vitric ash and absence of cross stratification are typical characteristic of the pyroclastic fall deposits (Cas & Wright, 1987; Fisher & Schmincke, 1984). Owing to interaction of water vapour and siliceous magma, such pumice and shard-bearing as well as fine-grained tuffaceous layers usually originate in response to phreatic eruptions (Self & sparks, 1978). Attributes of laterally

Fig. 46: Layered partially chertified pyroclastic flow deposits with chertified lapilli-bearing ash pockets, Tadpatri Formation.

continuous layered tuffaceous facies of the Tadpatri as event stratigraphic units were utilised as 'marker' levels to correlate the lower part of the facies in scattered outcrops over an area of 250 sq. km.

Pumiceous sandstone (F3)

The facies is well stratified (0.4 to 0.75m thick) showing sheet geometry and fining-upward trends. It includes well sorted very coarse to fine intercalated sand beds with pebble to cobble sized pumice and muddy ash clasts. Base of this facies is an irregular surface with sole marks upon which pebble- to cobble-sized ash clasts and pumice fragments are well concentrated. The pumice and ash clast-bearing beds are overlain by trough cross-stratified pumice and ash beds. The alignment of the pumice clasts is concordant with trough cross-stratification. In turn, these cross-stratified beds are overlain by mud-cracked fine ash beds with ripple cross-lamination or horizontal lamination.

This is a fluvial deposit attested by the attributes such as channelised disposition, well-sorted fining-upward sequences and sheet geometry due to accretion caused by shifting flow (Allen, 1965; Walker & Cant, 1984). Presence of mud-cracked surfaces lends support to occasional intermittent subaerial exposures. Pebble to cobble-sized pumice and ash clasts on erosional bases are suggestive of lag deposits. Fining-upward sequences are indicative of lateral migration of channels or deceleration of flow resulting from declined channel activity. Resedimentation of volcaniclastic elements comprising pebble- to cobble-sized pumices, lapilli and lithic chertified ash clasts was caused by fluvial action subsequent to or concomitant with subaerial volcanism (e.g.Kataoka & Nakazo, 2002).

Channelised resedimented volcaniclastics (F4)

Massive unstratified mixture of pebbly to cobbly pumice clasts, development of stratified sequences comprising subordiate amount of volcaniclastics, epiclastic and terrigenous clastics characterise the facies. Troughed channels with such infills also characterize the facies. There are discrete zones which show massive destruction of internal details, deformation in vertically changing patterns, microfaults, segmented zones passing upward into disrupted layers, thin micro-brecciated zones and diversely oriented fragments of thinly laminated sandstone-mudstone alternations, torn-out pieces of laminae, discrete zones with disturbed internal structures and at times occasional mosaics of hieroglyph-like deformed laminae. In some places convolute layers with complex geometry are also discernible.

The facies is developed peripherally to the subaerially emplaced felsic volcanics and volcaniclastics and its facies attributes indicate that the reworking of pyroclastic fall deposits occurred in proximity of the subaerially emplaced volcanic facies.

Dominance of channel facies with infills of tuffaceous material and its altered equivalents comprising smectitic clays and other epiclastics, intermixed with subordinate terrigeneous clastics indicate the role of fluvial processes as the main control of resedimentation (Kataoka & Nakajo, 2002; Cas & Wright, 1987; Reading, 1996; Fisher & Schmincke, 1984; Walker, 1984).

Attributes defined by discrete zones with disturbed internal structures at stratigraphic levels of the facies were matched with similar case histories and is concluded as seismic signatures (cf. Lowe, 1982; Plaziat & Purser, 1998). Seismically induced fluidized structures in discrete zones are produced owing to confined fluid pressure in sediment due to seismicity (Lowe, 1982; Bose et al., 1997). The pressure might exceed that of overburden and cause selective liquefaction (Price, 1973; Middleton, 1967; Lowe, 1982; Bose et al., 1997; Plaziat & Purser, 1998). Sedimentary layers with variable rheology under this situation suffered differential stretching and brittle failure with generation of signatures of extensional processes leading to formation of seismites.

The Tadpatri resedimented facies at discrete zones record such extensional features and has been identified as seismically triggered structures. Tadpatri paleoseismites in resedimented facies bear the evidences of frequent seismicity in the basin (cf. Price, 1973; Middleton, 1967; Bose et al., 1997; Plaziat & Purser, 1998).

Stratified clast- and granule bearing coarse- to medium sandstones (F5)

The reddish brown- to brown and yellow coloured sandy facies contains significantly high volume of volcanogenic sand and clasts. Laterally continuous beds are traceable for tens of meters. Sheet-like and lenticular bed geometries are discernible in the stratal assemblage. Sandbodies are simple to multistorey, 0.9 to 1.8m thick, lenticular ribbons (average width/depth ratio 9:1) which comprise primarily parallel stratifications and subordinate trough cross beds. Sharp, flat, gently undulatory or highly irregular lower bedding contacts are identifiable. Parallel stratifications are common.

Profuse disorganized volcanogenic and terrigenous clasts and granules form a significant facies attribute, which is particularly conspicuous over the top of volcaniclastic layers. Shale pebbles, rip-up clasts, broken volcanic pumice, lapilli-bearing acidic, volcanic rock fragments, very coarse- to coarse and medium volcanogenic terrigenous sand, clasts and granules constitute the lithology of the facies. Bed to bed distribution of sand, clasts and granules are variable. Size range of the clasts varies from 1.5cm to 12cm. Maximum size of the clasts derived from different beds (measured from average of 20 large clasts) is 8cm.

The clasts are ill-sorted, angular- to subangular and subrounded. Clast imbrication is indistinct. Low-angle stratifications and crude planar stratifications are common. Spatio-vertical relationships in sedimentary structures within the facies also include gradation of clast- and granule-rich sandstones into planar and trough cross-bedded sandstones. At places, lenticular pebbly sandstones containing both angular- to subrounded volcanic and sandstone clasts (2-5cm in long dimension) grade into epiclastic- and fairly well-rounded sandstone clast-bearing lenses and sheets.

Erosively based coarse sandstones with clasts and granules, lenticular and multistorey bed geometry and presence of parallel stratifications indicate important role of tractive currents for the deposition of the facies (e.g. Nemec & Steel, 1984; Bridge et al., 1986; Reading, 1996; Miall, 1996). Parallel laminations might have formed in fast and shallow flows across the lateral or transverse bar tops or in deeper- and faster flows where the dunes are washed out in favour of upper flow regime flat beds (cf. Karig, 1972). The facies resembles 'St' and 'Sp' faces of Miall (1996). Sheet-like clasts and granule-bearing sandy beds were probably deposited in wide shallow channels (e.g. Allen, 1983; Carling, 1999; Southard et al., 1984).

The sheet sandbodies might have probably been deposited by a system of aggrading and coalescing large-scale braided rivers which is analogous to Triassic Hawkesbury sandstone of Sydney basin (Conaghan & Jones, 1975; Jones & Rust, 1983) and Jurassic Westwater Canyon Member of the Morrison Formation, New Mexico (Campbell, 1976).

Low-angle stratifications and crude planar stratifications with coarse sand, clasts and granules and ill-sorted disorganized clasts point to its origin under an overall upper flow regime condition (Harms & Fahenstock, 1965; Stear, 1985). Variation in the texture and structures of the facies can be ascribed to a fluctuating flow condition and transport rates. Spatio-vertical association of the facies with trough cross beds point to changing water depth and velocity over the bar crests during the growth of active bars (e.g. Bridge & Tye, 2000; Crowley, 1983; Massari, 1983; Miall, 1996).

Trough and planar cross-bedded yellowish brown coarse sandstone (F6)

The facies is greyish white when fresh and yellowish brown wherever weathered with grain size varying from pebbles and granules to medium sand. Small and large trough- and planar cross-bedded sets with varying thickness are common. Locally, vertical transition between parallel laminated unit and trough cross- stratified unit is discernible. Some of the troughs show infills of ferricretes and volcaniclastic clasts. Planar cross-beds are closely associated with the trough cross-beds. Alignment of small granules, coarse clasts, volcaniclastics and ferricretes are found along the planar stratifications. Laterally extensive erosional surfaces separate individual cosets of the trough- and planar beds.

Troughs are interpreted as transverse features probably related to secondary flows generated by large-scale bedforms (Jones & Rust, 1983). Migration of transverse to sinuous crested out-of-crest subaqueous ripples and dunes deposit trough cross bedding (Rubin, 1987; Harms, 1975; Best et al, 2003). Planar cross-beds are products of migrating two-dimensional bedforms. Downflow migration of sinuous crested dunes are generated generally in the deeper parts of fluvial channels (Harms, 1975; Todd & Went, 1991; Ashworth et al., 1999; Leclair, 2002). Frequently deposited troughed and planar cross-bedded sands associated with three-dimensional dunes and two-dimensional bedforms respectively are common depositional styles in modern braided streams (Frazier & Osanik, 1961; Harms & Fahenstock, 1965; McGowen & Garner, 1970; Schwartz, 1978; Shelton & Noble, 1974; Cant & Walker, 1978; Bridge, 1993; Reading, 1996). Transitional growths of laminated facies from medium- to small troughs bear the imprints of waning energy conditions from upper- to lower flow regime within a single depositional phase (McKee et al., 1967; Nemec & Steel, 1988; Best, 1996; Miall, 1996; Lunt et al., 2004; Pierson & Scott, 1985; Smith, 1986; Turner & Monro, 1987). Spatio-vertically unimodal palaeocurrent vectors from troughs, preservation of erosional surfaces indicative of channels in profusion related to the final deposition of sequences and locally pebbly character of the lithology collectively can be ascribed to a high-energy braided fluvial. Fericrete infills of the troughs are probably reminiscences of paleosols. Fluvial successions at different stratigraphic levels of the Tadpatri attest to the importance of aggradational growth of the sediments in response to periodic exposures to subaerial conditions (cf. Jhonson, 1977).

Owing to repeated transgression and regression there were deposits of volcaniclastic bearing lagoonal and shallow-marine deposits interbedded with terrestrial fluvial and felsic volcaniclastics and ignimbrites. The lagoonal and shallow-marine deposits housing lapilli, lapilli tuffs and horizontally to subhorizontally arranged volcanic bombs suggest that there happened to be concomitant volcanism with shallow-water deposition.

Laminated Black Shale facies (F7)

The facies is characterised by millimeter-thin laminated grey to black coloured shales. Rippled units and micro-ripples are also prominent in the facies. Pyritic layers of ripple-drift laminations are main attributes of the facies. Pumice, lapilli, volcanic bombs and ash layers are commonly present within the beds. The facies is deposited in relatively small depressions within bed rocks of the Tadptri.

The facies attributes signify a low-energy lagoonal set-up where euxinic condition was prevalent and this led to the deposition of black shales and pyrites (cf. Reading, 1996; Middleton, 2003). The entire drainage basin registered smooth integration of catastrophic events. Thus major volcanic episodes, storm- or localised soil erosion events are represented in the lagoonal record. Presence of bedding parallel volcanic bombs, lapilli layers, glass shards and ash beds in association with thin black shales are reminiscents of major volcanic episode in the Tadpatri that was concomitant with lagoonal deposition.

Monotonous laminated mudstone-sandstone heterolith (F.8)

The facies is characterized in its lower part by the presence of monotonous laminated mudstone with existence of sporadic silt laminae, lenses and few sharp-based fine sandstone layers (ca. 10 cm thick). Compositionally the facies shows variation in clayey silt/ silty clay within parallel and mottled lenticular layers.

Rhythmic alternations, mudstone-sandstone layers, presence of basal lags, minor erosional contacts, structureless mud to very fine sandstone with thin mud interlaminae are additional facies attributes. In the upper part, laminated mudstone grades upward into interbedded heterolithic very fine sandstone (ca. 15 cm thick) and mudstone (ca. 2-4 cm thick). The succession is overlain by fining-upward cycles and truncated ultimately by an erosional surface. The individual fining-upward cycles (less than 25 cm thick) is in a continuum of flaser- and wavy lenticular layers (cf. Reineck & Wunderlich, 1968). The facies finally passes into rhythmite (mud and fine sand) dominated erosively based sandstones which have small lenticular concave-up fine sandy-muddy infills.

The spatio-vertically developed layers of the facies are suggestive of a tidal to subtidal impress and lateral migration of tidal channels during the closing-regressive phase. Presence of flaser-, wavy-, lenticular associations with heterolithic bedding (alternations of mudstone-sandstone layers), dominance of fining upward cycles, erosional bases and basal lags provide evidence for tidal channel deposition (Reineck and Singh, 1980).

The heterolithic bedding comprising thin mudstone and sandstone layers suggests that the succession has an imprint of tidal ridge progradation on a regressive shelf (e.g. Meckel, 1975; Dalrymple, 1992; Soria et al., 2003) as evidenced by occasional appearance of silt layers, lenses and sharp-based fine thin sandstone beds within the fining-upward cycles (e.g. Weimer et al, 1982). The presence of symmetric cycles is indicative of tidal action and these deposits may be considered as tidal rhythmites, with thicker sets of strata representing deposition during spring and thinner sets during neap tides (Kvale & Barnhill, 1994; Mazumder & Arima, 2005).

Interbedded wavy lenticular sandstone-mudstone (F.9)

The sandy-muddy layers (average sand/shale ratio 2:5) that characterize the depositional units of the facies are lithic bodies with a muddier upward trends, conglomeratic bodies with mud clasts and volcaniclastic fragments, cross-laminated very fine sandstone bodies, sandy- and muddy lenticles. Other salient facies attributes include monotonous, inter-stratified wavy lenticular sandstone-mudstone successions, heterolithic bedding, identifiable ripple cross-lamination with some bipolarity in very fine sandstones with thin mud drapes on sandier lithosomes, few sand lenticles in muddier layers (fining-upward) and sigmoidal foresets. A sigmoid set generally thickens in the down current direction and then thins towards the toe and climbs up the stoss side of the set in the down current direction. Low- to high angle planar cross-laminations and reactivation surfaces are not uncommon. Occasionally trough cross-bedded and rippled layers are also discernible.

layered-cake sequence. Internally the facies is thinly laminated in the range of 2-6mm, cumulatively forming an assemblage of layers, 30-140 cm thick. The bedding planes are sculptured by well-developed parting lineations trending 110º-290°.

Fig. 47: Laterally cutting braided-fluvial channels with infills of planar- and trough cross-stratifications and ripples in and around Gandikota and Nossam regions, Gandikota Formation.

Parallel or laminated cosets are either horizontal or dip at low angles of 3°-5°. Sometimes the lamina-sets get disturbed due to presence of dropstones. Ripple lamination and climbing ripple laminations are frequently preserved in sand and silt beds. Small fine-grained sandy ripples float within the muddy sheets. Occasionally bedding planar surfaces of parallel laminated facies are marked by current crescents. The flow vector as indicated by these current crescents is 310°. Locally parallel laminated facies shows small and large channel-like scours, filled with either laminae concordant with the lower bounding surface or cross-bedded. Flutes and load structures on the soles are locally found.

Lamination of mud, silt and very fine sand is common in overbank areas and represents deposition from suspension (Miall, 1996). Maximum thickness of the sets have been recorded about 140 cm, however, cumulative thickness of the facies can be much more as reported in other cases depending on sediment supply, fluvial style and basin subsidence rates (Miall, 1996).

Thin sheets of the muddy-sandy facies attest to its deposition from unidirectional waning currents, probably overbank sheet flows. Thin muddy drapes along sandy bedforms reflect suspension deposits formed during the low flow stage (cf. Nichols, 1987; Reading, 1996; Miall, 1996).

Channelised sand bodies inset in underlying mudstones, silty mudstones with centimeter thick beds of well-cemented siltstone and fine- to very fine sandstones are interpreted as fluvial sandstones. Scoured sandstone beds with lower nonerosional contacts have been interpreted as the product of overbank sedimentation during overbank flow from the river channels (cf. Elliot, 1981).

The close parallelism of the well developed parting lineation and flow vector from current crescents indicate that the mean flow direction was towards 310° in relation to deposition of overbank fines that were on the flanks of westerly flowing fluvial channels.

Sheet-form sandstone and siltstone (F2)

The facies is characterized by red-coloured sandstones and siltstones in variable proportions in different regions. Sandstones and siltstones are commonly interbedded with occasionally lendsoidal beds of matrix-supported intraformational clasts. These are poor to well sorted and frequently contain pebble-grade rip-up clasts. Beds are often amalgamated (upto 2.5 m thick) with centrimetric to decimetric scale horizontal stratifications. Individual beds range in thickness from 0.1 to 0.45 m.

The sheet-like geometry of the facies is suggestive of deposition from laterally unconfined flows. Variable sorting and presence of floating clasts are further suggestive of rapid deposition from sediment-rich fluid gravity flows or hyperconcentrated stream flows (Koster, 1978). The facies resembles D2 facies of Wells & Harvey (1987) and fluidal sediments of Nemec & Steel (1984). As the fine grain size and thinness of the bedding gradually becomes more pronounced at higher stratigraphic levels, it is probable that basinward shift of the fluvial sedimentation was gradually attended which is attested by such facies attributes.

Coarse- to medium grained planar cross-bedded reddish-yellow sandstone (F3)

Dispersed pebbles, subordinate mud clasts, volcaniclastic clasts, ferricrete clasts and quartz granules in association with coarse- to medium grained reddish yellow sands define the lithology of the facies. Planar cross beds, either solitary or in groups, with sharp, flat or slightly scoured bases and tops, laterally persistent over tens of meters, are prominent facies attributes. Individual sets are laterally persistent along downcurrent over few meters.

Downstream accreted macroforms are bounded on the top by convex-up bounding surfaces and bottom by flat surfaces. The well-bounded internal architecture consists of sets of well-bounded cross-laminae. The stoss side of the macroform is clearly erosional while the lee side is accretionary. Maximum height is nearly 152 cm, while the length is nearly 8m.

The erosional surfaces are carpeted by intraformational red mud clasts. Individual as well as set thickness of such structures in centrimetric- to meter scale declines along downcurrent directions. The exposed vertical section, oriented parallel to the regional palaeocurrent direction, indicates presence of compound large tabular cross-beds (0.4 m to 3 m) and smaller ones (0.3 to 0.6 m). Smaller units are discernible immediately above a large scale tabular cross strata. Compound cross beds and reactivation surfaces are common. Thinning- and fining-upward trends within the cosets of compound tabular cross-beds are observed. Thin-bedded sandstone-mudstone layers, at times, overlie the planar tabular and cross-bedded facies. The interbedded facies, at different levels also interfinger with horizontally laminated and trough-cross-stratified facies.

Laterally accreting bars and migration of two-dimensional bedforms promoted such growth of planar tabular cross beds (cf. Boothroyd & Ashley, 1975; Harms et al, 1982; Todd & Went, 1991; Allen, 1983; Anderson et al., 1999; Ashworth et al., 1999; Best, 1996; Collinson, 1996; Bridge & Tye, 2000; Bridge, 2003). Fairly deep channel configuration is attested by the height of the cross beds. Association of sheet flood deposits and cross-bedded sandstone facies in the present case history points to their formation as point bars in low-sinuosity streams under oscillatory energy condition (cf. Fedo & Cooper, 1990). Such association is also attainable in high-sinuosity stream systems under specific physiographic and depositional controls. Scouring by superimposed smaller bed forms or response to fluctuating flow stage can be attributed to reactivation surfaces (Collinson, 1970; McCabe & Jones, 1977). Sedimentation of overbank fines is precluded owing to the presence of thin-bedded sandstone-mudstone facies.

Preservation of ferricrete-, volcaniclastic- and mud clasts in the facies is suggestive of an overall short transport, reworking and quick burial leading to preservation of such clasts. The fact that mud and ferricrete clasts and scarcity of fines in the sandy fluvial deposits suggest their temporally overbank deposition or in protected channel milieu and subsequent reworking (cf. Long, 1978). At least formation of partly slip faced and outbuilding of bank-attached compound bars are indicated by the presence of large-scale tabular cross strata of the facies.

During the migration of smaller bedforms over the lee of the slower moving or stationary larger bed forms is further attested by compound cross-beds indicative of low flow condition during the sedimentation (Blodgett & Stanley, 1980). In addition, lateral persistence of the tabular cross strata over tens of meters can be ascribed to the stability of these bed forms over a considerable span of time. Subtle differences, however, in their depositional processes is reflected from variability of their geometry and orientation. Small repetitive sand waves, in contrast, appear to be responsible for the deposition of small-scale tabular cross-strata in the region. Formation of such small-scale structures was attended at the close of lower velocities than those responsible for the origin of the dunes. It is suggested that these small structures were formed during the extremely waning stage of the velocities that deposited dunes or very soon after the cessation of such velocity profiles. Such hydrodynamic conditions are reflected from deposition of small-scale tabular cross-strata on bar tops that are commonly deposited with shoaled waters (Todd & went, 1991; Allen, 1983).

Channels filled by simple vertical aggradation commonly show fining upward successions reflecting one of the two processes: progressive abandonment as a result of upstream avulsion or the plugging action of few dynamic events (e.g. flash floods). The thickness of such channels can not exceed the depth of the channel and is likely to be much less where the dynamic event strips away earlier deposits before depositing their sediment load. Such cycles in channel fills may represent flood events (Goodin, 1991; Miall, 1996).

In channels of great lateral continuity, flood events would leave their imprints in prominent fining upward trends and in case of successive floods there is a probability of superimposition of fining upward trends in multilateral and multistorey sand bodies (Selley, 1985). Such sporadic impress of fining upward trends in the tectonically sourced Tadpatri fluvial sandstones therefore stand for records of flood generation (cf. Cant & Walker, 1978; Ori, 1982; Arche, 1983; Crowley, 1983; Shepherd, 1987; Todd & Went, 1991; Bridge, 2003).

Interstratified black mud- and glauconitic sandstone (F.4)

The facies is characterized by dark grey to black mud and siltstones interbedded with glauconitic sandstone tongues. The mud and sand are loosely packed and separable into laminae. Subtle fining upward cycles (0.45 to 0.80 m thick) is one of the salient attributes of this facies. Occasionally coarsening-upward cycled sandy lithosomes abruptly culminate in current-rippled very fine sandstone lithosomes with small trough-cross-bedding.

In addition, plano-concave lenticular greenish-white and white- to yellow glauconite bearing well to poorly sorted sand form sandier lithosomes that are laterally continuous for fifty centimeter to hundreds of centimeters with variable relief. Laterally the plano-concave lenticular sandy lithosomes in association with heterolithic bedding show highly disturbed internal fabric. Associated lithosomes are black shaley layers. Prominent sedimentary structures discernible are trough-, low angle planar cross-stratifications and current ripple laminations. Concave-up erosively based sandy lensoidal bodies with small troughs, ripples, ripple-cross-laminations are discernible. The basal surfaces, sometimes, are marked by clay chips and clasts. Set boundaries are often marked by clay-chips and double mud drapes. Decreasing upward thickness is noticeable in sets. Herringbone cross stratification is common. The plano-concave sand lenses often record lateral accretion.

Herringbone structures indicating bidirectional currents; double mud drapes, rip-up clasts lend support to tidal origin for some of the channels (plano-concave sand-filled lenses). Double mud drapes along sets and foresets are characteristic features of tidal environment (Visser, 1980; Boersma & Terwindt, 1981; Smith, 1988). Abundant presence of large scale cross beds with alternating opposite current directions can be ascribed to the action of strong bidirectional currents as identifiable in tidal channel system (Reineck & Singh, 1980). The channels can be assumed to represent tidal creeks cutting across the marshy environment behind the barrier bar complex with sandy-muddy heteroliths. Glauconite in some of the sandy layers is a strong indicator of shallow-marine sedimentation (Pettijohn, 1975; Reading, 1996).

Hummocky and glauconitic trough-cross-stratified sandstone (F.5)

This facies consists of green, yellow, very fine- to fine grained hummocky- (HCS) and trough-cross-stratified glauconitic sandstone depicting presence of thickening upward lithosomes upto 3.5m thick (Fig.48). The facies is either thinly developed or absent in the pinch-out area of the sandstone tongues. Upwards, shale-encased single hummocky beds to amalgamated beds with occasional muddy stringers are discernible. Lower part of the stratigraphic section in lithosomal disposition of this facies shows prominent identifiable 4-12cm thick intervals of silty mudstones with disturbed bedding and inter-stratified hummocky cross-stratified sandstone lithosomes. Increase in the thickness and abundance of sandstone beds prominently define an upward-coarsening trend. Erosively based individual hummocky sets are prominent with sole marks and very low-angle (3°-15°), curved to undulating laminae that are often trough-shaped and/or forming broadly arching 'hummocks'. Intersecting broad troughs defining low-angle curved intersection of laminae are identifiable. Formation of ripple cross-laminations (2-5cm thick) with sharp basal contact over the top of this HCS lithosomes is a distinctive character.

Green and slightly altered minerals are identifiable as fresh and altered glauconitic grains. Glauconites originated from potassic rich volcanic fragments and pelleted glauconites from faecal pellets with micro-reducing environments. Glauconite formation was caused by adsorption of K^+ ions from the sea water by currents with variable directions at the sediment-water interface (McRae, 1972) which is quite common in agitated high-energy sea water. Glauconites so formed were subsequently redistributed in the infills of different structures and beds under current action.

Fig. 48: Hummocky and glauconitic trough-cross-stratified sandstones, Gandikota Formation.

The HCS lithosomes are interpreted as representing deposition above storm wave base in the lower shoreface environment within a storm-dominated inner shelf setting (Cheel & Leckie, 1993; Dott & Bourgeois, 1982; Leckie & walker, 1982; Walker, 1984). Variation in internal structures of the hummocks was caused by variable energy conditions and current direction. Moreover, rate of aggradation of the hummocky lithosomes also prompted specific form of internal stratification (Cheel & Leckie, 1993). Development of amalgamated hummocks and cross-stratified lithosomes in the facies along with coarsening upward trends is diagnostic of a shoaling upward regressive shoreface succession (Leckie & Walker, 1982; Pemberton et al., 1992; Soria et al., 2003).

The Nallamalai Group

The Nallamalai Group having faulted contact with the underlying Chitravati Group, comprise strata which are low grade metasediments or highly lithified volcaniclastic units. The tectonic setting of Nallamalai is complex and the intensity and stages of deformation increase gradually from northeast to southwest. The Thokapalle complex shows two-sets of folds : antiform and synform. The axial traces of the antiform and synform trend 035°-215° and truncate abruptly against the fault trending northeast-southwest. With the increase of intensity of deformation towards south-west in Vontimitta, the Nalamalai strata show even folds belonging to polyphase deformation. Fold styles vary from neutral (reclined) through series of antiforms and synforms, with axial traces more or less parallel trending NNW-SSE, to right superposed folds. Folding of Nallamalai Group of rocks, to some extent, took place before the onset of the succeeding basin namely Srisailam. Additional folding in the eastern part was later followed by major thrusting at the eastern margin and also some Nallamalai westwards over the Kurnool Formation. The folding of Nallamalai appears to have been accompanied by doming upwards involving or due to local fluidization of basement or diapiric granitic intrusion.

The Nallamalai contains the Bairenkonda Formation in its lower part and the Cumbum Formation in the upper part. Facies analysis of both the formations has been focussed.

The Bairenkonda Formation

The formation shows tectonic contact with the underlying Gandikota Formation. The formation is characterized by reddish-brown sandstones, conglomerates and occasional intervening mudstones with laterally cutting channelised disposition (Fig.49).

Trough cross-bedded yellowish brown coarse sandstone (F1)

The facies is yellowish brown coarse sandstones dominated by small and large trough cross-beds with profuse infills of ferricretes and having varying thickness. Grain size of the sandstone varies from pebbles and granules to medium sand. At places, vertical transition between ferricrete granule bearing parallel laminated and trough cross-stratified units are discernible.

Surface water wash outs under high-energy conditions are considered as casual phenomenon that was responsible for derivation of paleosols. With the decrease of erosion of soil cover, crude deficiency in supply of paleosol materials took place in the upper part of the sequence. Troughs are transverse features probably related to secondary flows generated by large-scale bedforms (Jones & Rust, 1983). Trough cross-beddings are generated by migration of three dimensional subaqueous ripples and dunes (Rubin, 1987; Harms, 1975; Best et al,

Fig. 49 : Reddish-brown sandstones, conglomerates and intervening mudstones with laterally cutting channelised disposition, Bairenkonda Formation.

2003). The parallel laminations with alignment of ferricretes and coarse clasts and granules are suggestive of higher flow regime structures (Harms, 1975; Todd & Went, 1991; Ashworth et al., 1999; Leclair, 2002; McKee et al., 1967; Nemec & Steel, 1988; Best, 1996; Miall, 1996; Lunt et al., 2004). Braided fluvial tracts display repeated channel cutting, initiation, filling and abandonment with development of different types of bedforms.

Clast-supported stratified gravel (F2)

Alternately arranged clast-supported, horizontally-stratified gravel and planar cross-bedded gravels have a clast framework and abundant sand matrix. Individual layers are 5-20cm thick with multistory units ranging in thickness reaching 10-16m. In some sections, haphazard arrangements of intermixed solitary troughs float with infills of pebbles and granules. Bed contacts are frequently obscure because of absence of well-defined bedding. Clast imbrications frequently float within the lithology.

It is evident that the hydraulics of flow was randomly varying and apparently there was switching back and forth between lamellar and shooting flows, what is quite expected in the Precambrian scenario. Textural variations within these layers are significant due to variations in sorting arising out of changing hydraulic conditions and gravel-clast overpassing. The troughs appear to be reminiscents of migrating three-dimensional bedforms (cf.Miall, 1996).

Cross-stratified granule conglomerate and very coarse grained pebbly sandstone (F3)

The facies comprises poorly-sorted, very coarse- to coarse grained, subarkosic pebbly sandstone and minor pebble and granule conglomerate. The sandstones are poorly-sorted, and consist primarily of angular to subrounded grains of quartz, feldspar and volcaniclastic rock fragments.

The sandstones occur either as isolated lenticular bodies cutting into the underlying beds, or as amalgamated lenticular bodies compounded along erosive contacts and forming a sheet-like unit upto 2m thick. The sandbodies are often bounded by planar to gently undulatory erosional surfaces. Small lenses or irregular patches of pebbles or granules often mark the erosional bedding surfaces. The sheets are often multistoried, and can be traced for more than 100m along the strike in suitable exposures. Individual lenses pinch out both across and along the strike within a few meters. Their maximum thickness is upto 2m, but most of the units are 0.5-1m thick. The frequency of pebbles generally increases upward in the sequence.

Majority of the sandstone lenses consists of trough and planar cross-sets which occur as cosets. The trough sets are 10-30cm thick, and occasionally displays a fining-upward trend within the lenses. The foresets are generally poorly defined, and is manifested by minor differences in grain size. In a few sets they are normally graded. Pebbles are concentrated along many foresets though more commonly, the pebbles occur along the set boundaries as thin pebble sheets.

Horizontal- to gently inclined beds in some sandstones range in thickness from 5-15cm. Many of the inclined beds are composite amalgamated small lenticular bodies. The facies units are multistorey packages developed through vertical superposition of tabular beds, and exhibit fining-upward or coarsening- to fining-upward trend. Pebbles are often concentrated along the interfaces of the beds, the bed sets or the lenses. The pebbles are either in contact with each other or are matrix-supported.

Broadly lenticular to sheet-like sandbodies with unidirectional cross-stratification points to their emplacement as channel-fill stream flow deposits (Miall, 1977; Ramos & Sopena, 1983; McCabe, 1977). Coarse grain size as well as spatio-vertical amalgamation of the lenticular bodies into sheets, often multistoried, indicate that the sediments were deposited in low sinuosity streams (Olsen, 1987; Kelly & Olsen, 1993; Shanley & McCabe, 1993). The lateral persistence of bedding surfaces, tabular nature of the beds with cosets of trough cross strata and intercalated sets of planar cross strata coupled with poorly developed fining-upward trend resemble shallow braided fluvial sheet-sandstones (Campbell, 1976; Cant & Walker, 1978; Smith & Smith, 1984). Dimension of the individual lenses suggests that the channels had high width/depth ratio. Preservation of lenticular sand body geometry further indicates that the channels were prone to episodic and avulsive shifting of their courses (cf. Dreyer, 1993) within the streams. Relative paucity of planar cross beds also points to constant shifting of the channels with consequent destruction of the channel bars (e.g. Campbell, 1976). Abundance of small planar erosional surfaces, small scours creating small scale irregularities as well as occurrence of small lenticular bodies of gravel and sand on bedding surface appear to be the manifestation of frequent shifting of shallow channels on the sand flats or emergent bars (cf. Allen, 1965; Bull, 1977).

The trough and planar cross beds may represent the migrating subaqueous dunes along the inter-bar channels (McGowen & Groat, 1971; Boothroyd, 1972). The conglomerate beds at the bases and also on the upper surfaces of the lenses record winnowing by strong currents, either at erosion surfaces related to migration of bars and dunes (cf. Bluck, 1971; Levey, 1978; Reading & Collinson, 1996) or in some cases on top of bars. Contrasting grainsize defining the foresets in trough cross-strata may as well suggest changing water stage over flood cycles, whereas the graded foresets appear to reflect the stage fluctuations over the crest of the bars (cf. Rust, 1984; Rust & Jones, 1987; Bluck, 1981; Steel & Thompson, 1983; Smith, 1990). Abundance of small laterally impersistent erosional surfaces record local striking of bars and subchannels (e.g. Smith, 1990). Shifting and filling of small channels is also indicated by small fining upward units (cf. Olsen, 1987).

Facies analysis of the Bairenkonda siliciclastics reveals a braided-fluvial set which overlie the fluvial- to shallow-marine deposits of the Gandikota Formation.

The Cumbum Formation

Each of the groups of the Cuddapah Supergroup contain signatures of significant volcanicity, out of which the Tadpatri and the Cumbum Formation are significant in recording emplacement of volcanics, volcaniclastics & epiclastics in varieties of depositional milieus. As regards facies and their associations in the Cumbum, little has been done todate. Keeping in view this inadequacy, sedimentology of Cumbum volcaniclastics has been addressed.

The 2000 meter thick Cumbum Formation records signatures of subaerial-, lacustrine and shallow-marine felsic- to acidic volcanics and volcaniclastics and lend support to their origin in an overall control of extensional regime.

The facies assemblage of Cumbum Formation of the Nallamalai subbasin includes conglomerate-sand facies (now conglomerate-quartzite association), sand-mud alternation facies (now quartzite-phyllite/schist alternation), mud facies (phyllite/schist) and carbonate-mud facies (metamorphosed into crystalline limestone/dolomite-phyllite/schist). Mafic and/or felsic to silicic volcanics are often associated with the facies assemblage (Figs. 50, 51 & 52). The sequences were involved in greenschist facies metamorphism and moderate to strong deformation. Mudstone and sandstone, unless otherwise mentioned, are written here instead of metapelites and metapssamites respectively.

Detailed sedimentological investigations of the Cumbum strata in and around Cheyuru river valley near Hostavaram records the following major facies associations :

- Tuffaceous interbedded rhyolitic layers and volcaniclastics with ignimbrites (Figs.53 & 54)

- Pyroclastic fall deposits-, subaerial emplacement of vertically aligned volcanic bombs, lapilli and pumice (Figs. 55 & 56).
- Channelised subaqueous deposition of ignimbrites with framboids, pumice, lapilli and bedding parallel volcanic bombs (Fig. 57).

The volcaniclastics are intimately associated with another significant facies which is laterally persistent for tens of kilometers. Lystric faults, sandstone dykes, graben-like structures, pseudonodules

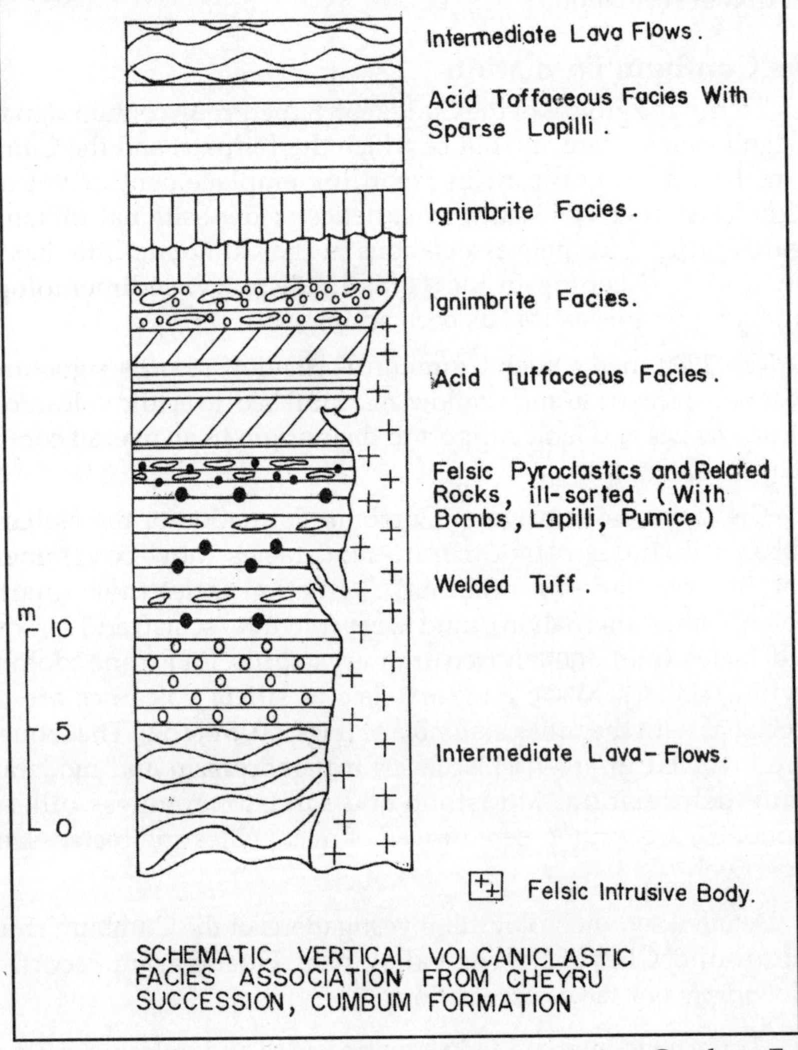

Fig. 50: Schematic vertical volcaniclastic facies association, Cumbum Formation near Cheuru dam, Cumbum Formation.

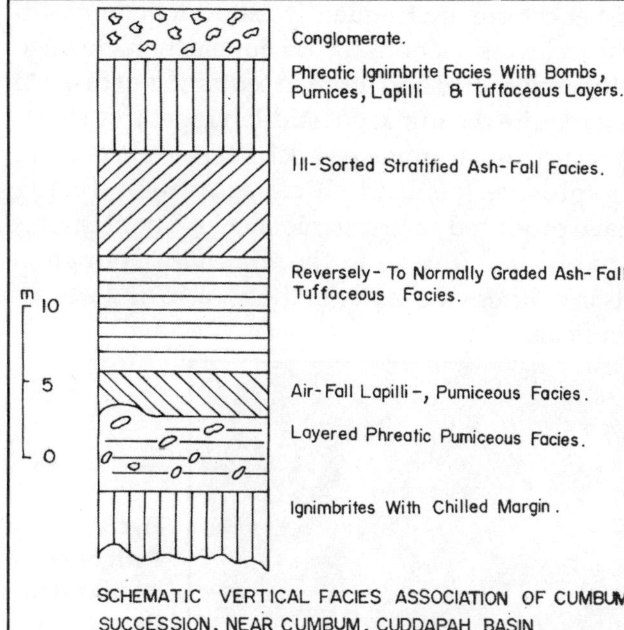

Fig. 51 : Schematic vertical facies association, near Cumbum, Cumbum Formation.

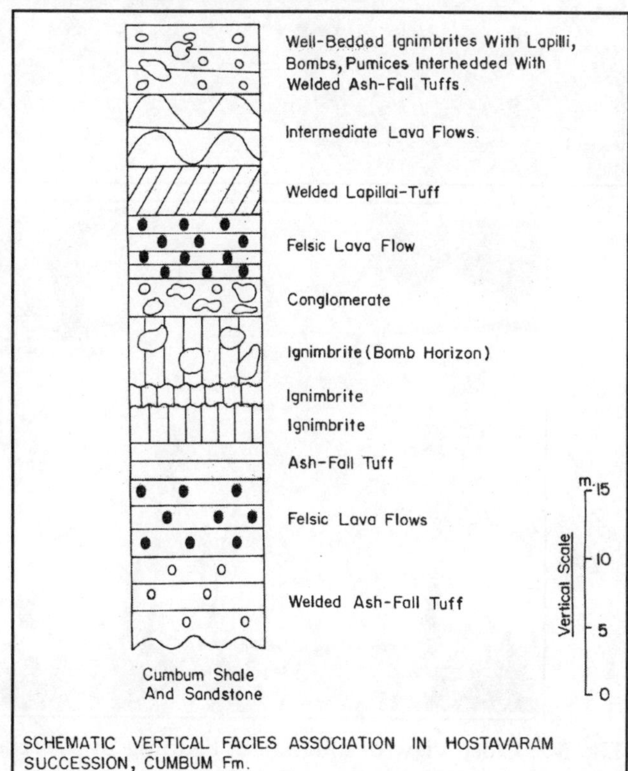

Fig. 52 : Schematic vertical facies association, near Hostavaram, Cumbum Formation.

and cyclic olistostromes et ceterra are frequently found together with the volcaniclastics. These facies associations record liquefaction, flowage and fluidization of soft to semilithified and lithified sediments at discrete yet distinct zones which require the impact of sudden repetitive short-lived high energy pulses such as earthquake shocks. It is expected that intermittent highly explosive felsic to silicic eruptions during sedimentation might have produced such seismic shocks. The seismites and associated facies, which are unique to the Cumbum Formation has been dealt in details in a different chapter of the book to add weight to its stratigraphic significance.

Fig. 53 : Tuffaceous interbedded rhyolitic layers and volcaniclastics with ignimbrites, Cumbum Formation.

Fig. 54 : Close view of allignment bedding in pyroclastic flow deposits with clusters of lapilli and pumice, developed from base surges, Cumbum Formation.

Fig. 55 : Pyroclastic flow deposits with lapilli and tubular flows, Cumbum Formation.

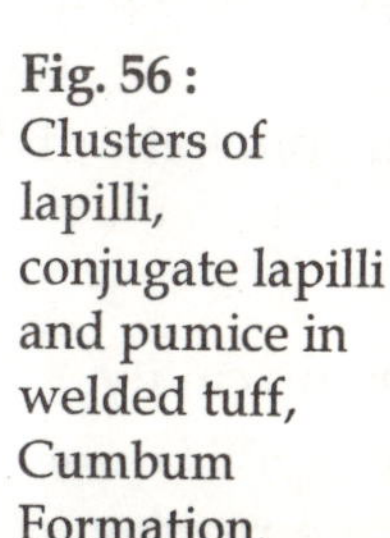

Fig. 56 : Clusters of lapilli, conjugate lapilli and pumice in welded tuff, Cumbum Formation.

Fig. 57 : Close view of a framboid from a channelised ignimbrite deposit, Cumbum Formation.

Facies attributes of the Cumbum succession in and around Cheyuru river valley are as follows :

17. Subaerially emplaced fine-to very fine serpentinised volcaniclastics. (Top)
16. Subaerially emplaced volcaniclastic facies dominated by tear-drop lapilli.
15. Low energy fine to very fine terrigeneous facies.
14. Subaerially emplaced volcaniclastic facies with subordinate lapilli tuff (tear drop).
13. Subaerially emplaced very fine volcaniclastic (ash fall) facies (chertified).
12. Subaerially emplaced coarse- to medium volcaniclastics (ash fall).
11. Low-energy, mainly fine- to very fine terrigenous facies (argillaceous) intermixed with carbonate facies.
10. Low energy terrigenous facies intermixed with volcaniclastics and epiclastics.
9. Subaerially emplaced volcaniclastic facies dominated by lapilli tuffs.
8. Subaerially emplaced volcaniclastic facies with pockets of chertified fine- to very fine tuffs.
7. Low-energy subaqueous terrigenous facies
6. Subaerially emplaced silicic lava with subvertical bomb cavity
5. Channelised silicic lava interbedded with pumice and lapilli layers. In the upper part of this layer development of wave ripples is discernible.
4. Ignimbrite flow; upper part semiconsolidated to consolidated, dragged forward by lower hot viscous layer responsible for the origin of vertical cracks in the upper layers. Lineation produced by stretching fiamme also observed.
3. Ash fall tuff layer with lapilli, both conjugate and simple, bombs and pumices.
2. Ignimbrite layers with bedding parallel disposition of volcanic bombs depicting subaqueous deposition of ignimbrites. Imbricate arrangements of volcanic bombs with concentric inner layers are discernible. Tbc Bouma divisions are often recorded here.
1. Subaqueously settled volcanic bombs with long axis subparallel to the bedding. Ripple phase discernible in the bottom, laminated chertified ash layers on top.

The upper part of the sequence is subaerially emplaced volcaniclastics, which in turn were emplaced as welded tuffs (Fig.58). The welded tuffs, however, show dominance of tear-drop lapilli. Tear drop shape of lapilli indicative of subaerial emplacement and free fall through atmosphere, earmarked subaerial emplacement of ash fall during the closing phase of phreatic volcanism.

Salient aspects of the sequence

1. Subaerially emplaced volcaniclastic rocks around Cheyuru valley is exposed for a strike length of about 14 km. Ash fall tuffs, lapilli tuffs or lithic tuffs are found as laterally continuous beds which are in most of the cases epidotised or serpentinised along certain preferential surfaces which define either successive curvilinear layers or climbing curvilinear bed forms. These are current ripples over the top of the subaerially emplaced rhyolitic flows. Ash, lapilli and/or lithic fragments might have formed as thin veneer on the top of the rippled surfaces.
2. Chertification is promoted by fine-grained ash layers or their patches. Types of chert are light grey to greyish brown in colour

Fig. 58: Welded ash fall tuff deposits with pumice and lapilli; slight dragging of lapilli suggestive from parallelism of their long axis with lamina, Cumbum Formation.

which are typically associated with chertfied ashes. This feature serves as a tool to identify thin layers or fine-grained ash and/or lapilli tuff over the top of rhyolitic flows.

3. This layering or flow foliation is frequently folded. Microscopically spherulitic growth has been identified. Vesiculated pumiceous layers may occur as bands with obsidian and spherulitic layers.
4. There are also different types of rhyolitic flow textures. It has been observed that at the flow fronts rhyolite lavas are steep and are typically terminated by tallas apron. Aprons of brecciated lavas and pumice blocks are built by the auto-brecciation of lava. Auto-brecciation is made possible due to differential stress field which is set up in the simple lava flow. The lower most part being concealed from the atmosphere, remains hot and fluid unlike the upper part which gets solidified and or semisolidified due to radiation under subaerial condition. Therefore relative motion is set in a single flow under a differential stress field. This could cause shattering into blocks in many portions of the upper part of the solidified flow causing autobrecciation. Other features of rhyolite lavas include great prevalence of pumice of different volcaniclastic facies with relatively high proportion of pumice blocks.
5. Association of excessive volumes of pumiceous material with rhyolitic lava is an evidence for great explosiveness of the eruption. Dynamic processes operating within the system itself did modify proportion of pumiceous material with the rhyolitic lava and sedimentary piles.
6. Volcaniclastic rich deep-water turbidites is a significant facies variant of the Cumbum (Fig. 59). The fine-grained turbidites are interbedded with insitu-bedded chert at different stratigraphic levels. Bouma sequences developed are T_{cd}, $T_{d/d}$, $T_{e/e}$ where modal Bouma sequence is represented by T_{de}. The low-energy turbidites, being the most dominant variants in Cumbum Formation got deposited in low-energy deep-water turbidite basin (Bouma, 1962; Mutti, 2000; Biswas, 2004).
7. Recent and ancient analogues suggest that the layered and non-layered disposition of the Cumbum volcaniclastics can be interpreted as products of explosive acid lavas with limited spread (Fisher & Schmincke, 1984). Presence of tear-drop lapilli,

subvertically emplaced volcanic bombs, vitric ash and overall absence of cross stratification are characteristic of pyroclastic fall deposits (Sparks & Wright, 1979). Presence of seismites is evidenced from discrete spatially continuous lithounits with vertically repetitive changing patterns, graded microfaults, disrupted layers, segmented thin microbrecciated zones, broken-up pieces of laminae and complex convoluted layers (Bose et al., 1997; Prat, 1992; Price & Cosgrove, 1990; Shiki & Yamazaki, 1996). Profuse presence of volcanic bombs with their axes parallel- to subparallel to the bedding is indicative of their transport and deposition in a subaqueous setting (Cas & Wright, 1997; Fisher & Schmincke, 1984).

Fig. 59: Volcaniclastic-rich deep-water calcareous turbidites; variation of energy condition is suggested from deep v-shaped turbidite channels to open flat-bottomed distal channels on top left. Variation of such channel configuration is indicative of instability of the basin. Modal Bouma sequence Tde, is indicative of an overall low-energy aspect of the turbidite deposition related to distant milieu, Naramrajupallae near Rajampet, Cuddapah district.

Summary of the Depositional Story of the Cumbum Formation

- Owing to rarity of clear-cut evidence of Precambrian ignimbrites, the present case history of Meso-Proterozoic Cumbum is of special significance.

- Recognition of ash-fall tuff is made with facies attributes related to pyroclastic character, degree of sorting, thickness, layering, aerial extent, gentle dip, welding and deformation of pumice, fused tuffs contrasted with welded tuffs, divitrification and vapour phase minerals, jointing erosion forms, fossil fumaroles, microscopically identified globules, bubbles, micro-fiamme and shards.
- Attributes such as pumice, tear-drop lapilli, vitric ash, overall absence of cross-stratification and vertical emplacement of bombs characterise subaerial pyroclastic fall deposits.
- Subaqueous- to resedimented pyroclastics and volcaniclastics are attested by bedding parallel bomb levels in tuffaceous facies (now chertified); chevron cross-bedding suggestive of tidalites.
- Scoria and steam-pipes can be ascribed to syndepositional gaseous activity.
- Phreatic eruptions in response to magma-water/vapour interaction is the main allocyclic genetic control for origin of silicic magmas, ash flows, welded- and non-welded ash fall tuffs of Cumbum felsic- to acidic volcaniclastics.
- Tectonic instability inherent in an extensional regime coupled with climatic variations, probably orbitally controlled, are among the important allocyclic elements that did have periodic impress on repetitive major volcanic phases.
- The lower one, the Cumbum felsic- to acidic effusives followed up stratigraphically higher Tadpatri silicic lavas, ash fall tuffs and volcaniclastics.
- Interaction between periodic percolated water vapour phase and silicic magma at intermediate- to shallow depths controlled violent phreatic eruptions of Cumbum and Tadpatri with limited spread.

The Srisailam Formation

The closing phase of Cuddapah sedimentation was marked by deposition of the Srisailam Formation. Facies analysis of the uppermost formation aims at an understanding of the depositional history during the closure of the basin. The three Members of the Srisailam Formation, namely the Kaklet Vagu, Tapasipenta and Egalapenta beds, arranged in ascending order, are characterised by frequent facies changes. The Egalapenta Member at the top, in particular, shows remarkable vertical and spatial facies variations and intertonguing of aeolian, fluviatile and shallow-marine facies. The Egalapenta facies mosaic is a token representation of a Proterzoic erg system and has been therefore dealt separately in a different chapter as a unique feature of the Cuddapah basin.

The Kaklet Vagu Member

Reconaissance traverses were undertaken along the Kaklet Vagu river section and it was found that best developed exposures of rocks of the Kaklet Vagu Member are exposed near the Kaklet Vagu and Bugga Vagu confluence, both being the tributaries of the river Krishna.

Facies analysis of the rocks reveals the following facies. Multi-directional planar cross-bedded reddish brown sandstones, channelised trough cross-bedded sandstones and disorganized sandstones mainly characterise the Kaklet Vagu rocks.

Multidirectional planar cross-bedded reddish brown sandstone facies (F1)

The facies comprises medium- to coarse-grained reddish brown sandstone with subordinate mudclasts, quartz granules and dispersed pebbles (Fig.60). It is characterised by large multidirectional planar cross-beds at an angle of 16°-30°. Tabular cross-beds at low angles (4°-10°) have also been recorded. They have sharp, flat or slightly scoured bases and tops. They are found both in solitary form as well as in groups. Set thickness ranges from a few centimeters to a few meters and descends in a down-palaeoflow direction. Both large and small scale tabular cross-strata are discernible. Volcaniclastics comprising glass shards, pumice and lapilli are also found within the facies. Mean palaeoflow direction of this facies measured from particular levels is towards northwest and west. Laterally cutting lenticular sandstones with concave-up erosional surfaces with infills of ripples and cross-stratifications also form a major part in the facies.

Multidirectional planar tabular cross-beds with coarse clasts and granules were formed by migrating two dimensional bedforms and laterally accreting bars under an overall high-energy condition (cf. Allen 1983, Harms et al. 1982, Todd and Went 1991). Their deposition in fairly deep channels can be deciphered from the height of the cross-beds. Laterally cutting lenticular sandstones with concave-up erosional surfaces are channelised sandstones with infills characteristic of fluvial deposits. Deposition of the facies under a pulsatory high-energy condition in embayments and protected channel environments is indicated by the presence of mudclasts, volcaniclastic clasts, clasts of ferricretes and insignificant fines. The volcaniclastics in the channels have come as wash outs from a nearby volcanically affected terrain. Volcaniclastic dominated Tadpatri and stratigraphically lower Cuddapah strata seem to be the probable provenance.

Channelised trough cross-stratified facies (F2)

The facies comprises medium- to large channels with profuse troughs and ripples (Fig.61). Both small to large troughs are preserved. The cross-strata are markedly concave-up and tangentially or concordantly related to the underlying bounding surfaces. Lithologically, the facies ranges from medium-grained red sandstones to sandstones with scattered pebbles and granules.

Moderate to large channels with trough cross-beddings are products of migrating subaqueous dunes and ripples mainly in the deeper portion of fluvial channels (Rubin 1987; Harms et al. 1975; Todd and Went 1991, Harms et al. 1975).

Disorganised Sandstone (F3)

The facies includes very poorly sorted fine- to very coarse-grained sandstones with abrupt vertical and lateral variations of grain size and internal structures. Lower part of the facies is conglomeratic with nearly massive beds and the upper part bears shallow trough cross-beds and planar cross-beds.

Generally, very coarse to pebbly nature of the facies and local occurrence of torrential cross-beds with sigmoidal foresets designate high energy flow conditions (Roe 1987). Massive beds and poorly differentiated foresets indicate rapid sedimentation from highly sediment-charged flow (Smith 1986). The attributes probably point to the origin of the facies from sediment gravity flow during flood stage. Ill-sorted, structureless massive sand bodies support their deposition from unconfined flows during overbank flooding.

The three major facies documented from the Kaklet Vagu Member points to a braided-fluvial environment which grades up into the Tapasipenta siliciclastics.

The Tapasipenta Member

Reconaissance traverses along the Tapasipenta exposures led to the identification of ideally preserved sequence of some beds of

Tapasipenta sequence, with gradational relations with the lower Kaklet Vagu and upper Egalapenta Member.

Facies analysis of the lower part of the Tapasipenta Member has been done along a well exposed section along the Krishna gorge (Fig.62). The section comprises few couples of sedimentary units, of which the first couple starts with a 2m thick fine-grained sandstone horizon which has gradational contact with the underlying Kaklet Vagu Member. The unit is a channel facies with large scale cross-beddings at the bottom and rippled unit on its top. Laterally cutting channels is a characteristic of the facies. Volcaniclastics also form a significant proportion in the unit. The second unit of the first couple is a 16cm thick tuffaceous mudstone facies. The lower part of the unit is characterised by small ripple-drift laminations. The middle part contains horizontal

Fig.60 : Multidirectional planar cross-bedded reddish brown sandstones, Kaklet Vagu Member, Srisailam Formation.

laminations with partings of coarse sand and white mica. Festoons with development of foresets are a characteristic feature. Ash and glass shards are also discernible. The upper part of the facies comprises laterally cutting channels with infills of cross-beds, troughs and ripples. Interlacing channels are common.

The tuffaceous mudstone facies is overlain by the first unit of the second couple which is a 20cm thick sandstone facies. The lower part of the facies is dominated by laterally cutting channels with pebbles of variable dimensions. These layers of pebbles are discontinuous and thin-bedded.

The sandstone facies is overlain by the second unit of the second couple which is a 15cm thick mudstone facies. The facies is characterized by climbing ripple-drift laminations.

The 15cm thick rippled mudstone facies is overlain by first unit of the third couple which is 55cm thick sandstone. The basal part comprises unassorted shale pebbles of variable dimensions forming the channel lag deposits. This grade up into ripple-drift laminations.

The sandstone facies is in turn overlain by the second unit of the third couple, comprising a 14cm thick mudstone facies, lower part of which is rippled, upper part is laminated with micaceous partings along the bedding planes.

The different facies attributes within the three main sandstone-mudstone couples (Fig.62) in the well exposed Tapasipenta section in the lower part of the member characterizes a braided-fluvial deposit with fluctuating energy condition. The laterally cutting and interlacing

Fig. 61: Channels with troughs and ripples, Kaklet Vagu Member, Srisailam Formation.

concave-up scoured surfaces enclosing lensoidal bodies of trough- and planar-cross-beds might be interpreted as laterally cutting braided channels. Association of the channelised deposits with climbing ripple-laminations and planar-cross-beds on top in the section can be attributed to the existence of bar deposits (cf. Reading, 1996; Miall,1996; Allen,1983). Discontinuous layers of thin, unassorted shale pebble layers form the channel lag deposits and they occupy the lower part of the channels or the bars. The shale pebbles of channel lag deposits reflect higher energy pulses and the grading of the pebbles to fine-grained rippled sandy-muddy beds point to waning of energy conditions.

The fluvial siliciclastics of the lower Tapasipenta is onlapped by glauconite- and felsic volcaniclastic-bearing sandstone-mudstone sequences in the upper part of the member (Fig.63). Facies analysis of the fine-grained sandstone-mudstone sequence reveals the following attributes.

Large-scale planar cross-stratified glauconitic acid volcaniclastic-bearing sandstones, heterolithic sandstone-mudstone facies and trough-cross-stratified glauconitic sandstones are characteristic features in the upper part of the Tapasipenta Member.

Large-scale planar cross-stratified glauconitic sandstone facies comprises sigmoidal sets of cross-stratifications up to 60cm thick and extend approximately 10m. The toesets of the sigmoidal beds have a tangential contact with the underlying beds and thin and fine rapidly in the downdip direction. The upper part of the sigmoidal sets grade up into planar cross-stratifications.

Large scale cross-stratified glauconitic sandstone of the facies are products of migration of sandwaves in subtidal environment with intermittent high energy waves. Presence of sigmoidal bundles within the facies corroborate tidal origin (cf. Kreisa and Moiola 1986). The attributes of this facies therefore points to a shallow marine origin. The shallow-marine facies onlapped the braided-fluvial siliciclastics of the lower part of the Tapasipenta Member owing to transgression of the sea.

Heterolithic sandstone-mudstone facies comprises millimetre-thick sand-mud interlaminations. Horizontally stratified mudstones and alternating fine-grained sandstones with cross-beds also characterise the facies. The planar cross-bedded fine-grained sandstones display a channelised disposition. Wave ripples and wavy bedding occur abundantly. The sand-mud alternations are laterally persistent.

Millimetre-thick interlaminations of sand and mud are reminiscents of tidal bedding (cf. Reineck and Singh 1980). Local occurrences of wavy bedding, symmetrical ripples, also support a tidal flat origin. Channelised fine-grained cross-bedded sandstones might be interpreted as tidal channels. The facies attributes are peculiar to and characteristic of shallow-marine environment.

The trough cross-stratified glauconitic medium-grained sandstones display grouped sets of troughs with infills of planar cross-beds. The tops of the beds contain wavy ripples, laminations, and mud-draped sandy ripples.

Presence of wavy bedding and ripples in association with the troughs are due to reworking of the sediments by oscillating wave action. This medium-grained facies was deposited in a subtidal environment following its similarity with other shallow marine sand deposits (Narayan 1971, Michelson and Dott 1973, McCave 1973).

Facies analysis reveals that the Tapasipenta Member bears impress of braided fluvial environment in its lower part and onlap of shallow-marine deposits in the upper part. The upper part grades into the Egalapenta siliciclastics which is another mosaic of different environments owing to pulsatory condition of the basin.

The Egalapenta Member

The Egalapenta Member of the Srisailam Formation marks the last phase of the Cuddapah sedimentation. Cratonic quartz sandstones

Fig.62 : Sandstone-Mudstone couples, Tapasipenta Member, Srisailam Formation.

of the Egalapenta Member, besides their extreme compositional maturity and paucity of shale are particularly notable for their sheet-like geometry. If the sandstones were entirely marine, as long supposed in Indian stratigraphy, such geometry is difficult to explain in terms of modern shelfal analogues. The modern shelfal sediments are generally held to be either relict or only slightly reworked by transgression (palimpsest) (Dott et al. 1986, Swift 1970, Swift and Thorne 1991, Emery 1952). This speaks of the significance and necessity of detailed facies analysis for understanding the controls of sedimentation during the closure of the basin.

Fig. 63: Glauconite and felsic volcaniclastic-bearing sandstone-mudstone sequence, upper Tapasipenta Member, Srisailam Formation.

Approximately 238m thick Egalapenta Member of the Srisailam Formation studied at different localities in the Srisailam hills in and around Krishna Gorge West region covers an area of 490sq.km. The Egalapenta is a reddish brown quartz-cemented quartz arenite. The ridge-forming Egalapenta sandstones exposed at somewhat inaccessible terrains form a conspicuous topographic feature in the Srisailam region.

The Egalapenta Member is divisible into three sequences (Fig.64):

a) sequence 1: it is gradational to the underlying Tapasipenta Member as well as to the base of the overlying sequence 2 (c.94m);
b) sequence 2: it includes impure carbonate-sandstone-mudstone interbedded with low angle stratified sandstones (c.20m to 32m) and grades to the base of sequence 3;

c) sequence 3: it is the upper sequence which extends from top of the sequence 2 and is thickest (c.112m) in the southeastern flank of the study area.

The terrigenous clastics of sequence 1 consist of fine- to very coarse, plane-bedded, planar- and trough cross-bedded sandstones with dispersed granules and clasts or combinations of these. Laterally extensive sandstone sheets, several metres thick, generally concave-up and intersecting erosional surfaces are notable. Lenticular or multistorey style of internal geometry is discernible.

Eight facies identified from sequence 1 are planar/horizontally stratified medium to coarse reddish brown sandstone (F1), red mudstone-sandstone (F2), trough cross-bedded, reddish brown coarse sandstone (F3), planar cross-bedded coarse-grained reddish brown sandstone (F4), disorganised fine-grained to pebbly sandstone (F5), convolute bedded reddish brown coarse sandstone with deformation structures (F6), stratified clast- and granule-bearing sandstone (F7) and small trough and planar cross-bedded coarse to pebbly sandstone (F8) (Figs.65 & 66).

Three facies, discernible in lithosomes of glauconitic sandstone, are rhythmic cross-stratified sandstone (M1), trough cross-stratified medium-grained glauconitic sandstone (M2) and mudstone with intercalated sandstone (M3) (Figs.67, 68 & 69).

Sequence 1 of Egalapenta sandstone is gradational to the underlying marine Tapasipenta deposits. The transition consists of c.3m thick interval in which grain-size gradually increased from fine sands to coarse sands. With the exception of minor development of thin (few millimetres to few centimetres) red mudstone sheets in association with fine-grained sandy ripples, ripple-drifts and parallel laminations, sandstones of the facies assemblage are reddish brown and medium- to coarse-grained.

Attributes of the facies complexes are: planar stratification, trough cross-stratification, interdigitated trough cross-stratifications with scattered pebbles, granules and ferricrete, planar (large and small scale) tabular cross-stratification, locally developed conglomerates spatially and vertically grading into disorganised fine to pebbly internal fabric of shallow troughs and planar cross-beds, locally developed deformational fabric, parallel to low angle stratifications with profuse clasts of quartz and ferricrete granules showing random and frequent variations in texture, sheet-like geometry in sandstones with cosets of small troughs and planar cross-beds, several deeply scoured concave-up large scale erosional surfaces often lined with clasts of pebbles and granules, overlain by multistorey fills of trough cross-stratified sands and stacked extensive tabular sand bodies bounded by planar or broadly undulated erosional surfaces.

The eight facies (F1 to F8), their facies associations and palaeocurrent patterns reflect a braided fluvial depositional environment. Dominant palaeoflow directions of the braided fluvial system were consistently westerly and northwesterly. Gradational contact of sequence 1 with the underlying shallow-marine Tapasipenta beds point towards increasing supply of land-derived sediments which led to rapid progradation and formation of high energy braided fluvial systems.

Westerly and northwesterly palaeoflows support the derivation of Egalapenta sediments from terrigenous clastics of lower Cuddapah and granitic gneissic basement complex. Almost total withdrawal of feldspar from the Egalapenta sands is ascribed to aggressive weathering during the Proterozoic under the influence of carbon dioxide rich atmosphere (Eriksson et al. 1998) and or to sediment recycling.

Facies attributes of the three facies M1, M2 and M3 point to a shallow-marine depositional condition. Terrigenous clastics of sequence 1 (c.94m thick), intertongues with glauconitic sandstone wedges at two stratigraphic levels. The lower (LS) and upper (US) tongues are 4m to 9m and 3.5m to 12m thick respectively.

Sequence 2 : Approximately 2m to 2.5m above the upper shallow-marine tongue (US), the sequence 2 is a facies association of two facies (E2 and E3). Patches of impure carbonate-sandstone-mudstone (E3) is repetitively interstratified with horizontal to low angle stratified reddish brown medium- to coarse-grained sandstone (E2). The sequence is laterally extensive for tens of kilometres. An upsection rapid increase in the bedding thickness of E2 is discernible. Impure carbonates of the sequence 2 is ascribed to an arid coastal sabkha origin from the facies attributes such as

- presence of bird's eye structures indicative of supratidal origin (Friedman and Sanders 1978),
- laminated dolomites characteristic of low ridges and natural levees bordering tidal channels (Friedman and Sanders 1978),

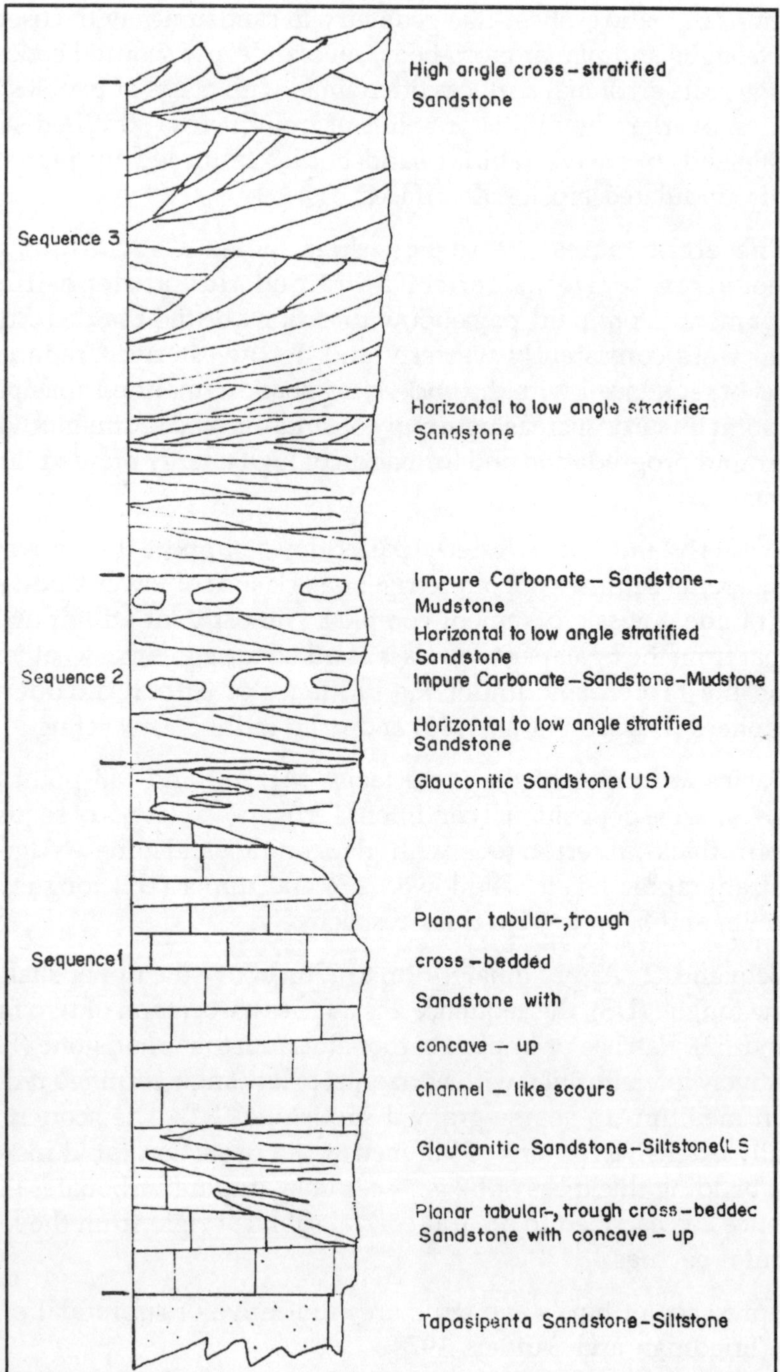

Fig.64 : Schematic representation of the different depositional sequences of the Egalapenta Member.

Fig. 65: Trough cross-stratified sandstones, sequence 1, Egalapenta Member.

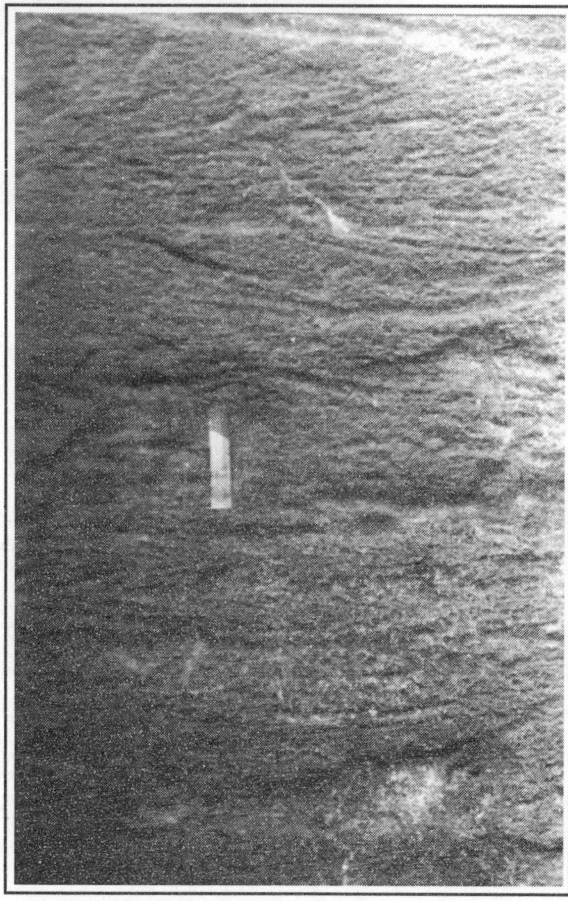

Fig.66 : Trough cross-stratified sandstones, sequence 1, Egalapenta Member.

Fig.67 : Heterolithic sandstone-mudstone laminations, sequence 1, Egalapenta Member.

Fig.68 : Glauconitic sandstone-mudstone deposits with sygmoidal bundles, sequence 1, Egalapenta Member.

Fig.69 : Glauconitic sandstone-mudstone deposits, sequence 1, Egalapenta Member.

- nodular anhydrites (chicken-wires) and 'enterolithic structures' in dolomites indicative of penecontemporaneous deformation due to forceful growth of nodular anhydrites under complex geochemistry of pore fluid in the carbonate mud of the arid sea-marginal sabkha deposits (Wood and Wolfe 1969, Friedman and Sanders 1978).

Intimately associated sandstone-mudstone layers with facies attributes such as wind ripples, small dunes, aeolian sand sheets, coarse-grained sand lags (Chan 1989, Fryberger et al. 1979), small ephemeral channels and troughs (Rust 1978, Glennie 1987) with infills of horizontal stratifications, trough and tabular cross-beds and aeolian stratifications namely wind-ripple strata and small dune foresets reflect frequent interference by aeolian processes and intermittent ephemeral stream action. Arid coastal sabkha deposits are repetitively interbedded with low angle stratified medium- to coarse-grained sand sheets.

Rapidly deposited significant sediment volume of aeolian sand sheets during the closing phase of sedimentation of sequence 2 were modified by alluvial incursions and wind action in the coastal scenario. The set-up is comparable to an erg-margin.

Arid coastal sabkha deposits repetitively interbedded with aeolian sand sheets attest to rapid progradation of terrigenous sands under the influence of wind action with occasional ephemeral episodes which alternated with marine transgression over coastal dune fields causing deposition of thin carbonate layers (cf. Fairbridge and Bourgeois 1978). The Devonian Helderberg Group is among the best analogues for mixed quartz sand and carbonate sediments (Driese and Dott 1984). Basinward prograding aeolian sand sheets of sequence 2 developed as repetitive coastal offlaps interbedded with arid sea-marginal sabkhas represented by impure carbonate-sandstone-mudstone assemblage.

Sequence 3 : Besides extreme compositional maturity and paucity of shale, the lensoidal, high angle cross-stratified and horizontal to low angle stratified sandstones of sequence 3 exhibit sheet-like geometry. Remarkable facies variations are documented in the Egalapenta Member during the closing phase of the Cuddapah sedimentation. The supermature quartz arenites of sequence 3 are reddish brown, medium- to fine-grained mostly with lensoidal and sheet geometry. Coarse- to very coarse-grained sandstone is an important facies variant of sequence 3. Seven facies (E1, E2 and E4 to E8) are discernible in sequence 3. These are large scale cross-bedded sandstone (E1), horizontal to low angle cross-stratified sandstone (E2), mudstone-sandstone (E4), flat parallel laminated sandstone (E5), large scale low angle cross-bedded sandstone (E6), trough cross-laminated sandstone (E7) and ripple laminated (E8) facies (Fig.70).

The facies attributes are :

E1 Cosets or isolated sets of laterally persistent steeply dipping (25°-32°), large tabular or wedge-shaped cross-beds, grainflow-grainfall layers and translatent strata;

E2 Sheet-like geometry, wind ripples, granule ripples, low angle to nearly horizontal stratifications, coarse-grained sand lags, coarse-grained low amplitude bedforms lacking well-developed slip faces, bimodality of coarse to very coarse (0.71mm to 1.92mm), rounded to well-rounded sands, broken roundness and fine to very fine sand and coarse silt (0.18mm to 0.05mm);

E4 Thinly bedded mudstone to fine-grained muddy sandstone with ripples and ripple-drifts, wind ripple strata, low angle stratifications and adhesion cross-laminae along with adhesion laminae and adhesion warts, shale-pebbles and mudcrack surfaces;

E5 Horizontal to subhorizontal cosets of parallel laminations, small and large channel-like scours with infills of translatent strata and wind ripples, mudcracks and shale-pebbles;

E6 Laterally extensive thick tabular tangential cross-beds with low foreset inclination;

E7 Closely interlocked trough sets;

E8 Slightly sinuous and locally linguoid ripples and ripple-laminated sets.

The large tabular cross-beds are preserved remnants of migrating large dunes, either straight-crested or crescentic (Kocurek and Dott 1981, Hunter 1981). Superimposed sets of large scale high angle cross-beds associated with grainflow, grainfall and translatent strata are the remnant deposits of downwind migrating dune bedforms or draas – sensu Wilson (1972). High angle cross-strata, abundant grainflow, wind ripple strata, medium to fine grain size and excellent sorting within the facies are comparable to those found in modern aeolian dunes with well-developed slip faces (Hunter 1977, McKee 1966, Inman et al. 1966). Wedge-shaped cross-beds of the facies suggest the presence of elongate oblique dunes. Such association is quite common in dune fields (Tsoar 1982, Rubin and Hunter 1983).

Sheet-like geometry, dominance of horizontal to low angle stratifications, presence of wind-ripple laminae, small dune cross-sets represent an aeolian sand sheet deposit. Packages of low angle wind-ripple strata of fine to medium sand are interpreted to represent zibars which are low-amplitude aeolian bedforms that lack slipfaces and developed in the marginal parts of dune field or in sand sheets (cf. Nielson and Kocurek 1986). Wind-ripple laminae of medium to coarse sand represent the amalgamated interzibar deposits (cf. Nielson and Kocurek 1986). Presence of distinct coarse mode of sand within a fine to very fine mode and absence of medium sand suggest an aeolian bimodality which is a characteristic feature of deflationary flats.

Mudstone to fine-grained muddy sandstone with ripples, ripple-drifts and thin laminations suggest their deposition in a quiet subaqueous environment such as shallow pool or lake within the sand sea. Presence of adhesion structures, wind-ripple strata and dessication features like mudcracks, shale-pebbles are indicative of drying up sequences.

Fig.70 : High-angled cross-stratified reddish-brown medium-grained sandstones, indicative of aeolian dune, overlying the interdune sandstones and mudstone deposits with shale pebbles and mud curls near the scale, sequence 3, Egalapenta Member. Distance between two scales=3.2m; height of scale=16cm.

Extensively developed parallel laminated sandstones with randomly scattered mudclasts probably represent high energy ephemeral flash floods. Meager bank stability in absence of vegetation during the Proterozoic might have promoted unconfined flows during the floods leading to formation of flash flood facies. Development of major non-aeolian sequence of flash flood facies within an aeolian regime is indicative of deposition owing to nonconfined flows. In-channel deposits with preserved channel forms suggest limited migration of ephemeral streams as a result of intense localised surface run-off (cf. Glennie 1987). Infills of aeolian strata within such channels indicate that those channels must have been quickly filled by aeolian sand under sand saturated subaerial condition dominated by wind action.

Tabular tangential cross-bedded sandstone resulted from migration of linguoid bars (cf. Smith 1970). Trough cross-laminated sandstones were generated from vertically stacked small dunes. Presence of ripple laminations and absence of reverse grading suggest their formation in subaqueous environment. Uncommon presence of lateral accretion bedding, tabular geometry of the sandstone bodies, frequent dessication features, mud draping at the top of thin waning flow sequence, erosional surfaces topped by red shale-pebbles are common attributes of E5 to E8 ephemeral facies association and ephemeral fluvial deposits of present and past (Rust 1978, Sneh 1983, Steel et al. 1977, Allen 1983, Cant and Walker 1976). The ephemeral fluvial facies (E5 to E8) are identifiable in association with dune and sand sheet deposits.

Spatial and vertical facies associations reveal ubiquitous presence of aeolian stratification types and bedforms. These are translatent strata, grainflow and grainfall layers, adhesion structures, wind ripples, high angle large scale cross-strata representing aeolian dunes, superimposed aeolian dunes or draas (sensu Wilson 1972), low angle stratified aeolian sand sheets (Chan 1989, Fryberger et al. 1979) and zibar and interzibar deposits (Neilson and Kocurek 1986). Associated nonaeolian facies includes braided ephemeral fluvial deposits and mudstone-sandstone of inland sabkha deposit. Vertical change from amalgamated interdune deposits to dunes and draas records dune-field initiation or lateral migration of peripheral to more centralised portion within the dune-field. Signatures of aeolian environments within the sequence 3 are comparable to those in a modern erg. Transition from braided fluvial deposits to a repetitive sequence of aeolian sandsheets alternating with coastal sabkha carbonates, followed by vertically arranged dunes, interdunes and ending in draas reflects alternation of humid and arid conditions which might have been orbitally-controlled. Eustatic fluctuations account for repetitive occurrence of environments and palaeogeographic shifts.

Salient Aspects of Basin Modelling

Intracratonic basins are important features of all continental tectonics from the Precambrian to the present time. Many contain important accumulations of hydrocarbons and many more are rich in ore deposits. The processes of formation and evolution of intracratonic basins are some of the unresolved problems in geodynamics.

The period between the Eparchaean interval and the Cambrian period is marked throughout the world by vast sedimentary accumulations in continental depressions. These sedimentary accumulations give the paleoenvironment of deposition and record the history of subsidence; the associated igneous materials help to unravel the tectonomagmatic events and mantle activity in basin evolution on the Archaean crust. Role of mantle upwelling and magmatism in basin evolution, crustal uplift, lineament reactivation and in crustal rifting is still poorly understood. General mechanisms proposed for sedimentary basin evolution also remain conjectural. No single mechanism has yet appeared to be fully satisfactory. Varieties of structural settings and tectono-magmatic events associated with intracratonic basins have made it difficult to arrive at a single satisfactory explanation (Bhattacharji, 1987).

With increased exploration of deeper levels of well known intracratonic basins from all over the globe, it has become clear that most intracratonic basins overlie fossil rifts. The geology and geodynamics of intracratonic basins such as Illinois, Williston, Michigan, Amadeus and Parana basins demonstrate that they were formed by a similar sequence of processes. These processes, in order of occurrence, were lithospheric stretching, mechanical fault-controlled subsidence, thermal subsidence and contraction and margin of slower thermal sequence with reactivated subsidence due to uncompensated excess mass (Sleep et al., 1980; Aheren & Dekeou, 1989). These events were followed by flexural fold and subsidence in some of the basins (Leighton & Kolata, 1990). Moreover, timing of initiation of thermal subsidence in some of the basins is remarkably close (530-500Ma). Some of the intracratonic basins of Western Europe and India also share a common timing of initiation of thermal subsidence.

To explain basin forming processes, concurrent initition of thermal subsidence of Mesozoic-Cenozoic intracratonic basins of Eurasia, several common characteristics of the basins should be considered. First and foremost are the cause and localization of rifting and reactivation of the basins (Aheren & Dekeou, 1989). Secondly, major Meso-Neoproterozoic and Paleozoic intracratonic basins of North America, Europe, Africa, South America and India appeared to have been initiated during breakup and assembly of major supercontinents either during the latest Proterozoic/Early Cambrian (Bond et al., 1984; Biswas, 2005). Thirdly, correlations of interregional unconformities that separate intracratonic sedimentary sequences have not been addressed with certainty. Over last 2 billion years, four major episodes of supercontinent accretion have been followed by supercontinent breakup (Condie, 1982). During periods, when continents were dispersed, heat loss from the earth's interior was dissipated through the mid-oceanic ridges whereas during times when continents were merged into a supercontinent, ridge volcanism and associated heat loss diminished considerably, partially because spreading rates may have been reduced as well. Consequently the supercontinents might have acted as heat lenses.

The evolutionary history of the Proterozoic platform basins of the Indian shield, some of which, experienced several cycles of uplift and subsidence, episodic magmatism prior to and during the sedimentary accumulations such as the lower Cuddapah depositional setting and syn- to postdepositional deformation as in the case of the upper Cuddapah Nallamalai Fold Belt are complex. The igneous episodes during the lower Cuddapah are intense and include mantle xenolith bearing ultramafic-, mafic- and felsic volcanism while the Nallamalai in the upper Cuddapah recorded dwindling stages of volcanism with Kimberlites, Nepheline-Syenites, Lamproites (Chelima) and Sodalites (Markapur) (Dasgupta, 2000; Dasgupta, 1986). The Cuddapah basin hosts such evidences that dictate explanation on role of mantle upwelling and magmatism, lineament reactivation, crustal uplift and assemblage of Rodinia supercontinent in basin evolution with some degree of certainty in the case of the Proterozoic Cuddapah basin.

Presence of garnet-spinel-palgioclase lherzolite xenoliths in picrites and basalts occurred in clusters in the first phase of mafic intrusives in the Tadpatri shales of the lower Cuddapah. Nodular xenoliths are deformed and contain thin veins and patches of calcite-rich carbonatites. The presence of garnet lherzolite indicates the maximum depth of derivation of these nodules to be 200km : Moho. The earlier picrite magma during ascent through the mantle possibly halted at an intermediate magma chamber near Moho to form cumulates before resumption of the journey to the upper levels of the crust (Dasgupta, 1986). Association of modified and unmodified mantle xenoliths in ultramafic sills and dykes of the lower Cuddapah strata provides a strong evidence of reactivation of deep-seated faults leading upto upper mantle and role of mantle perturbation in basin dynamics of the lower Cuddapah. Occurrence of round garnet grains in the core of euhedral olivine crystals embedded in orthopyroxone intercumulous raises the possibility that the garnet is a residual refractory material of the mantle source which underwent partial melting and the melt ascended into the lower Cuddapah platformal crust in two stages which does indicate also reactivation of deep-seated faults leading upto upper mantle (Dasgupta, 1986; Kaila et al,1987).

Owing to contrasting depositional styles and basin dynamics with the lower Cuddpah formations, the basinal history of Nallamalai Fold Belt of the upper Cuddapah appears to have been generated under different tectonosedimentary framework that makes understanding of the Nallamalai Fold Belt more complex. However, the salient aspects of sedimentation, igneous episodes and tectonic styles have been covered on the basis of available data.

The Nallamalai Group : a link between India and Antarctica in the Rodinia supercontinent

The southern extension of the Eastern Ghats Granulite Belt EGGB) occurs to the east of the Nallamalai Fold Belt (NFB) with an intervening 50 km wide tract of the Nellore schist Belt. The latter is thought to be equivalent to the Dharwar schists of the Karnataka craton. Model melting reactions and geochemical studies have already correlated high-Mg metapetalites of the Eastern Ghats Granulite Belt (EGGB) with those in the Rayner Complex, East Antarctica.

The Rodinia supercontinent was assembled by late Mesoproterozoic (1300-1000 Ma) orogenies that produced two major mobile belts in India : the EGGB and the Central Indian Tectonic Zone in eastern and central India, respectively. Connection between the Dharwar craton of Peninsular India and Archean rocks of the Napier Complex of Antarctica was established which did exist possibly earlier than 1.0 Ga. The Proterozoic Rayner Complex surrounds the Archean Napier Complex and contains reworked Napier Complex crust. The EGGB and southern granulite terranes of Tamilnadu and Kerala are mainly composed of Palaeoproterozoic to Archean protoliths affected by phases of high-grade metamorphism. The suture linking India with Antarctica is represented by the EGGB-Circum East Antarctic Mobile Belt. Mesoproterozoic suturing of the Circum East Antarctic Mobile Belt and the EGGB provide a geodynamic and tectonostratigraphic framework to study relations between Antarctica, Australia, India and Sri lanka within Rodinia.

Deformed Meso- to Neoproterozoic Nallamalai Group of mainly metasedimentary rocks with deep-water turbidite-contourite sequences and associated shelfal facies in a fold and thrust belt are truncated in the east by the high-grade Nellore schist belt of the Eastern Ghats Granulite Belt. The Nallamalai basin can therefore be considered as a peripheral foreland basin that formed as a result of west-directed thrusting in the Eastern Ghats Granulite Belt and got gradually modified into a foreland basin.

Diverse lithotectonic ensembles of the belt include huge volcaniclastic-rich flysch deposits that now include greenschist facies rocks, metapelites, marbles and calcsilicates. The deep-water

Nallamalai flysch basin was also associated with episodic intrusive phases of kimberlites and lamproites (most about 1.2-1.1 Ga) and also generation of felsic plutons of the Ishwarakuppam Complex. Nepheline-bearing gneisses and sodalite bearing metasediments may be remnants of fossil plate boundaries, reflecting destruction of an ocean basin during plate collision.

The Proterozoic cratonic basins situated along the eastern margin of the eastern Dharwar and Bastar cratons in India occur immediately west of the high-grade metamorphic belt of the Eastern Ghats. The spatial association has led to the plausible explanation of the NFB as an evolved back-arc basin or one in a retro-arc setting behind the collisional belt i.e. the Eastern Ghats. In this scenario, deformation in arcuate NFB is prior to the main E-W directed compression during the (latest Mesoproterozoic-early Neoproterozoic) Eastern Ghats orogeny (cf. Saha, 2002).

Considering effect of the tectonic convergence of India and Antarctica along the Eastern Ghats collisional zone (e.g., Naqvi and Rogers, 1987) on the adjoining intracratonic basin, it has been considered that the folding and thrusting in the Nallamalai Fold Belt has occurred between the latest Mesoproterozoic and early Neoproterozoic i.e. prior to the main deformation of the Eastern Ghats orogeny at ~1000 Ma (Grenvillian orogeny). Hence Nallamalai can be considered as a part of the Rodinia supercontinent, when India and East Antarctica were joined together (Figs.71,71a,71b & 71c).

Megashear couples near the border of Trans Himalaya with related transported basins, south of Tarim is cited as an analogue for drifting of the major part of the Nallamalai basinal piles and ultimately its collision with the lower Cuddapah strata (Dasgupta, 2005).

The overlying platformal Srisailam Formation, Palnad and Kurnool strata, separated by unconformity, were deposited on the cratonised basement of the Nallamalai subbasin during the closing phase of the Cuddapah sedimentation. However, Kurnool and Palnad basins are not included in the Cuddapah Supergroup. Their category according to basin dynamics should be included as successor basins of the Cuddapah (Dasgupta, 2000).

History of the Cuddapah sedimentation is thus represented by deposition of platformal lower Cuddapah strata unconformably over the cratonic Archaean Dharwar basement. The lower Cuddapah, is

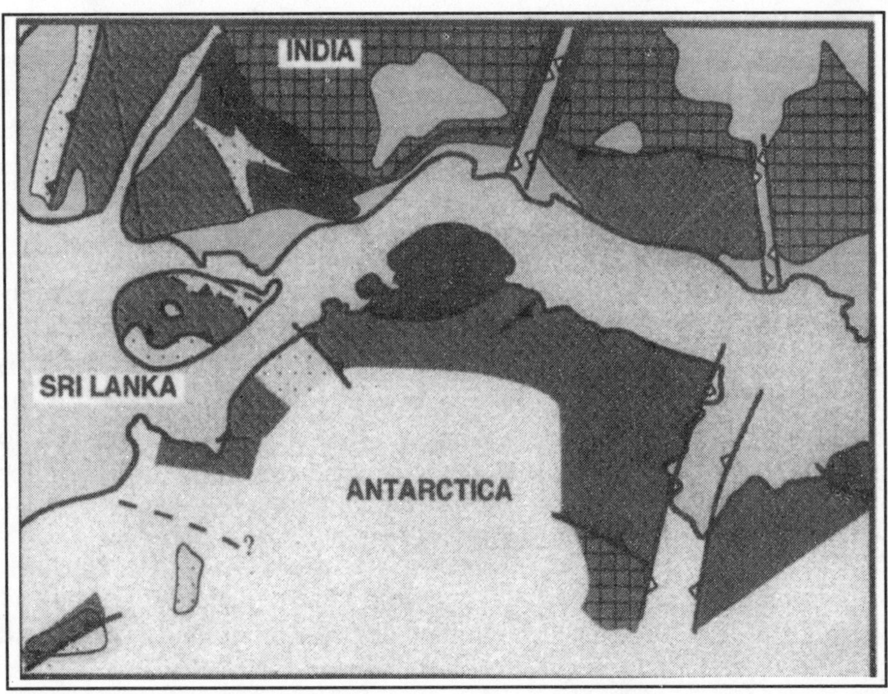

Fig.71 : Closeness of India and East Antarctica in Rodinia supercontinent (modified after GSI, 1995).

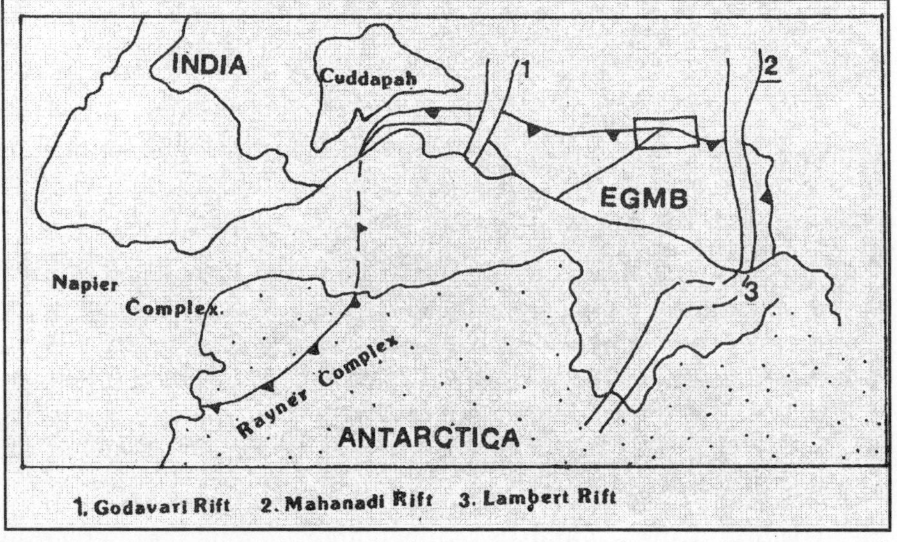

Fig.71a : Schematic diagram showing Closeness of India and East Antarctica in Rodinia supercontinent (modified after GSI, 1995).

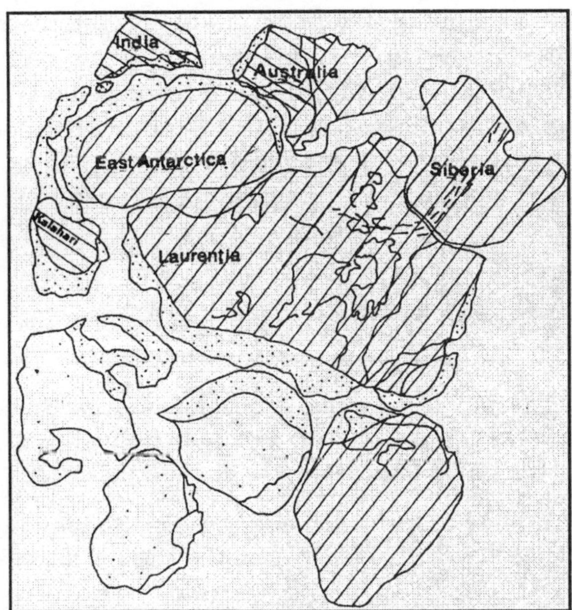

Fig.71b : Schematic diagram showing closeness of India, Australia and Antarctica (modified after GSI, 1995).

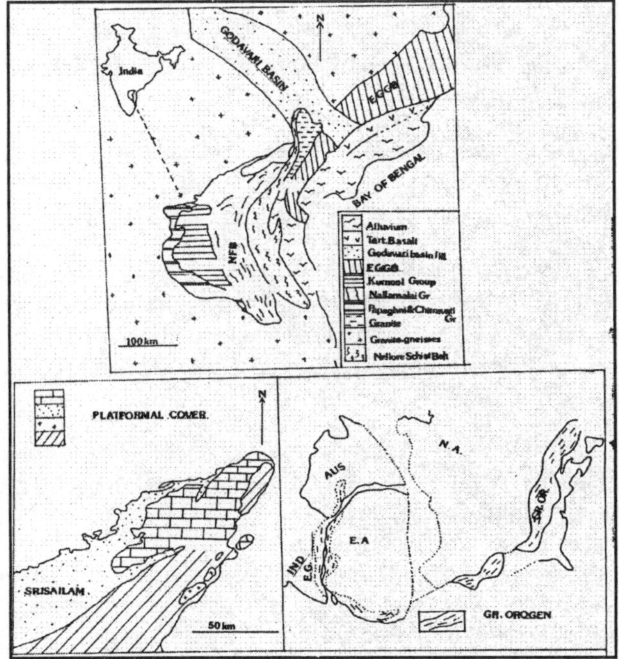

Fig.71c : Relationship of the Cuddapah basin with Nellore Schist Belt and Eastern Ghats Granulite Belt (EGGB) and its closeness with Grenville orogen (after Saha,2002).

inturn, tectonically separated from the gradually cratonised Nallamalai Fold Belt, a product of complex interplay of geodynamic processes during the assemblage of Rodinia supercontinent. Ultimately the platformal sediments of Srisailam, Palnad and the Kurnool blanketed the underlying formations. Deep seated weathering through course of time has exposed the Cuddapah rocks in the basin.

REFERENCES

Aheren, J.L. and Dekeou, P.J., 1989. Evolution of the lithosphere beneath the Michigan basin : Earth and Planetary Science Letters,v.95,pp.73-84.

Allen,J.R.L.,1983. Studies in fluviatile sedimentation : bars, bar complexes and sandstone sheets (low-sinuosity braided streams) in the Brownstone (L.Devonian), Welsh Borders. Sed.Geol., v.33, pp.237-293.

Allen,J.R.L.,1965. The sedimentation and palaeogeography of the Old Red Sandstone of Anglesey, North Wales. Yorkshire. Geol.Soc.Proc, v.35, pp.139-185.

Allen,J.R.L.,1982. Developments in sedimentology: Sedimentary structures. Their character and physical basis. Elsevier (Amsterdam). v.30A and B.

Anderson, M.P., Aiken, J.S., Webb, E.K. and Mickelson, D.M., 1999. Sedimentology and hydrogeology of two braided stream deposits. Sed. Geol., v. 129, pp. 189-199.

Arche,A.,1983. Coarse-grain meander lobe deposits in the Jarama River, Madrid, Spain. In: Modern and ancient fluvial systems. (J.D.Collinson and J.Lewin, eds.). Int.Ass.Sediment.Spec.Publ., v.6, pp.313-321

Ashworth, P.J., Best, J.L., Peakall, J. and Lorsong, J.A., 1999. The influence of aggradation rate on braided alluvial architecture: field study and physical scale modelling of the Ashburton River gravels, Canterbury Plains, New Zealand. In: N.D. Smith and J. Rodgers (Eds.), Fluvial Sedimentology VI, IAS Sp. Publ., v. 28, pp. 333-346.

Beaty,C.B.,1990. Anatomy of a White Mountain's debris flow – the making of an alluvial fan. In : Alluvial Fans : A field approach (Ed. by A.H.Rachocki and M.Church),John Wiley,pp.69-89.

Best, J.L., Ashworth, P.J., Bristow, C. and Roden, J., 2003. Three-dimensional sedimentary structure of a large, mid-channel sand braided bar, Jamuna River, Bangladesh. Jour. Sed. Res., v. 73, pp. 516-530.

Best, J.L., 1996. The fluid dynamics of small-scale alluvial bedforms. In: P.A. Carling and M.R. Dawson (Eds.), Advances in Fluvial Stratigraphy, John Wiley and Sons, New York, pp. 67-125.

Bhattacharji,S.,1987. Lineaments and igneous episodes in the evolution of Intracratonic Proterozoic basin on the Indian Shield. S. Sen. Comm. V. Hindusthan Pub. Co., New Delhi, India, 13, 1-15.

Biswas, A., 2004 : Mesoproterozoic Turbidite Systems From Cuddapah Basin, India : Records Of Deep-Water Clastics, Nallamalai Group, Congress vol. of 32nd International Geological Congress, Florence Session, Italy.

Biswas, A.,2005 : Mesoproterozoic Nallamalai Sedimentation In The Cuddapah Basin : A Link Between India And East Antarctica In Rodinia Supercontinent : Symposium volume in 'Supercontinents and Earth Evolution Symposium', Freemantle, Australia.

Blair, T.C., 1985. Depositional chronology, sedimentary processes and the resulting vertical stratification sequences in the Roaring river alluvial fan, Rocky Mountain National Park, Colorado. In: Flores, R.M. and Harvey, M.D. (Eds.), Field guide book to modern and ancient fluvial systems in the United States, Fort Collins, Colorado, Proceed. 3rd Int. Fluvial Conf., pp. 96-101.

Blair, T.C., 1987. Tectonic and hydrologic controls on cyclic alluvial fan, fluvial and lacustrine rift-basin sedimentation, Jurassic- lowermost Cretaceous Todos Santos Formation, Chiapas, Mexico, Jour. Sed. Pet., v. 57, pp. 845-862.

Blair, T.C. and McPherson, J.G., 1994. Alluvial fans and their natural distinction from rivers based on morphology, hydraulic processes, sedimentary processes and facies assemblages. Jour.Sed.Res., v. A64, pp.450-489.

Blodgett,R.H. and Stanley,K.O.,1980. Stratification, bedforms and discharge relations of the Platte braided river system, Nebraska. J.Sedim.Petrol., v.50, pp.139-148.

Bluck,B.J.,1971. Sedimentation in the meandering river Endrick. Scott.J.Geol., v.7, pp.93-138.

Bluck,B.J.,1981. Upper Old Red Sandstone, Firt of Clyde. In: Field Guides to Modern and Ancient Fluvial Systems in Britain and Spain (T.Elliot, ed.). Proc.2nd Cong.Fluvial Seds.Univ.Keele, U.K., Int.Ass.Sediment., 5.14-5.20.

Boersma,J.R. and Terwindt,J.H.J.,1981. Neap-spring sequences in intertidal shoal deposits in a mesotidal estuary. Sedimentology, v.28, pp.151-170.

Bond,G.C., Nickeson,P.A. and Kominz,M.A., 1984. Breakup of a supercontinent between 625 Ma and 555ma : new evidence and implications for continental histories : Earth and Planetary Science Letters,v.70,pp.325-345.

Boothroyd,J.C. and Ashley,G.M.,1975. Process, bar morphology, and sedimentary structures on braided outwash fans, northeastern Gulf of Alaska. In: Glacio-Fluvial And Glacio-Lacustrine Sedimentation (A.V.Jopling and B.C.McDonald, eds.). Soc.Econ.Paleontol.Mineral Spec. Paper, v.23, pp.193-222.

Boothroyd, J.C., 1972. Coarse-grained sedimentation on a braided Clutwacsh Ran, Northeast Gulf of Alaska. Tech. Rep. No. 6, Coastal Research Division, University of South California, Columbia, 127p.

Bose,P.K., Banerjee,S. and Sarkar,S.,1997. Slope-controlled seismic deformation and tectonic framework of deposition : Koldaha Shale, India. Tectonophysics, v.269, pp.151-169.

Bouma, A.H., 1962. Sedimentology of some flysch deposits, a graphic approach to facies interpretation. Elsevier, Amsterdam, 168p.

Bridge, J.S., 1993. Description and interpretation of fluvial deposits: a critical perspective. Sedimentology, v. 40, pp. 801-810.

Bridge, J.S., 2003. Rivers and flood plains. Balckwell Scientific, Oxford, 504p.

Bridge, J.S. and Tye, R.S., 2000. Interpreting the dimensions of ancient fluvial channel bars, channels and channel belts from wireline logs and cores. Am. Assoc. Petrol. Geol. Bull., v. 84, pp. 1205-1228.

Bridge, J.S., Smith, N.D., Trent, F., Gabel, S.L. and Bernstein, P., 1986. Sedimentology and morphology of a low-sinuosity river: Calamus River, Nebraska Sand Hills. Sedimentology, v. 33, pp. 851-870.

Brierley, G.J., 1991. Bar sedimentology of the Squamish river, British Columbia : definition and application of morphostratigraphic units.J. Sedim. Petrol., v.61., pp.211-225.

Buatois, L.A., Mangano, M.G. and Car, T.R., 1999. Sedimentology and ichnology of the Palaeozoic estuarine and shoreface reservoir, Morrow Sandstone, Lower Pennsylvanian of southwest Kansas, USA. Kansas Geol. Surv. Current Res. In Earth Sc., v. 243(1), pp. 1-35.

Bull,W.B.,1977. The alluvial fan environment. Prog.Phys.Geogr., v.1, pp.222-270.

Campbell, C.V., 1976. Reservoir geometry of a fluvial sheet sandstone. AAPG Bull., v. 60, pp. 1009-1020.

Cant, D.J. and Walker, R.G., 1978. Fluvial processes and facies sequences in the sandy braided South Saskatchewan River, Canada. Sedimentology, v.25, pp.625-648.

Cant,D.J. and Walker,R.G.,1976. Development of a braided-fluvial facies model for the Devonian Battery Point Sandstone, Quebec. Can.J.Earth.Sci., v.13, pp.102-119.

Carling, P.A., 1999. Subaqueous gravel dunes. Jour. Sed. Res., v. 69, pp. 534-545.

Cas, R.A.F. & Wright, J.V., 1987 : Volcanic successions- modern and ancient., Allen & Unwin London, 528p.

Chan,M.A.,1989. Erg margin of the Permian White Rim Sandstone, SE Utah. Sedimentology, v.36, pp.235-251.

Chapin, C.E. and Elston, W.E., 1979. Ash-Flow Tuffs., Sp. Paper, Geol. Soc. Am., 211p

Cheel, R.J. and Leckie, D.A., 1993. Hummocky cross stratification. In: V.P. Wright (Ed.), Sedimentology Review 1, Blackwell Scientific, pp. 103-122.

Cloud, P.E. and Semikatov, M.A., 1969. Proterozoic stromatolitic zonation.Amer.Jour.Sci., pp.1017-1061.

Coleman, J.M., 1969. Brahmaputra River : Channel processes and sedimentation. Sedimentary Geology, v.3, pp.129-239.

Collinson,J.D.,1970. Bedforms of the Tana river, Norway. Geogr.Ann., v.52A, pp.31-55.

Collinson, J.D., 1996. Alluvial sediments. In: H.G. Reading (Ed.), Sedimentary Environments: Processes, Facies and Stratigraphy, Blackwell Scientific, Oxford, pp. 37-82.

Conaghan, P.J. and Jones, J.G., 1975. The Hawkesbury Sandstone and the Brahmaputra: A depositional model for continental sandstones. Jour. Geol. Soc. Aust., v. 22, pp. 275-283.

Condie,K.C.,1982. Early and middle Proterozoic supracrustal successions and their tectonic settings. Amer.J.Sci., v.282, pp.341-357.

Condie,K.C.,1992. Proterozoic crustal evolution. Developments in Precambrian Geology, 10, Elsevier, p.1-54.

Crowley,K.D.,1983. Large-scale bed configurations (macroforms), Platte River basin, Colorado and Nebraska: Primary structures and formative processes. Geol.Soc.Am.Bull., v.94, pp.117-133.

Dalrymple, R.W., 1992. Tidal depositional systems. In: R.G. Walker and S.P. James (Eds.). Facies models : responses to sea-level change. Geol. Assoc. Can., Ontarip. pp. 195-218.

Dasgupta, P.K.,1986. Picrite borne mantle xenoliths from Cuddapah Basin, South India. Ind. J. Earth Sciences 13(4), 333-338.

Dasgupta, P.K., 2000. The Neoproterozoic sequences from the Kurnool Supergroup and its significance in Meso-Neoproterozoic boundary in the global scenario. Abs. Vol. International Geological Congress, Rio-de-Janeiro session, Brazil, Aug. 2000.

Dasgupta, P.K., 2005. The Pan-African Orogeny in Northwestern Himalaya : The missing link in the transition from Rodinia to Gondwana. Symposium volume in 'Supercontinents and Earth Evolution Symposium', Freemantle, Australia.

De Raff, J.F.M. and Boersma, J.R., 1971. Tidal deposits and their sedimentary structures. Geol. Mijnbouw, v. 50, pp. 479-504.

De Raff, J.F.M., Boersma, J.R., Gelder, A., 1977. Wave-generated structures and sequences from a shallow-marine succession, Lower Carboniferous, County Cork, Ireland. Sedimentology, v. 24, pp. 451-483.

DeCelles,P.G., Gray,M.B., Ridgway,K.D., Cole,R.B., Pivnik,D.A., Pequera,N. and Srivastava,P., 1991. Controls on synorogenic alluvial-fan architecture, Beartooth Conglomerate (Paleocene), Wyoming and Montana, Sedimentology,v.38, pp.567-590.

Dott,R.H. and Bourgeois,J.,1982. Hummocky stratification: Significance of its variable bedding sequences. Geol.Soc.Amer.Bull., v.93, pp.663-680.

Dott,R.H.Jr., Byers,C.W., Fielder,G.W., Stenzel,S.R. and Winfree,K.E.,1986. Aeolian to marine transition in Cambro-Ordovician cratonic sheet sandstone of the northern Mississippi Valley, USA. Sedimentology, v.33, pp.345-367.

Dreyer, T., 1993. Quantified fluvial architecture in ephemeral stream deposits of Esplugafreda Formation (Palaeocene), Tremp-Graus Basin, northern Spain. In: M. Mazo, and C. Puidefabregas (Eds.), Alluvial sedimentation, Int. Assoc. Sed. Sp. Publ., v. 17, pp. 337-362.

Driese,S.G. and Dott,R.H.Jr.,1984. Model for sandstone-carbonate "cyclothems" based on upper member of Morgan Formation (Middle Pennsylvanian) of northern Utah and Colorado. Am.Assoc.Petrol.Geol.Bull., v.68(5), pp.574-597.

Dunham, R.J., 1962. Classification of carbonate rocks according to depositional texture in classificationof carbonate rocks (Ham,W.E., Ed.). Tulsa, Okla.,Amer.Assoc. Petrol.Geol., v.43., pp108-121.

Elliot, T. (Ed.), 1981. Field guide to modern and ancient fluvial systems in Britain and Spain. Proceed. 2ng Int. Conf. On fluvial systems: sedimentology and processes, Int. Assoc. Sed. Sp. Publ., 193p.

Emery,K.O.,1952. Continental shelf sediments off southern California. Bull.Geol.Soc.Am., v.63, pp.1105-1108.

Eriksson,P.G.; Condie,K.C.; Tirsgaard,H.; Mueller,W.U; Altermann,W.; Miall,A.D.; Aspler,L.B.; Catuneanu,O. and Chiarenzelli,J.R.,1998. Precambrian clastic sedimentation systems. Sedimentary Geology., v.120, pp.5-53.

Fairbridge,R.W. and Bourgeois,J.,1978. The encyclopedia of sedimentology, encyclopedia of earth sciences. Dowden,Hutchinson and Ross Inc., Stroudsburg. v.6, 901p.

Fedo,C.M. and Cooper,J.D.,1990. Braided fluvial to marine transition : The basal lower Cambrian Wood Canyon Formation, Southern Marble Mountains, Mojave Desert, California. J.Sedim.Petrol., v.60(2), pp.220-234.

Fisher, R.V. & Schmincke, H.U., 1984. Pyroclastic rocks, Springer-Verlag, 472p.

Folk, R.L., 1959. Practical petrographic classification of limestones. Bull.Amer.Assoc.Petrol.Geol.,v.43.,pp.1-38

Frazier,D.E. and Osanik,A.,1961. Point-bar deposits; Old River Locksite, Louisiana. Trans.Gulf.Coast.Assoc.Geol.Socs., v.11, pp.121-137.

Friedman,G.M. and Sanders,J.E.,1978. Principles of sedimentology. John Wiley and Sons. 792p.

Fryberger,S.G., Ahlbrandt,T.S. and Andrews,S.,1979. Origin, sedimentary features and significance of low angle aeolian sand sheet deposits, Great Sand Dunes National Monument and vicinity, Colorado. J.Sedim.Petrol., v.49, pp.733-746.

Furnes, H.I.B., Friedliefsson,A. and Atkins, F.B., 1980. Subglacial volcanics – on the fprmation of acid hyaloclastiles.Jour.Volc.Geoltherm.,Res.,v.8., pp.95-110.

Glennie,K.W.,1987. Desert sedimentary environment present and past – a summary. Sediment.Geol., v.50, pp.135-165.

Godin, P.D., 1991. Fining-upward cycles in the sandy braided-river deposits of the Westwater Canyon Member (Upper Juarassic), Morrison Formation, New Mexico. Sed. Geol., v. 70, pp. 61-82.

Goodwin, A.M., 1981. Archaean Plates and greenstone sets.In: A. Kroner (ed) Precambrian Plate Tectonics, Elsevier, Amsterdam, pp.105-135.

Greb, S.F. and Archer, A.W., 1995. Rhythmic sedimentation in a mixed tide and wave deposit, Hazel Patch Sandstone (Pennsylvanian), Eatern Kentucky Coal Field. Jour. Sed. Res., v. B65(1), pp. 96-106.

Grotzinger,J.P.,1986. Cyclicity and palaeoenvironmental dynamics, Rocknest Platform, Northwest Canada, Geol. Soc. of Amer. Bull., v.97, pp.1208-1231.

Harms,J.C. and Fahnestock,R.K.,1965. Stratification, bedforms and flow phenomena (with an example from the Rio Grande). Soc.Econ. Palaeontologists Mineralogists Spec. Publ., v.12, pp.84-115.

Harms,J.C., Southard,J.B., Spearing,D.R., Walker,R.G.,1975. Depositional environments as interpreted from primary sedimentaryy structures and stratification sequences. Soc.Econ.Paleontol.Mineral., Short Course, v.2, p.161.

Harms,J.C.; Southard,J.B. and Walker,R.G.,1982. Structures and sequences in clastic rocks. Soc.Econ.Paleontol.Mineral., Short Course, 9.

Harms, J.C., 1975. Stratification and sequence in prograding shoreline deposits. In: Depositional environments as interpreted from primary sedimentary structures and stratification sequences, SEPM short course 2, Dallas, pp. 81-102.

Hunter,R.E.,1977. Basic types of stratification in small aeolian dunes. Sedimentology, v.24, pp.361-387.

Hunter,R.E.,1981. Stratification styles in eolian sandstones : some Pennsylvanian to Jurassic examples from the western interior USA. In : Recent and Ancient nonmarine depositional environments : Models for exploration (F.G.Ethridge and R.M.Flores, eds.), Spec.Publ.Soc.Econ.Paleont.Miner. Tulsa, v.31, pp.315-329.

Inman,D.L., Ewing,G.C. and Corliss,J.B.,1966. Coastal sand dunes of Guerrero Negro, Baja California, Mexico.Bull.Geol.Soc.Amer., v.77, pp.787-802.

Jhonson, H.D., 1977. Shallow-marine sand bar sequences: an example from the late Precambrian of North Norway. Sedimentology, v. 24, pp. 245-270.

Jones, B.G. and Rust, B.R., 1983. Massive sandstone facies in the Hawkesbury Sandstone, a Triassic fluvial deposit near Sydney, Australia. Jour. Sed. Petrol., v. 53, pp. 1249-1259.

Kaila,K.L., Tewari,H.C., Roy Chowdhury,K., Rao,V.K., Sridhar,A.R. and Mall,D.M.,1987. Crustal structure of the Proterozoic Cuddapah Basin of India from deep seismi soundings and gravity data. In : S. Asano and W.D.Mooney (editors), Seismic studies of the continental lithosphere. Tectonophysics, v.140, pp.1-12.

Karig, D.E., 1972. Remnant arcs. Geol. Soc. Am. Bull., v. 83, pp. 1057-1068.

Kataoka, K. and Nakajo, T., 2002. Volcaniclastic resedimentation in distal fluvial basins induced by large-volume explosive volcanism: the Ebisutoge-Fukuda tephra, Plio-Pleistocene boundary, Central Japan. Sedimentology, v. 49, pp. 319-334.

Kelley, S. and Olsen, H., 1993. Terminal fans with reference to Devonian examples. In: A.J. Fleet, K. Kelts and M.R. Talbot (Eds), Current research in fluvial sedimentology, Geol. Soc. Lond. Sp. Publ., v. 40, pp. 3-326.

Klein, G. DeV., 1971. A sedimentary model for determining palaeotidal range. Geol. Soc. Am. Bull., v. 82, pp. 2585-2592.

Kocurek,G. and Dott,R.H.,Jr.,1981. Distinction and uses of stratification types in the interpretation of aeolian sand. J.Sedim.Petrol., v.51, pp.579-595.

Koster, E.H., 1978. Transverse ribs: their characteristics, origin and palaeohydraulic significance. Can. Soc. Petrol. Geol. Mem. 5. Pp. 161-186.

Kreisa, R.D. and Moiola, R.J., 1986. Sigmoidal tidal bundles and other tide generated sedimentary structures of the Curtis Formation, Utah. Geol. Soc. Am. Bull., v. 97, pp. 381-387.

Krumbein, W.C. and Sloss, L.L., 1963. Stratigraphy and sedimentation. 2nd Edn., Freeman and Co., 660p.

Kvale, E.P. and Barnhill, M.L., 1994. Evolution of Lower Pennsylvanian estuarine facies within two adjacent palaeovalleys, Illinois basin, Indiana. In : R.W. Dalrymple, R. Boyd and B.A. Zaitlin (Eds.0, Incised-valley systems : origin and sedimentary sequences. Sp.Publ. Econ. Paleont. Miner. Soc. Sedim. Geol., Tulsa, v. 52, pp. 191-207.

Leckie, D.A. and Walker, R.G., 1982. Storm- and tide-dominated shorelines in Cretaceous Moosebar-Lower Gates interval-outcrop equivalents of deep basin gas trap in Western Canada. AAPG Bull., v. 66, pp. 138-157.

Leclair, S.F., 2002. Preservation of cross-strata due to the migration of subaqueous dunes: an experimental investigation. Sedimentology, v. 49, pp. 1157-1180.

Leighton,M.W. and Kolata,D.R. 1990. Selected interior cratonic basins and their place in the scheme of global tectonics : a synthesis : American Association of Petroleum Geologists memoir.v.51,pp.729-797.

Levey, R.A., 1978. Bed-form distribution and internal stratification of coarse-grained point bars, upper Congaree River. In: A.D. Miall (Ed.), Fluvial sedimentology, Can. Soc. Petrol. Geol., Mem., No. 5, pp. 105-127.

Long,D.G.F.,1978. Proterozoic stream deposits : some problems of recognition and interpretation of ancient sandy fluvial systems. In: A.D.Miall (Ed.),. Fluvial Sedimentology, Can. Soc. Pet. Geol. Mem., v. 5, pp. 313-342.

Lowe, D.R., 1982. Sediment gravity flows: Depositional models with special reference to the deposits of high-density turbidity currents. Jour. Sed. Pet., v.52, pp. 279-297.

Lunt, I.A., Bridge, J.S. and Tye, R.S., 2004. Development of a 3-D depositional model of braided river gravels and sands to improve aquifer characterisation. In: J.S. Bridge and D.W. Hyndman (Eds.), Aquifer characterisation, SEPM Sp. Publ., v. 80.

Massari,F.,1983. Tabular cross-bedding in Messinian fluvial channel conglomerates, Southern Alps, Italy. In: J.D. Collinson and J. Lewin (Eds.), Modern and Ancient Fluvial Systems, Int. Assoc. Sed. Sp. Publ., pp. 287-300.

Mazumder, R. and Arima, M., 2005. Tidal rhythmites and their implications. Earth Sc. Rev., v. 69, pp. 79-95.

McCabe,P.J. and Jones,C.M.,1977. Formation of reactivation surfaces within superimposed deltas and bedforms. J.Sedim.Petrol., v.47, pp.707-715.

McCabe, P.J., 1977. Deep distributory channels and giant bedforms in the Upper Carboniferous of the Central Pennines northern England. Sedimentology, v. 24, pp. 271-290.

McCave,I.N.,1973. The sedimentology of a transgression : Portland Point and Cooksburg Members (Middle Devonian), New York State. J.Sedim.Petrol., v.43, pp.484-504.

McGowen,J.H. and Garner,L.E.,1970. Physiographic features and stratification types of coarse-grained point bars: modern and ancient examples. Sedimentology, v.14, pp.77-111.

McGowen, J.H. and Groat, C.G., 1971. Van Horn Sandstone, west Texas: an alluvial fan model for mineral exploration. Bur. Econ. Geol. Report of Investigations, v. 72, Austin, Texas, 57p.

McKee,E.D.,1966. Structures of dunes at White Sand National Monument, New Mexico (and a comparison with structures of dunes from other selected areas). Sedimentology, v.7, pp.1-69.

McKee,E.D., Crosby,E.J. and Berryhill,H.L.Jr.,1967. Flood deposits, Bijou Creek, Colorado, June (1965). J.Sedim.Petrol., v.37, pp.829-851.

McRae,S.G.,1972. Glauconite. Earth Science Review., v.8, pp.367-440.

Meckel, L.D., 1975. Holocene sand bodies in the Colorado delta area, northern Gulf of California. In: M.L. Broussard (Ed.), Deltas: Models for Exploration, Houston Geol. Soc., Houston, pp. 239-265.

Miall, A.D., 1977. A review of the braided river depositional environment. Earth Sc. Rev., v. 13, pp. 1-62.

Miall,A.D.,1996. The Geology of Fluvial Deposits: Sedimentary Facies, Basin Analysis and Petroleum Geology. Springer-Verlag, Berlin, 582p.

Michelson,P.C. and Dott,R.H.,Jr.,1973. Orientation analysis of trough cross-stratification in Upper Cambrian sandstones of Western Wisconsin. J.Sedim.Petrol., v.43, pp.784-794.

Middleton, G.V.,1967. Experiments on density and turbidity currents: III. Deposition of sediment. Can. Jour. Earth Sc., v. 4, pp. 475-505.

Middleton, G.V., 2003 (Ed.). Encyclopedia of sediments and sedimentary rocks. Kluwer Academic Press, 871p.

Mutti, E., 2000. An introduction to the analysis of Ancient Turbidite Basins from an outcrop perspective. AAPG Continuing Education Course Note Series, No. 39, 98p.

Naqvi,S.M. and Rogers,J.G. 1987. Precambrian geology of India. Oxford University Press, New York,223p.

Narayan,J.,1971. Sedimentary structures in the Lower Greensand of the Weald, England and the Bas-Boulonnaise, France. Sed.Geol, v.6, pp.73-109.

Neilson,J. and Kocurek,G.,1986. Climbing zibars of the Algodones. Sedimentary Geology, v.48, pp.1-15.

Nemec,W. and Steel,R.J.,1984. Alluvial and coastal conglomerates: their significant and some comments on gravelly mass flow deposits. In: Sedimentology Of Gravels And Conglomerates (E.H.Koster and R.J.Steel, eds.). Mem.Can.Soc.Petrol.Geol.Calgary, v.10, pp.1-31.

Nichols, G.J., 1987. Structural controls on fluvial distributory systems- the Luna system, northern Spain. In: F.G. Ethridge, R.M. Flores, M.D. Harvey (Eds.), Recent developments in fluvial sedimentology. Soc. Econ. Paleontol. Miner. Sp. Publ., v. 39, pp. 269-277.

Olsen, H., 1987. Ancient ephemeral stream deposits: a local terminal fan model from the Bunter Sandstone Formation (L. Triassic) in the Tonder-3, - 4 and-5 wales, Denmark. In: L. Frostick and I. Reid (Eds.), Desert sediments: ancient and modern. Geol. Soc. Sp. Publ., v. 35, pp. 69-86.

Ori,G.G.,1982. Braided to meandering channel patterns in humid-region alluvial fan deposits, River Reno, Po Plain (northern Italy). Sed. Geol., v. 31, pp. 231-248.

Park, R., 1976. A note on the significance of laminations in stromatolites. Sedimentology., v.23.,pp.379-393.

Pemberton, S.G., Van Wagonar, J.C. and Wach, G.D., 1992. Ichnofacies of a wave-dominated shore line. In: S.G. Pemberton (Ed.), Applications of Ichnology to petroleum Exploration-A core workshop. Soc. Econ. Min. Palaeon., v.17, pp. 339-382.

Pettijohn, F.J., 1975. Sedimentary rocks. Harper, New York, 718p.

Pierson, T.C. and Scott, K.M., 1985. Downstream dilution of a lahar: transition from debris flow to hyperconcentrated streamflow. Water Resour. Res., v. 21, pp. 1511-1524.

Pierson,T.C. 1981. Dominant particle support mechanisms in debris flows at Mt. Thomas, New Zealand and implications for flow mobility. Sedimentology,v.28.,pp.49-60.

Plaziat,J.C. & Purser,B.H.,1998: The tectonic significance of seismic sedimentary deformations within the syn- and post-rift deposits of the north-western (Egyptian) Red Sea coast and Gulf of Suez. In : Sedimentation and Tectonics of Rift Basins : Red Sea Gulf of Aden. (B.H.Purser and D.W.Bosence, eds.), Chapman & Hall, London. pp.347-652.

Pratt,B.R.,1992: Shrinkage features ("molartooth" structure) in Proterozoic limestones – New model for their origin through synsedimentary earthquake-induced dewatering: Geol.Soc.Amer.Abstracts with Programmes, v.24(7), 53p.

Price,N.J. & Cosgrove,J.W.,1990: Analysis of geological structures. Cambridge University Press. pp.502.

Price,R.A.,1973: Large-scale gravitational flow of supra-crustal rocks, Southern Canadian Rockies. In : Gravity and Tectonics (ed. by K.de Jong & R.Scholten), Wiley, New York. pp. 491-502.

Ramos, A. and Sopena, A., 1983. Gravel bars in low sinuosity streams (Permian and Triassic, Central Spain). In: J.D. Collinson and J. Lewin (Eds.), Modern and ancient fluvial systems, Int. Assoc. Sedim. Sp. Publ., v. 6, pp. 301-312.

Reading, H.G. and Collinson, J.D., 1996. Clastic coasts. In: H.G. Reading (Ed.), Sedimentary environment: processes, facies and stratigraphy, 3rd Edn., Blackwell, Oxford, pp. 154-231.

Reading, H.G., 1996. Sedimentary environments: Processes, facies and stratigraphy. 3rd Edn., Blackwell, Oxford, 688p.

Reineck, H.E. and Wunderlich, F., 1968. Classification and origin of flaser and lenticular bedding. Sedimentology, v. 11, pp. 99-104.

Reineck, H.E. and Singh, I.B., 1980. Depositional sedimentary environments. 2nd Edn., Springer-Verlag, New York, 549p.

Roe,S.L.,1987. Cross-strata and bedforms of probable transitional dune to upper-stage plane-bed origin from a late Precambrian fluvial sandstone, northern Norway. Sedimentology, v.34, pp.89-101.

Ross,C.S. and Smith,R.L. 1961. Ash-flow tuffs: their origin, geologic relations and identification.U.S.Geol.SurvetProf.Pap.,v.366,pp.1-77.

Rubin,D.M.,1987. Cross-bedding, bedforms and paleocurrents. Soc.Econ.Paleontol.Mineral. Concepts in Sedimentology and Paleontology, v.1, 188p.

Rubin,D.M. and Hunter,R.E.,1983. Reconstructing bedform assemblages from compound cross-bedding In: Brookfield,M.E. and Ahlbrandt,T.S. (Eds.) Eolian Sediments and Processes. Elsevier, Amsterdam, pp.407-427.

Rust, B.R. and Jones, B.G., 1987. The Hawkesbury Sandstone, south of Sydney, Australia: Triassic analog for the deposit of a large, braided river. Jour. Sed. Pet., v. 57, pp. 222-233.

Rust,B.R.,1978. Depositional models for braided alluvium. In: Fluvial Sedimentology (A.D.Miall, ed.). Can.Soc.Petrol.Geol.Mem., v.5, pp.605-625.

Rust, B.R., 1984. Proximal braid-plain deposits in the Middle Devonian Malbaia Formation of eastern Gaspe, Quebec, Canada. Sedimentology, v. 31, pp. 675-695.

Saha, D., 2002. Multi-stage deformation in the Nallamalai Fold Belt, Cuddapah Basin, South India-Implications for Mesoproterozoic tectonism along southeastern margin of India., Gondwana Research.,v.5.,pp.701-719.

Schmincke, H.U., 1974. Volcanological aspects of peralkaline silicic welded ash flow tuffs. Volcanology, v. 38, pp. 594-636.

Schwartz,D.E.,1978. Hydrology and current orientation analysis of a braided to meandering transition; the Red River in Oklahoma and Texas, USA. In: A.D.Miall (Ed.), Fluvial Sedimentology, Can.Soc.Petrol.Geol.Mem., v.5, pp.231-256.

Self, S. and Sparks, R.S.J., 1978. Characteristics of widespread pyroclastic deposits formed by the interaction of silicic magma and water. Bull. Volcanol, v. 41, pp. 196-212.

Selley, R.C., 1985. Ancient sedimentary environments, 3rd. edn., Chapman and Hall, 317p.

Shanley, K.W. and McCabe, P.J., 1993. Alluvial architecture in a sequence stratigraphic framework: a case history from the upper Cretaceous Utah. In: S.S. Flint and I.D. Bryant (Eds.), The geological modeling in hydrocarbon reservoirs and outcrop analogs. Int. Assoc. Sed. Sp. Publ., v. 15, pp. 21-56.

Shanmugam, G., Poffenberger, M., and Toro Alava, J., 2000, Tide-dominated estuarine facies in the Hollin and Napo ("T" and "U") Formations (Cretaceous), Sacha Field, Oriente Basin, Ecuador: AAPG Bulletin, v. 84, pp. 652-682-.

Shelton,J.W. and Noble,R.L.,1974. Depositional features of braided-meandering stream. Amer.Assoc. Petroleum Geologists Bull., v.58, pp.742-749.

Shepherd,R.G.,1987. Lateral accretion surfaces in ephemeral stream point bars, Rio Puerco, New Mexico. In: Recent developments in fluvial sedimentology (F.G.Ethridge, R.M.Flores and M.D.Harvey, eds.). Spec. Publ. Soc. Econ. Paleont. Miner.Tulsa, v.39, pp.93-98

Shiki,T. & Yamazaki,T.,1996: Tsunami-induced conglomerates in Miocene upper bathyal deposits, Chita Peninsula, Central Japan. Sedimentary Geology, v.104, pp.175-188.

Sleep,N.H., Nunn,J.A. and Chou,L., 1980. Platform Basins : Annual review of Earth and Planetary Sciences,v.8,pp.17-34.

Smith, A.L. and Roobol, M.J., 1982. Andesitic pyroclastic flows. In: R.S. Thorpe (Ed.), Orogenic andesites, Wiley, pp. 415-433.

Smith, N.D. and Smith D.G., 1984. William River: an outstanding example of channel widening and braiding caused by bed-load addition. Geology, v. 12, pp. 78-82.

Smith,N.D.,1970. The braided stream depositional environment : Comparison of the Platte river with some Silurian clastic rocks, North-Central Appalachians. Geol.Soc.Amer.Bull., v.81, pp.2993-3014.

Smith,G.A.,1986. Coarse-grained nonmarine volcaniclastic sediment: terminology and depositional process. Geol.Soc.Am.Bull., v.97, pp.1-10.

Smith, G.A., 1988. Sedimentology of proximal to distal volcaniclastics dispersed across an active foldbelt; Ellensberg Formation (late Miocene), central Washington, Sedimentology, v. 35, pp. 953-977.

Smith, S.A, 1990. The sedimentology and accretionary style of an ancient gravel-braid stream: the Buddleigu Salterton pebble beds (Lower Traissic), SW England. Sed. Geol., v. 67, pp. 199-219.

Sneh, A., 1983. Desert stream sequences in the Sinai Peninsula. J.Sedim.Petrol., v.53, pp1271-1280.

Soria, J.M., Fernandez, J., Garcia, F. and Viseras, C., 2003. Correlative lowstand deltaic and shelf systems in Guadix Basin (Late Miocene Beltic Cordillera, Spain): the stratigraphic record of forced and natural regressions. Jour. Sed. Res., v. 73(6), pp. 912-925.

Southard, J.B., Smith, N.D. and Kuhnle, R.A., 1984. Chutes and lobes: newly identified elements of braiding in shallow gravelly streams. In: E.H.Koster and R.J.Steel Eds.), Sedimentology of Gravels and Conglomerates, Can.Soc.Petrol.Geol.Mem., v. 10, pp. 51-59.

Sparks, R.S.J. and Wright, J.V., 1979. Welded air-fall tuffs. In: C.E. Chapin and W.E. Elston (Eds.), Ash-flow tuffs, Geol. Soc. Am. Sp. Paper, pp. 155-166.

Stear, W.M., 1985. Comparison of the bedform distribution and dynamics of modern and ancient ephemeral flood deposits in the southwestern Karoo region, South Africa. Sed. Geol., v. 45, pp. 209-230.

Steel, R.J. and Thompson, D.B., 1983. Structures and textures in Triassic braided stream conglomerates ('Bunter" Pebble Beds) in the Sherwood Sandstone Group, North Staffordshire, England. Sedimentology, v. 30, pp. 341-367.

Steel, R.J., Maehle, S., Nilsen, H., Roe.S.L. and Spinnanger, A., 1977. Coarsening-upward cycles in the alluvial of Hornelen Basin (Devonian), Norway: sedimentary response to tectonic events. Geol. Soc.Am.Bull., v.88, pp.1124-1134.

Swift, D.J.P., 1970. Quaternary shelves and the return to grade. Mar.Geol., v.8, pp.5-30.

Swift, D.J.P. and Thorne, J.A., 1991. Sedimentation on continental margins, I: a general model for shelf sedimentation. In: Shelf Sand and Sandstone Bodies: Geometry, Facies and Sequence Stratigraphy (D.J.P.Swift, G.F.Oertel, R.W.Tillman and J.A.Thorne, eds.). Int.Ass.Sediment. Spec.Publ., pp.3-31.

Todd, S.P. and Went, D.J., 1991. Lateral migration of sand-bed rivers : examples from the Devonian Glashabeg Formation, South West Ireland and the Cambrian Alderney Sandstone Formation Channel Islands. Sedimentology, v.38, pp.997-1020.

Tsoar, H., 1982. Internal structure and surface geometry of longitudinal (seif) dunes. J.Sedim.Petrol., v.52, pp.823-831.

Turner, B.R. and Monro, M., 1987. Channel formation and migration by mass-flow processes in the Lower Carboniferous fluviatile Fell Sandstone Group, northeast England. Sedimentology, v.34, pp.1107-1122.

Visser,M.J.,1980. Neap-spring cycles reflected in Holocene subtidal large-scale bedform deposits: a preliminary note. Geology, v.8, pp543-546.

Walker,R.G. and Cant,D.J.,1984. Sandy flow systems. In: Facies models, 2nd edition (R.G.Walker, ed.). Geosci.Can., Reprint Series, v.1, pp.71-89.

Walker, R.G., 1984. General introduction: facies, facies sequences and facies models. In: R.G. Walker (Ed.), Facies models, 2nd ed., Geosci. Can. Repr. Ser., v.1, pp. 1-9.

Weimer, R.J., Howard, J.D. & Lindsay, D.R., 1982. Tidal flats. In: P.A. Scholle and Spearing, D. (Eds.), Sandstone depositional environments. Mem. Am. Assoc. Petrol. Geolg.v. 31, pp. 191-245.

Wells, S.G. and Harvey, A.M., 1987. Sedimentologic and geomorphologic variations in storm-generated alluvial fans, Howgill fells, northwest England. Geol. Soc. Am. Bull., v. 98, pp. 182-198.

Williams,G.E.,1969. Characteristics and origin of a Precambrian pediment. J.Geol., v.77, pp.185-207.

Williams,G.E.,1971. Flood deposits of the sand-bed ephemeral streams of Central Australia. Sedimentology, v.17, pp.1-40.

Wilson,I.G.,1972. Aeolian bedforms: their development and origin. Sedimentology, v.19, pp.173-210.

Windley,B.F., 1984. The Evolving continents. 2nd Edition. John Wiley and Sons. pp.399.

Wohletz, K.H. and Sheridan, M.F., 1979. A model of pyroclastic surge. Sp. Pap. Geol. Soc. Am., v. 180, pp. 177-194.

Wood,G.V. and Wolfe,M.J.,1969. Sabkha cycles in the Arab/Darb Formation off the Trucial coast of Arabia. Sedimentology, v.12, pp.165-191.

3

Hydrocarbon Possibilities in Frontier Basins : A case history from the Proterozoic Cuddapah Basin

Introduction

Widespread shallow water sedimentary basins became a dominant feature of the earth's surface during the Proterozoic (Rogers, 1993, Goodwin, 1996). Under favourable life bearing conditions as well as for photosynthesis, primitive life forms including cyanobacteria evolved at a faster rate and produced enormous amount of organic matter which could not be conceptualised till mid fifties as source of petroleum. However, a qualitative change emerged in this frame with the oil production from fractured Precambrian basement rock in Central Kansas (Walter, 1953). Since then Precambrian organic matter has been extensively investigated for evidence of the nature of early life (Schopf, 1983), but it has only rarely been considered as a source for oil and gas as Precambrian sedimentary strata has undergone extensive burial, thermal alteration and tectonic movement, such that any petroleum formed was considered to have been lost or destroyed (Hobson & Tiratsoo, 1981). In addition, doubts existed about the availability of suitable organic material due to the presumed primitive nature of the Precambrian biota. However, commercial accumulation of indigenous petroleum in Siberia (Meyerhoff, 1980) and Oman (Al-Marjeby & Nash, 1986) and prospective hydrocarbon basins of Australia (Murray et al, 1980, Jackson et al, 1988) have altogether changed the approach for exploration of oil and gas in Precambrian terrain.

Despite the existence of large number of Proterozoic basins in India, unfortunately exploration for hydrocarbon is still in its infancy as the source rock potentials has not been ascertained in these basins owing to non-availability of stratigraphic and sedimentological details.

Commercial production of Proterozoic sourced-oil from some basins of Russia, China and Oman have drawn attention to Proterozoic oil exploration in regions including India. Cuddapah Basin (1.9 Ga- 0.6 Ga) evolved under an overall control of extensional regime. Aeolian, fluvial, sabkha, lagoonal, shallow marine, slope apron- to deep water turbidite facies are identifiable in Cuddapah rock rocords. Cyclicities in clastic-carbonate couples, carbonaceous shales and volcaniclastics, and imprints of transgressive-regressive pulses characterise Cuddapah Supergroup. Vempallae-, Tadpatri- and Cumbum carbonate-, and carbonaceous argillite facies have yielded 36 different Acritrachous and other microbiotic forms, indicative of immense biotic activities during sedimentation. Lithounits record extensive reserves of microbial mats, their eroded fragments, layered and domal stromatolites, pisolites and composite oolites controlled by the growth of bacterial colonies. The microbiota include identifiable filamentous algae, desycladacean forms and cyanobacteria to which origin of many Proterozoic oil deposits have been attributed.

Carbon, derived from reduction of atmospheric carbon dioxide by hydrogen from frequent volcanic emanations during sedimentation was an additive source to the carbon load produced by prolific photosynthesis by microbiota. Redistribution of carbon from carbonaceous sources during upward migration of quartz veins seems to have contributed to the growth of lenses and pockets of carbon in rocks. Such carbon content through further organic and/or other processes might have produced oil. Hydrocarbon accumulations specially in permeable fractured source/ reservoir zones of lower Cuddapah rocks in carbon-rich and biotic activity-dominated Cuddapah basin might have escaped destructive effects of metamorphism and deformation. The Cuddapah scenario in ancient stable cratonic set-up did not witness disruptions such as trap breaching, imperfect sealing or thermal obliteration. This highlights the possibilities of finding valuable hydrocabon resources specially in the lower Cuddapah facies.

Proterozoic petroliferous Dripping Spring Quartzite Sequence of USA, Officer- and MacArthur basins, Australia are analogous to Cuddapah basinal set-up.

This chapter discusses salient aspects of facies and depositional milieus, occurrence of sediments rich in organic matter, occurrence of source rock and reservoir potentials and hydrocarbon biomarkers in the Palaeo- to Neo Proterozoic (ca. 1.9-0.6 Ga) Cuddapah basin of Peninsular India.

Salient Aspects of Facies and Depositional Milieu

The Cuddapah Basin refers to an area about 44,000 square km of unmetamorphosed and relatively undeformed Proterozoic sedimentary rocks of Andhra Pradesh, India (King, 1872, Murthy, 1982). This region is not a basin in the strict sense but represents possibly only a part of a much more extensive area of Proterozoic sedimentary rocks spreading far beyond the present day limit of the basin in the north, south and west. Sedimentary rocks of the Cuddapah basin form stratigraphically simple succession (Murthy, 1979). The salient aspects of the facies, their transition and depositional milieus are mentioned below:

1. Cyclicity in all the Formations including the transition facies of different Groups such as the Papaghni, the Chitravati, the Nallamalai and Srisailam Sequence of the Proterozoic Cuddapah Supergroup has been established for the first time in the Indian Stratigraphy.
2. Contrary to the accepted views of dominance of shallow-marine milieu, discovery of aeolian-, fluvial-, sabkha- and deep-water turbidite facies are considered as original contribution in the Cuddapah scenario.
3. A strong correlation has been established among the different stages of basinal evolution, role of oscillatory movements, volcanism and extensional tectonics.
4. Examples of such correlations are:

a) Facies relations in the Gulcheru Formation of the Papaghni Group are indicative of recession of a piedmont-like source and shift of terrestrial fan set-up away from the basin. The extensional regime probably accounts for a progressive increase in the domain of the protobasinal phase during the Gulcheru sedimentation.

b) Cyclicities in Vempallae and transition facies are related to the superposition of periodic pulses of volcanism on an oscillatory set-up and concomitant faulting in an extensional regime. Global sea-level changes seem to have little contribution towards the formation of the cyclic associations.

c) The Hostavaram – Naramrajupallae tectofacies of the Nallamalai Group, comprises chaotic facies gradational to impure carbonate turbidites with a volumetrically important volcaniclastic element in each cycle and indicated deposition under the overall control of syndepositional faulting in an extensional regime7

d) The Egalapenta Member of the Srisailam Sequence is marked by accumulation of cycles of shallow-marine facies which might be ascribed to eustatic oscillatory changes in sea-level

5. Eustatic changes in sea-level as substantiated by shallow-marine signatures appear to have played a subordinate role in the evolution of the Cuddapah basin.

6. Tectonic controls of sedimentation indicate that the basinal evolution took place under an overall influence of extensional regime where intermittent volcanism and concomitant faulting on an oscillatory set-up played a dominant role.

Occurrence of sediments rich in organic matter

The Cuddapah Supergroup records prolific growth of stromatolites in the Vempallae–, Tadpatri–, and Cumbum facies in association with ubiquitous development of carbonaceous argillite facies housing significant microbiotic forms (Figs. 1, 2, 3, 4 and 5). Salient aspects of the Vempallae stromatolites have been highlighted showing the importance of the similar stromatolitic forms associated with other Cuddapah Formations.

Vempallae Stromatolites

Twenty one lithofacies with strong volcaniclastic element in a shelfal set-up, often forming assymetric to fining upward cycles and punctuated by non-cyclic phases both in northern and southern sectors attest to an overall pulsating basin dynamics.

Noncyclic carbonates include volcaniclastic and genetically associated chert as two dominant elements. Dolostones spatially interfingering with and vertically grading into volcanogenic chert suggest syndepositional origin for the chert, epiclastics and carbonates. Generation, transportation and deposition of the intraclasts and the growth patterns of the long axis of the stromatolites indicate well defined current directions even under the influence of wave action in the Upper Vempallae Sequences (cf. Grotzinger, 1988) while mudstone dominated Lower Vempallae Sequences with apparent sheet geometry characterising low energy shallow marine southern facies (cf. Aitken, 1967).

Vempallae biosedimentary forms such as cryptalgaminites, pseudocoloumnar, cumulate, domal and columnar stromatolites delineate three informal assemblage biozones. The Vempallae stromatolites mostly reflect ecologically controlled morphology as suggested from great variations in their architectural patterns (Figs. 1, 2 and 3) (cf. Black, 1933; Ginsburg et al, 1954, Hoffman, 1967, Hoffman, 1973, Logan, 1961, Logan et al, 1999).

Responses of subtle ecological imbalance as suggested from spatial and temporal changes in microfacies was duly reflected in subsequent changes in architectural patterns of stromatolites. Strong positive correlation between the intraclastic carbonates and relatively simple growth patterns of the stromatolites corroborates such ecological and morphological tuning.

Facies modeling reveals that the Vempallae formation was deposited in a shelfal set-up with complex biogeochemical variations when pulses of subaerial volcanism were generated in response to thermal regime (Dasgupta, 1986; Dasgupta et al, 1990).

Microbial Mats

Ancient microbial mudchips have been described from shallow-marine Lower Cambrian of Jordan (Pfulger, 1999), Neoproterozoic shallow-marine strata of the Nama Group, Namibia and South Africa (Pfulger and Greese,1996). Some argillaceous horizons in Vempallae, Tadpatri as well as Cumbum are packed with microbial muddy chips in the shallow-marine facies. The chips are usually fine grained. Microscopically irregularly shaped coaly patches are circular, elliptical or triangular or bit irregular in shape in association with mud and very fine sand. The carbonaceous patches in microbial mats show angular to subrounded corner and up to a length of 4.5 cm. Petrologically silt comprises clastic quartz, feldspar and mica flakes. Microbial mat chips are in association with curled mudflakes. On some bed surfaces of marine facies of rock concentrically rolled up sily laminae, 2 to 3 mm thick analogous to previously described 'roll ups' biogenic structures are generally ascribed to soft sediment deformation or desiccation of microbial mats (Dasgupta et al, 1988; Eriksson et al, 2000). It appears that possibly microbial mats imparted the cohesiveness to the muddy layers which curled up and formed flakes owing to desiccation under subaerial conditions in shallow-marine to lagoonal set-up. Ancient and modern analogues support the conclusion (Gerdes et al, 2000, and Pfulger and Greese, 1996). The mudflakes in association with many carbonaceous microbial elements from Vempallae - Tadpatri- and Cumbum marine facies are strongly

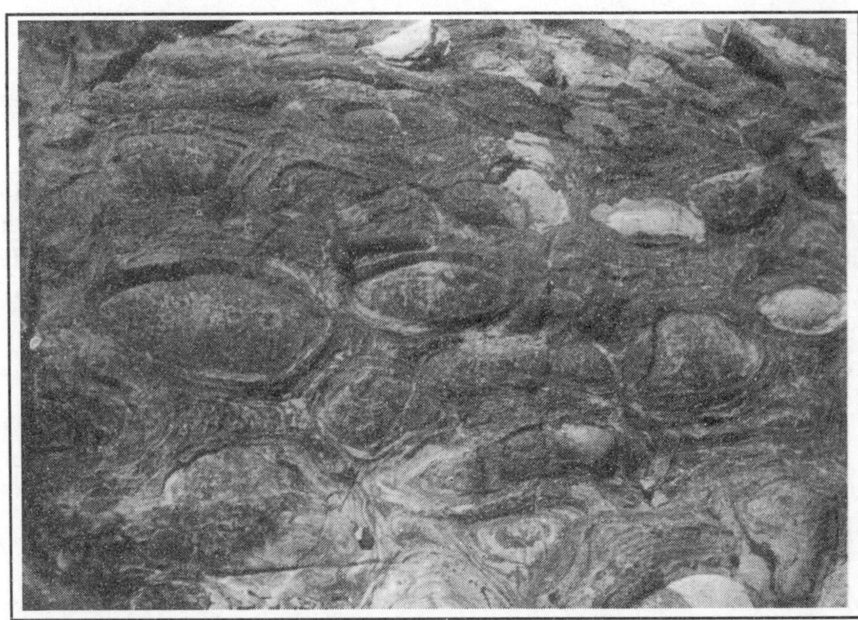

Fig.1 : Colony of elongated, oval-shaped stromatolites with occasionally preserved carbon specks, near Tumlapallae, Vempallae Formation.

Fig.2 : Laterally linked pseudocolumnar stromatolites with gently convex lamina in the vicinity of oval-shaped stromatolites, Vempallae Formation.

Fig.3 : Pseudocolumnar stromatolites with circular heads, Vempallae Formation.

oriented in a particular direction and indicates presence of strong currents. The flakes were cohesive enough to retain their shape possibly owing to the sticky polysaccharides and proteins of the microbial mats that glued non-cohesive silty grains.

Acritarchs and related microbiota from Pulivendla, Tadpatri and Gandikota Formations:

Palynological work on Pulivendla Quartzites, Tadpatri Shales and Gandikota Quartzites of the Chitravati Group has yielded different palynomorphs classifiable under specific and non-specific taxonomic categories which have been monographically described and illustrated (Viswanathiah et al, 1982). Based on these palynomorphs an attempts has been made to give stratigraphical remarks to the rocks with which they are associated. Besides, the presence of abundant life in these rocks is stressed. The microbiota recognised indicates that the microbiota in the Pulivendla Quartzites and Tadpatri Shales relatively have a lesser concentration than those in Gandikota Quartzites. The rock types have yielded thirty six different Acritarchous and other microbiotic forms classifiable under various taxonomic categories

Genus Protosphaeridium (Timofeev, 1966) from Acritarcha Group Evitt identified are subspherical to oval, dark brown, 15-25 m in diameter, wall single layered, stout smooth, dense opaque surface. These forms differ from most other species in possessing a smooth and faintly sculptured surface and the absence of nucleus.

Protosphaeridium densum Timofeev are spherical, yellowish brown to dark brown dense, psilate texture, 10 - 15 m in diameter, wall single layered, subopaque. This form bears close similarity to Protosphaeridium laccatum and Protosphaeridium parvulum (Timofeev, 1974) in colour and size, but it differs from them in having smaller dimensions and much thicker envelope.

Protosphaeridium pusillum, Granomerginata, Leiosphaeridia, Stictosphaeridium implexum Timofeev, Kiddivella Polysphaeritae, Tasmanites sp, Chlorophytes, Rhodophytes, Rymnesiophytes, and/ Dinoflagellates or their precursors are most dominant among other identifiable acritrachs (Viswanathiah et al,1982).

Other microfossils identified with acritarchs are two distinct types of segmented Algal Filaments. The Algal Filaments of first major types are spindle shaped, elongated, yellowish brown, wall single layered thick segmentation distinct surface translucent.

Segment Algal Filaments of the other type are lenticular, brown colour, single layered wall, thick segmentation distinct, clear constriction from one segment to the other, segments vary from 3 to 5 in number; the lower and the upper segments are blunt, a medium furrow is visible, a small fold in the lowermost segment is distinct.

Based on these palynomorphy and algal remains stratigraphic remarks to the rocks with which these are associated are drawn. Besides, presence of abundant life in the rocks strongly points to the possibilities of hydrocarbon generation from these strata.

Stratigraphic Implications

Deciphering specific stratigraphic implications of the microbiota have not been attempted since the forms presently identified show varied stratigraphical ranges. Efforts are been made therefore to analyse the characterisation of stratigraphic impact of specific forms. In addition, more microbiotic finding are being examined from several other rock types. However, the total micro organic assemblage in its entirety has certain taxonomic and evolutionary manifestations worthy of further probe noted here under.

The forms that have been already recognised in the underlying Papaghni Group earlier on comparison with the forms presently figured suggest that the forms of the Chitravati Group have greater taxonomic diversity and relatively higher evolutionary status. The higher evolutionary status, recognisable is being manifested by typical characters related to the nature of the wall, peripheral processes, ornamental features and size. The aspects outlined could ultimately throw considerable amount of light while analysing specific stratigraphic value on one hand and evolutionary status on the other. From this point of view, the microforms recognised are of particular significance for the study of microbiotic life of Chitravati Group, in particular and Cuddapah Super Group in general. Middle Precambrian age element is significant based on the microbiotic characteristics. Further, from a correlation study of microbiota, it appears that a major component of the organic content bears affinity to the microbiota that has been earlier described from the upper Lokapur Formation of the Kaladgi Group, South India.

Hydrocarbon Biomarkers and Source Rocks

Cuddapah lithounits record modest to extensive reserves of microbial mats, their eroded fragments, layered and domal stromatolites, pisolites and composite oolites controlled by the growth of bacterial colonies (cf. Arouri et al, 2000). Significant volume of carbonaceous material from primitive organic source appears to have an additional inorganic counterpart. Additive carbon was probably derived owing to reduction of atmospheric carbon dioxide by hydrogen from frequent volcanic emanation during sedimentation. In addition, Acritrachous and other microbiota such as identifiable filamentous algae, desycladacean forms and cyanobacteria in association with ubiquitous carbonaceous black shales define, prominent biomarkers at stratigraphic levels. Exposures of such shales are well preserved in Vempallae–, Cumbum, Pullampet–Bhakrapet outcrops on the Cuddapah–Vontimitta Road. However, biomarker distributions of different types of kerogen should be addressed in order to ascertain the prognostic hydrocarbon reserves and potentials of Cuddapah source rocks.

Reservoirs and Traps

Geological studies of Cuddapah rocks till date is based on outcrops only. Owing to non-availability of outcrop-to-subcrop relations overall facies variations suggest general characteristics of presumed reservoirs and traps.

Sandstones are widespread at the base of several sedimentary cycles of the Cuddapah Supergroup of Peninsular India. Qualitatively high fracture porosities as discernible from Cuddapah facies complexes appear to be formidable factor for oil and gas.

Cyclical sedimentation, particularly in transition facies of Papaghni - Chitravati - Nallamalai and Srisailam Formation resulted in alternation of sandstone and clay and/or rocks of sabkha origin. The clay members form seals for underlying sandstones reservoirs. One of the regional seals are argillite in association with carbonaceous shales of the Vempallae with sparingly occurring stromatolite lamina (Fig.6). Thicknesses of the seals are variable within the range between 20 meters and 55 meters. Owing to the low extent of study of these ancient rocks, not a single local trap has been delineated. General characteristics of these rocks, however, suggest that a large number of traps of various kinds are present. The oil-gas zone of Riphean and Vendian of the Siberian craton may be an analogue of graben-horst traps in several aulacogens of Russian Platform (Fedorov, 1994). Evaporite sequences of the cyclic sabkha facies comprising impure carbonates, halites, anhydrite, gypsum with chickenwire structures and other associated facies attributes are frequently well developed in the Vempallae–, Tadpatri –, Cumbum –, Pullampet –, Srisailam Formations. Such potentially effective units with halite, claystone and siltstone, more than 10 meter thickness seems to form favourable traps in Meso– to Neoproterozoic clastic and carbonate subbasin of the Cuddapah which resembles in its hydrocarbon potential with that of McArthur - Officer - and Amadeus Basins and Adelaide Rift Complex of Australia (Crick et al, 1988; Carlsen et al, 1999).

Discussion and Conclusion

Frequently occurring carbonaceous shales with strong microbiotic content at different stratigraphic levels (Vempallae–, Tadpatri–, Cumbum–, Pullempet Formations) indicate petroleum source rock potential. The sediments contain well-preserved microbial mats. They occur variously as (1) autochthonous mats within epiclastic silts in the peritidal platform carbonates, and in slope-controlled deposits within the lowermost reaches of the photic zones, (2)allochthonous structurally dismembered mats within debris-flow deposits, (3) tempestite deposits within the peritidal carbonates that contain large intraclasts of biolaminated siltstones, and (4) detrital fragments and small intraclasts of mat-bound subtrate that occur in profusion within low-

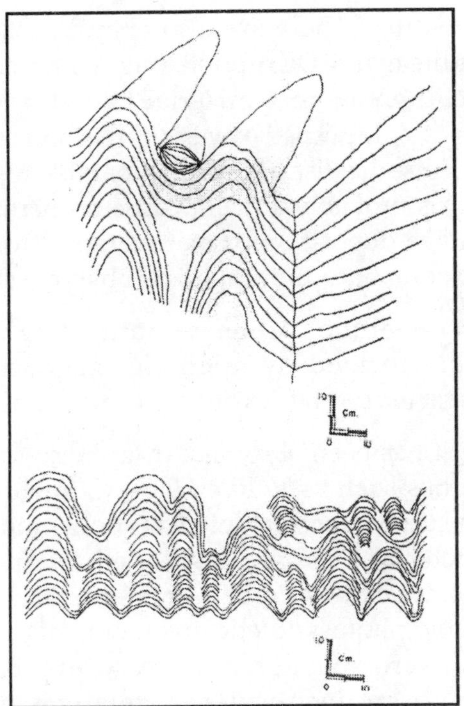

Fig.4 : Diagrametic sketch showing an order of pseudocolumnar stromatolites, Vempallae Formation.

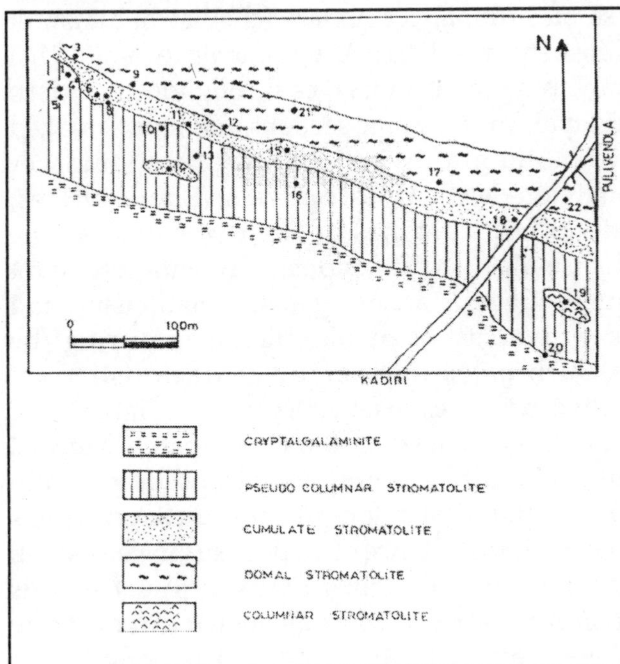

Fig.5 : Spatial distribution of different types of stromatolites, Vempallae ormation.

concentration turbidites. There are also abundant acritarchs within fine-grained turbidite marls. Disruption, fragmentation, dislodgement, removal of autochthonous mat material from lagoonal to shallow marine milieus was accomplished by various mechanisms, both ambient and catastrophic. Episodic disturbance comparable to those associated with pycnocline was probably a major cause of disruption and down slope removal and re-deposition of mat material from the shallower part of the photic zone onto the slope and abyssal setups.

Very finely crystalline carbonate, intimately associated with coccoids, probably formed by microbial mediation in peritidal environments characterised by warm to elevated water temperatures.

Furthermore, it has been observed, from Meso to Neoproterozoic Cuddapah analogues such as Pulivendla –, Cumbum –, Srisailam –, and still younger late Neoproterozoic Kurnool – and Palnad – facies that the cyanobacterial mats are associated alternately with non-cyanobacterial microbial mats. Probably non-cyanobacterial microbiota included anoxygenic photosynthetic green non-sulpher bacteria with minor terminal anaerobic sulphates – reducers. Some turbidite – siltstone samples from deep-water correlatives of the Cumbum containing only alochthonous mat detritus, have a cyanobacterial biomarker signature that is strongly overprinted by other bacterial signals, including those of terminal consumers (cf. Viswanathiah et al,1982; Aitken, 1967; Ginsberg et al, 1954). Abundant algae (acritarchs) are present in many palynological assemblages throughout the Meso to Neoproterozoic Cuddapah stratigraphic sections and show marked upward increase in taxonomic diversity. Sterane Biomarker distributions are consistent with Chlorophytes, Rhodophytes, Rymnesiophytes, and / Dinoflagellates or their precursors. The association of such diverse algal sterane biomarkers with abundant acanthomorphs acritarchs suggests a close genetic relationship and demonstrates the great diversity of planktonic algae during the terminal Proterozoic. Such paleo-microbiotic characteristics are discernible from conclusive evidences so far collected, highlight strong probability for hydrocarbon generation in the Cuddapah – Kurnool scenario. Despite non-availability of kerogen data predominantly significant cynobacterial input at stratigraphically different levels furthermore attest to the possible existance of oil deposits in Cuddapah basinal complex. Precambrian analogues lends support to this conclusion as many Proterozoic oil reserves are attributed to the activity of cyanobacteria (Taylor & Taylor, 1983).

Fig.6 : Argillite in association with carbonaceous shales of the Vempallae Formation with sparingly occurring stromatolite lamina.

REFERENCES

Aitken, J.D., 1967. Classification and environmental significance of cryptalgal limestones and dolomites with illustrations from the Cambrian and the Orodovician of southwestern Alberta, Journal of Sedimentary Petrology,v.37, pp.1163-1178.

Al-Marjeby, A. & Nash,D., 1986. A summary of the geology and oil habitat of the eastern flank hydrocarbon province of south Oman : Marine and Petroleum Geology, v.3, pp.306-315.

Arouri, K., Conaghan,P.J., Walter,M.R., Bischoff,G.C.O. & Grey,K., 2000. Reconnaissance sedimentology and hydrocarbon biomarkers of Ediacarian microbial mats and acritarchs, lower Ungoolya Group, Officer basin, Precambrian Research, v. 100, Issues 1-3, pp.235-280.

Balck, M., 1933. The algal sediments of Andros Island, Bahamas: Royal Society (London) Philos.Trans.Ser.B, v.222, pp.165-192.

Carlsen, G.M., Apak,S.N., Ghori,K.A.R., Grey,K. & Stevens,M.K., 1999. Petroleum potential of the Neoproterozoic western Officer basin, western Australia, based on a source-rock model from Empress-IA, APPEA Journal, pp. 322-341.

Crick,I.H., Boreham,C.J., Cook,A.C. & Powell,T.G., 1988. Petroleum geology and geochemistry of middle Proterozoic McArthur basin, northern Australia II: Assessment of source rock potential, AAPG Bulletin, v.72, pp.1495-1514.

Dasgupta,P.K., Dey,S. and De,S.,1990. Facies variations and different stages of basin evolution in the Proterozoic Cuddapah Basin. India, Abstract volume on Evolution of Precambrian Crust in India, Geological Survey of India, pp.5-6.

Dasgupta,P.K.,1986. The sedimentation model of the Papaghni Group (The Middle Proterozoic), the Cuddapah Basin. Peninsular India. 12th Int.Cong.Sedimentology, Canberra Session, pp.211-212.

Eriksson,P.G., Simpson,E.L., Eriksson,K.A., Bumby,A.J., Steyh,G.L. & Sarkar,S, 2000. Muddy Roll-up structures in siliciclastic interdune beds of the ca.1.8 Ga. Waterberg Group, South Africa, Palaios, v.15, pp.177-183.

Fedorov,S.G., 1994. Petroleum potential of Proterozoic rocks of Russian platform, Petroleum Geology, v.29, no.1/2.

Gerdes,G., Klenke,T. & Noffke,N., 2000. Microbial signatures in peritidal siliciclastic sediments: A catalogue: Sedimentology,v.47. pp.279-308.

Ginsburg,R.N., Isham,L.B., Bein,S.J., Kuperberg,J., 1954. Laminated algal sediments of South Florida and recognition in the fossil record, in : Aitken,J.D. 1967, Journal of Sedimentary Petrology,v.37, pp.1163-1178.

Goodwin,A.M.,1996. Principles of Precambrian Geology. Academic Press, London. 327p.

Grotzinger,J.P., 1988. Introduction to Precambrian Reef rocks. In : Geldster,H., James,N.P., Tebbut,G., eds. Reefs-Canada and adjacent areas, Canadian Soc. Petrol.Geology.

Hobson,G.D. & Tiratsoo,E.N., 1981. Introduction to petroleum geology : Houston, Gulf Publishing Co., 352p.

Hoffman,H.J., 1973. Stromatolytes: characteristics and utility, Earth Science Review,v.9, pp.339-373.

Hoffman,P.F., 1967. Algal stromatolites: use in stratigraphic coorelation and palaeocurrent determination, Science, v.157, pp. 1043-1045.

Jackson,I., Sweet,P. & Powell,T.G., 1988. Studies on petroleum geology and geochemistry of the middle Proterozoic McArthur basin, northern Australia 1: petroleum potential : Australian Petroleum Association Journal,v.28, pt,1, pp.283-302.

King,W.,1872. The Kadapah and Karnul Formations in the Madras Presidency. Mem.Geol.Serv.India, v.8, Pt.1, p.346.

Logan,B.W., 1961. Cryptozoon and associated stromatolytes from the recent of Shark Bay, W.Australia: Journal of Geology,v.69, pp.517-533.

Logan,G.A., Calver,C.R., Gorjan,P., Summons,R.E., Hayes,J.M. & Walter,M.R., 1999. Terminal Proterozoic benthic microbial mats and their environmental significance, Geochim. Cosmochim. Acta, v.63, pp.613-621.

Meyerhoff,A.A.,1980. Geology and petroleum fields in Proterozoic and Lower Cambrian strata, Lena-Tunguska Petroleum province, Eastern Siberia, USSR, in M.T. Halbouty, ed., Giant oil and gas fields of the decade, 1969-1978: AAPG memoir, 30pp.225-252.

Murray,G.E., Kaczor,M.J. & McArthur,R.E., 1980. Indigenous Precambrian petroleum revisited, AAPG Bulletin, v.64, pp.1681-1700.

Murthy,Y.G.K.,1979. Salient features of the revised geological map of the Cuddapah basin. 3rd Workshop on Status, Problems and Programmes in Indian Peninsular Shield. Institute of Indian Peninsular Geology, Hyderabad, pp.1-21.

Murthy,Y.G.K.,1981. The Cuddapah Basin – A review of basin development and basement framework relations: 4th workshop on status problem in the Cuddapah Basin. Institute of Indian Peninsular Geology, Hyderabad, pp.51-72.

Pfulger,F. & Greese,P.G., 1996. Anon actualistic sedimentary structures: Sedimentary Gelogy,v.102, pp.263-274.

Pfulger,F., 1999. Matground structures and redox facies: Palaios, v.14, pp.25-39.

Rogers,J.J.W.,1993. A History of the Earth. Cambridge Univ.Press, 312p.

Schopf,J.W., ed., 1983. Earth's earliest biosphere : Princeton, Princeton University Press, 543p.

Taylor,T.N. & Taylor,E.L., 1983. The biology and evolution of fossil plants, Prentice Hall, New Jersey, 378p.

Timofeev,B.V. 1966.Mikropaleogitologicheskoe issledovanie drevnikh Svit : Izdat "Nauka" Moskva.

Timofeev,B.V. 1974. Microfossils of Proterozoic and Early Palaeozoic of USSR, Acad. Nauk. USSR.III.

Viswanathiah,M.N., Venkatachalapathy,V. & Sekhar,R.G.C., 1982. Acritarchs and related microbiota from Pulivendla Tadpatri and Gandikota Formations of the Chitravati Group, Cuddapah Supergroup, pp. 91-106., in : Bhattacharji,S & Balakrishna,S (eds.) Monograph 2: Evolution of the intracratonic Cuddapah basin, south India, Institute of Indian Peninsular Geology, Hyderabad.

Walter,R.F., 1953. Oil production fractured Precambrian basement in Central Kansas, AAPG Bulletin,v.37, pp.300-313.

4

Picrite Borne Mantle Xenoliths From Cuddapah Basin

Introduction

The role of the mafic magmatic elements in the Cuddapah basinal fill has engaged the attention of workers studying dynamics of the Proterozoic basin evolution (Sen and Rao 1967 ; Bhattacharji 1981; Bhattacharji and Singh 1984). It is well known that the Cuddapah magmatic activity peaked during the Tadpatri time when thick composite picrite-dolerite sills, flows, ash flows and tuffs were formed. With thermal cooling thereafter, the magmatic activity dwindled rapidly and basinal configuration changed in stages.

Interestingly, clusters of peridotite xenoliths have been found in the picrite intrusives in the basal Tadpatri Shales, especially, in their northwestern-most exposures situated northeast of Pulivendla.

Garnet-, spinel-, and plagioclase Iherzolite xenoliths occur in clusters in the first phase of mafic intrusives in the Tadpatri Shales of the Cuddapah basin. The nodular xenoliths are deformed and contain thin veins and patches of calcite-rich carbonatites. The presence of garnet Iherzolite indicates the maximum depth of derivation of these nodules to be 200 km. The carrier picrite magma, during ascent through the mantle, possibly halted at an intermediate magma chamber to form cumulates before resumption of its journey to the upper levels of the crust.

The heterogeneity of these xenoliths in both mineralogy and texture is significant and its bearing on the Proterozoic upper mantle beneath this segment of the south Indian shield is obvious.

Geological Background

The Proterozoic stratotype crescent-shaped Cuddapah basin consists in its lower parts mainly of an assemblage of easterly low dipping multilateral and multistorey conglomerate and sandstone bodies within mega-channels of the basal Papaghni (Gulcheru clastics) indicative of a fan complex and fluvial deposition, followed by fine-grained elastics, the Vempalle intrabasinals of the upper Papaghni Group and a cyclic sedimentation of sandstone (orthoquarzite)-shale-sandstone of the Chitravati Group (Dasgupta 1986). Magmatic activity started at the Vempalle stage and became intense during Tadpatri Shale sedimentation (upper stage of the Chitravati). This was followed by a break in sedimentation and erosion, and mafic, syenite and granitic intrusive activity occurred on a limited scale during the subsequent Naliamalai time and possibly also during the still later Srisailam sedimentation.

Two distinct phases of mafic intrusive activity accompanied the Tadpatri sedimentation. The first phase is associated with the basal Tadpatri argilites and comprises olivine/pyroxene rich picrite and dolerite. Two pyroxenes (opx and cpx) occur in the picrite which shows a general coarse-grained poikilitic fabric. The dolerite, with the somewhat coarse subophitic fabric, also contains two pyroxenes. The second phase of intrusive activity is associated with the upper part of the Tadpatri sequence. Despite their natural tendency to follow the bedding planes, the intrusives, especially those in the first phase, are seen to cut across the associated sedimentaries at low angles. There are also interconnecting feeder dykes.

Peridotite xenoliths occur as nodules in the first phase in exposures situated north-east of Pulivendla town. The nodules are generally ellipsoidal in shape and vary in dimension, the largest measuring 1.32 m across. They occur in clusters, although haphazard in distribution, and are fairly conspicuous in the exposures. Some of the nodules are cut by thin veins and patches of calcite-rich carbonatites (Gold,1966; Heinrich,1966).

Petrology of Xenoliths

The xenoliths are all peridotite and comprise olivine, enstatite (opx) and diposide (cpx) together with a few aluminous phases, namely garnet (ga), spinel (sp), plagioclase (plag) and phlogopite (phlog) (Fig.1). Petrological studies reveal the following three distinct mineralogical assemblages :

Fig.1 : Olivine and garnet (dark) poikilitically enclosed within orthopyroxene (Garnet lherzolite), Pulivendla Formation. Cross polars.x18.

1. ol - opx - cpx - ga - phlog - carb.
2. ol - opx - cpx - sp - piag - phlog -carb - biot.
3. ol - opx - cpx - plag - phlog - carb - biot.

Under the IUGS classification system (0'Hara in Wyliie, 1967; Streckeisen,1976) these assemblages are assigned to garnet - spinel and plagioclase Iherzolite respectively.

Transition between spinel Iherzolite and plagioclase Iherzolite is indicated by the presence of plagioclase rims around subrounded spinel grains in some xenoliths. The average volume percentages of the constituent mineral phases of the three types of Iherzolites have been obtained by modal analysis (Table.1).

Table 1. Volume Percent Modal Analysis of Three Lherzolites Types

Type	ol	opx	cpx	ga	Sp	plag	phlog	carb	biot
Garnet Lherzolite	61.2	12.5	14.6	3.2	tr.	–	0.8	2.4	tr.
Spinel Lherzolite	55.2	10.1	19.1	–.	7.9	3.4	0.4	2.2	1.0
Plagioclase Lherzolite	53.7	13.8	16.7	–	–	10.2	0.9	3.8	1.1

A characteristic feature of the Tadpatri nodules is the presence of porphyroclastic texture indicative of deformation (Mercier and Nicolas 1975). Porphyroclasts are made up mostly of large, elongate, strained intratelluric olivine with two modes (6 mm and 0.45 mm), while neoblasts are small, often polygonal and strain-free. Protogranular textural features such as marginal granulation, reduction in grain size and stretching by gliding with preferred orientation in case of enstatite porphyroclasts suggest a considerable degree of deformation. Partially annealed textures in plagioclase Iherzolite also retain former record of deformation. The annealed texture shows straight grain boundaries, 120° triple junctions and diminished undulose exctinction together with a few kinks.

Chemistry of Xenoliths and Host Picrite

Analyses of two samples of garnet Iherzolite, and average of three carbonatites are presented here (Table-2). Mean analysis of eight samples of picrites around Pulivendla (Sinha and Rao,1968) is also shown in the table.

Table 2: Chemical Compositions of Garnet Lherzolite, Carbonatite and Picrite, Cuddapah Basin, India.

	1	2	3	4	
SiO_2	45.86	40.64	36.82	44.34	(1.59)
TiO_2	0.42	0.19	0.42	0.62	(0.07)
Al_2O_3	2.21	8.52	6.35	8.57	(1.12)
Fe_2O_3	0.86	7.06	8.66	2.52	(0.55)
FeO	7.22	5.22	3.05	8.44	(0.38)
MnO	0.04	0.04	0.03	0.15	(0.02)
CaO	3.81	5.65	11.37	5.40	(0.33)
MgO	35.60	23.57	18.04	24.45	(2.18)
Na_2O	0.23	3.10	0.96	1.46	(1.11)
K_2O	0.62	0.60	0.22	0.26	(0.11)
P_2O_5	0.13	0.13	0.11	0.12	(0.04)
H_2O+	0.23	0.16	0.24	3.36	(0.44)
H_2O-	1.54	4.20	6.61		
CO_2	1.04	1.04	6.77		
	99.91	100.12	99.65	99.69	

Analyst: B.P.Gupta, Classical method

1. Garnet Iherzolite, Pulivendla, Cuddapah Basin
2. Garnet lherzolite, Pulivendla, Cuddapah Basin
3. Average of three analyses of carbonatites, Pulivendla, Cuddapah Basin
4. Average of eight analyses of picrite; Sinha and Rao 1968; the figures in bracket

The two analyses of garnet Iherzolite show differences in $Si0_2$, Al_20_3, MgO and Na_20. Compared to Lesotho garnet Iherzolite xenoliths (Cox *et al.* 1973 ; EVSP 1981), the Cuddapah varieties come under the *sensu stricto* category, although their $Ti0_2$ and $C0_2$ contents are a little on the higher side.

Compared to granular undeformed lherzolites (Nixon *et al.*, 1973) these deformed Iherzolites are less depleted in Na_20, $Ti0_2$, CaO and AI_20_3.

Discussion

Studies on mineralogical, textural and geochemical characteristics of garnet-, spinel- and plagioclase Iherzolite xenoliths in lavas and intrusives have led to a general consensus that they have been plucked from the upper mantle.

The presence of xenoliths of garnet-, spinel-, and plagioclase Iherzolite in the Tadpatri may be ascribed to considerable heterogeneity in the mantle source. Alternatively, the three varieties may suggest consecutive phase transition stages of one and the same parent material. The texture of the Iherzolites, however, does not give any indication of phase transition phenomena.

From experimental and theoretical studies (0' Hara in Wyllie 1967; Green 1963, 1964; Onuki and Tiba 1965, Mercier and Nicolas 1975; Ganguly etal 1983), it is generally concluded that garnet Iherzolite equilibrated at tempera-tures between 900°C and 1400°C and at pressure from 1.7 to 2.6 GPa, that is equivalent to depths of between 125 km and 200 km. Spinel Iherzolites are considered to have formed at lower pressure of 0.8 to 1.6 GPa and plagioclase Iherzolites at pressures less than 0.8 GPa. Garnet peridotite nodules were recovered from the drill core samples of the kimberlite pipes (1100 m.y.) at Wazrakarur, S. India, which are intrusive into the Archaean granites, and these nodules range from garnet harzburgite to Iherzolite and wehrlite, mostly having granular fabric and are reported to be derived from various depths between 125 and 200 km (Ganguly et. al. 1983).

The occurrence of round garnet grains in the core of euhedral olivine crystals (Fig.2) embedded in orthopyroxene intercumulus raises the possibility that garnet is the residual refractory material of the mantle source which underwent partial melting and the melt ascended into the Cuddapah continental crust in two stages :

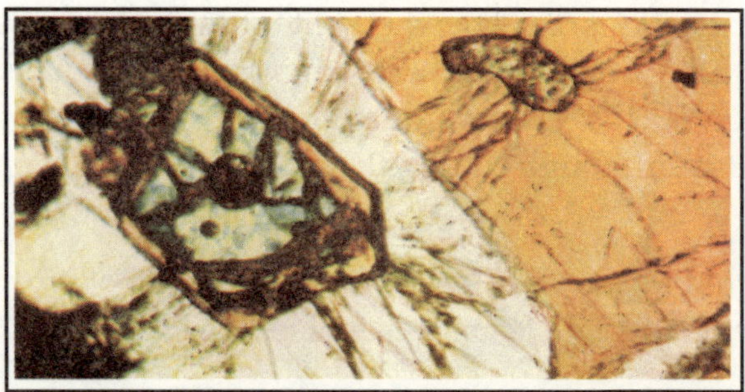

Fig.2 : Round garnet grains (dark) in core of euhedral olivine embedded in orthopyroxene (Garnet lherzolite), Pulivendla Formation. Cross polars.x18.

1. The magma came to a halt in a chamber beneath or above the Moho and olivine cumulates along with opx-cpx cumulates and intercumulates crystallized in this chamber.
2. The journey was resumed and the crystal-laden magma intruded the Tadpatri Shales.

One bearing of the Iherzolite xenoliths on the dynamics of the Cuddapah basinal evolution is the possible role of a mantle-derived magma cupola to initiate the basin depression.

REFERENCES

Basaltic Volcanism Study Project 1981 : Basaltic Volcanism on the terrestrial planets. Pergamon Press, Inc. New York, U. S. A.

Bhattacharji, S. 1981 : Evolution of an intrac-ratonic basin, in Evolution of an Intracratonic Cuddapah Basin. Balakrishna, S, (ed.) Inst. Ind. Penn. Geol. PubL, Hyderabad, A.P., Monograph 1, 7-28.

Bhattachaiji, S. and Singh, R. N., 1984 : 7her-momechanical structure of the Southern part of the Indian Shield and its relevance to Precambrian basin evolution. Tectonophys. 105, 103-120.

Cox, K. G., Gurney, J. J. and Harte, D. 1 973 : Xenoliths from the Matsoku Pipe, 76-92 *In* : Nixon, P. H.(ed.) Lesotho Kimberlites, Lesotho Nat. Devel. Crop. Maseru, Lesotho.

Dasgupta, P. K. 1986: Sedimentation Model of the Papaghni Group (Middle Proterozoic) the Cuddapah Basin, Peninsular India, Proceedings 12th Int. Sed. Cong. Canberra Session, Australia, Sym. Precamb. Sedimentology.

Ganguly, J. and Murty M. K. and Bhattacharya, P. K., 1983 : Garnet peridotite nodules from the Proterozoic krmberlite pipes of south India (Abstract), EOS, V, 64, No 45. 1983.

Gold, D. P. 1966: The average and typical chemical composition of Carbonatites. Miner, Soc. Ind. I. M. A. Vol. 83-91.

Green, D. H. 1963 : Alumina content of enstatite in a Venezuelan High temperature perio-dotite. Bull Geol. Soc. Amer. 74, 1397-1402.

Green, D. H. 1964 : The petrogenesis of the high-temperature peridotite intrusion in the Lizard Area, Cornwall. J. Petrol. 5, 134-188.

Harte, B. 1983 : Mantle peridotites and processes - the Kimberlite Sample, *In* Continental Basalts and Mantle Xenoliths, C. J. Haw-keswarth and M.J. Norry, eds,, 46-71. Shiva.

Heinrich, E. W. 1966 : The Geology of Carbonatites. Rand McNally & Co. Chicago, U.S.A.

Mercier, J. C. C, and Nicolas, A., 1975: Textures and fabrics of upper-mantle peridotites as illustrated by xeoliths from basalts. J. Petrology 16,454-87.

Nixon, P. H. and Boyd, F. R. 1973 : Petrogenesis of the granular and sheared ultrabasic nodule suite in Kimberlite,in Lesotho Kimberlite, P. H. Nixon, ed., 67-75. Lesotho Nat. Dev. Corpn. Meseru.

Wyllie, P. J. (ed) 1967 : Ultramafic and related rocks. Wiley. New York.

Onuki, H. and Tiba, T 1965: Notes on petrochemistry of ultrabasic Intrusives - specially, Alumium Distribution in co-existing pyroxenes, J. Jap. Assn Petr. Econ. Geol. 3, 215-227.

Sen, S. N, and Narasimha Rao, Ch. 1967: Igneous activity in Cuddapah Basin and adjacent areas and suggestions on the pal-eogeography of the basin *in* Proc. Symp. UMP, GRB and NGRI. Publ. Hydrabad, A. P. India, 8, 261-285.

Sinḥa, R. C. and M. Krishna Rao 1968: Geochemistry of the basic rocks around Pulivendla, Andhra Pradesh. Jour. Geochem. Soc. Ind. April 1968,25-52.

Streckeisen, A. 1976: To each plutonic rock its proper name. Earth Sci. Review 12 (1976), 1-33.

5

Proterozoic Erg System from the Egalapenta Member, Srisailam Formation

Introduction

Facies associations in the Egalapenta Member are suggestive of erg environment in all its three divisions, namely erg-margin, fore-erg and central-erg where detritus was supplied from earlier fluvial deposits.

Aeolian processes operate at a range of spatial and temporal scales, from the microscopic to the global and from microseconds to millennia. Sand and dust "storms", surface deflation and abrasion, as well as sand dune formation and migration are all part of the aeolian process spectrum.

Physiographically great sand seas or ergs cover areas of tens of thousands of square kilometers in sand hundreds of meters thick. The erg shows a hierarchy of bedforms or landforms; the largest forms are the draa which are large mound-clusters of smaller forms called dunes; some of the larger complex forms appear to be composed of smaller, simpler dunes. Dunes, in turn, are covered over by ripples. All these forms move with the wind, the smaller ones much faster than the larger ones. The smaller forms can respond to short term changes in wind direction; the larger ones respond far more sluggishly.

Ancient aeolian deposits are laterally extensive, homogenous bodies of fine-grained, well-sorted sandstones with large sweeping cross-bed sets; thick cross-bed cosets may be set apart by lenticular interdune deposits. Pebbles and granules are usually lacking except may be at the base of some cosets and on the perimeters of the aeolian deposits. Aeolian sandstone bodies may pass laterally into sabkha or braided-fluvial deposits or contrasting shallow-marine beds. Aeolian deposits probably are not transitional with anything; aeolian sands may interfinger with other associated environments such as shallow-marine or fluvial deposits, but do not grade into such sediments. Among the ancient sandstones, considered to be aeolian are the Coconino (Permian), Navajo (Jurassic) of USA and New Red sandstones (Permian) in England. This is to be noted that examples of Precambrian aeolianites and erg are few and far in literature owing to extreme cases of lithification, diagenesis and deformation either stratal characteristics are mixed up or crushed partially. Effectively such intermixed and deformed bodies form collage that is difficult to interpret.

The Egalapenta aeolian facies represents a Proterozoic wide sandy plain extended over a large area. Vertical and spatial facies associations suggest that stratigraphically at different levels, the sandy plain, a mosaic of dunes, inland sabkha, ephemeral fluvial deposits and sand sheets interspersed with isolated dunes, zibars and interzibars were deposited. It is evident that the superimposed dunes of Egalapenta were spatially separated by interdune areas which were occupied by deflationary desert flats and shallow inland sabkhas. The dominant assemblage of aeolian sand sheets with zibars and interzibars is rather uncommon in recent and ancient analogues. Despite many common aspects with facies assemblages of ancient and modern ergs (Kocurek 1981, Porter 1987, Sweet et al. 1988, Mountney & Thompson, 2002, Mountney & Jagger, 2004), the Egalapenta erg includes :

- coarse-grained sand sheets mainly without dunes, inland sabkha and ephemeral fluvial;
- bedforms without slip faces (zibars) and interbedded medium- to coarse-grained sandstones with bundles of wind-ripple laminae (interzibars);
- dunes with slip faces mainly formed by fine to medium sand.

Different types of aeolianites associated with the ancient Egalapenta sand sea, details of their depositional styles, facies attributes, associations and interpretation are focused in this chapter.

Facies Analysis of the Egalapenta Member

The Egalapenta beds constitute the youngest Member of the Srisailam Formation of the Cuddapah basin, India. Three sequences (designated 1, 2 and 3) gradational to one another are distinguishable within the Egalapenta Member. The Egalapenta Member mainly comprising red sandstone is gradational to the underlying Tapasipenta marine beds and displays complex interplay of fluvial, shallow-marine, arid sea-marginal sabkha and aeolian processes.

Fluvial facies attributes of sequence 1 comprise cosets of planar- and trough cross-strata with clasts, granules and ferricrete grains, rippled and laminated mudstone-sandstone and disorganised or deformed fabric. Associated shallow-marine facies with tidal imprint is characterised by glauconitic sandstone intercalated at two stratigraphic levels. Its salient attributes include sets of planar-, trough cross-strata, sigmoidal beds and mudstone-glauconitic sandstone interlaminations. Facies interrelations, regional correlations and palaeocurrent trends indicate westerly basinward migration of a depositional front of braided fluvial system(s) which was interfered by shallow-marine incursions.

Sequence 2 is marked by repetitive interbedding of anhydrite-bearing carbonate-sandstone-mudstone (coastal sabkha) with aeolian sand sheets. The facies associations attest to further basinward progradation of land alternating with coastal sabkha sedimentation and subsequent modifications of aeolian sands by alluvial incursions and wind action. The set-up is comparable to an erg-margin.

Aeolian and associated facies are characteristic of sequence 3. Dominant facies attributes include cosets of cross-strata of dune origin and horizontal to low angle strata either indicative of aeolian sand sheets or amalgamated interdune deposits. Aeolian facies complex and associated facies of sequence 3 are comparable to those in a modern erg.

Transition from braided fluvial deposits to a repetitive sequence of aeolian sand sheets alternating with coastal sabkha carbonate, followed by vertically arranged dunes and interdunes and ending in draas, reflects alternation of humid and arid conditions which might have been orbitally-controlled. Eustatic fluctuations account for repetitive occurrence of environments and palaeogeographic shifts.

FLUVIAL FACIES

Introduction

Interpreting depositional environment of thick Proterozoic arenaceous sequences with little intercalated fines is difficult and there are cases where depositional environments have been misread from rock records. The causes of misinterpretation are absence of body fossils and vegetation (Long 1978). The possible lines of evidence which may help to discriminate between different environments are grain size, petrography, sand body geometry, lithofacies associations and the type, scale, abundance and directional attributes of sedimentary structures.

Facies

Attempts have been made to interpret the depositional environment of the arenaceous sediments of sequence 1 of the Egalapenta Member. The rocks have been divided into eight facies on the basis of lithology, grain size, sedimentary structures and sand body geometry. They have been numbered as F1 to F8.

Planar/Horizontally Stratified Medium To Coarse Reddish Brown Sandstone (F1)

The facies consists of medium- to coarse-grained sandstone with subhorizontal, subparallel to low angle stratifications (Fig.1). Stacked extensive tabular sand bodies bounded by planar or broadly undulated erosional surfaces are discernible. Individual strata are few millimetres to two centimetres thick and occur in sets 0.2m to 1m thick. They have a sharp, relatively flat lower contact and a gradational upper contact with cross-stratified sandstones at places. Alternating light and dark coloured

Fig.1 : Partially visible long profile of a braided fluvial channel with scoured base, sequence 1 of Egalapenta Member. Horizontally stratified medium to coarse sandstones are discernible as channel-fills. Scale: length of the hammer = 42cm.

(commonly black), horizontal to slightly wavy layers, ranging between 0.5cm to 20cm in thickness, characterise the facies. Individual layers or beds, consist of medium- to coarse-grained, moderately well-sorted sediment, locally organised into very crude, normally graded sequences. These beds are commonly laterally persistent, but in places, pinch out abruptly. Parallel, stratified beds, often seem to change downcurrent into gently dipping cross-strata and are frequently associated with small trough cross-beds. The strata on closer inspection, reveals presence of internal ripple cross-laminations at places.

The planar-stratified sandstone developed most likely as the product of upper plane-bed sheet flood conditions in a distal alluvial plain setting. Although horizontal stratification can result from either low or high-energy conditions, accumulations of thick, locally pinching beds indicate high-velocity depositional conditions (Picard and High 1973). Sharp lower contact and millimetre- to centimetre-thick, locally pinching horizontal beds suggested that high-energy conditions existed during sheet flooding. Presence of internal cross-laminae indicates that these strata were generated by ripple migrations and are similar to the water-laid pseudoplanar cross-strata (Smith 1971). Gradation of these strata into low angle cross-stratification suggests that some of them may also be upper flow regime plane beds (Allen 1983, Kirk 1983, Haszeldine 1983). The plane-bedded sands occur widely amongst the deposits of modern low-sinuosity streams (Harms and Fahnestock 1965, Coleman 1969, Smith 1970, Williams 1970, Cant and Walker 1978). Plane-bedded sandy facies occurs only on the tops of transverse or longitudinal bars that were mainly cross-bedded internally. McKee and others (1967) reported that 3.5m thick horizontally stratified sand was deposited in the modern Bijou Creek during a single flooding event. This points to the fact that the planar stratified sandstones of the study area possibly developed during individual catastrophic episode. Occurrence of planar stratified sands is suggestive of uncontrolled flood conditions (Vos and Tankard 1981, Cudzil and Driese 1987).

Proterozoic was a time, devoid of vegetation and overbank mud, hence the river should have had a braided character (Schumm 1968, Cotter 1978, Fuller 1985). Presence of stacked extensive tabular sand bodies bounded by planar or broadly undulated erosional surfaces within the arenaceous sediments of sequence 1 of Egalapenta are likely the products of a sandy braided river system. Case histories from ancient and modern (Long 1978, Schumm 1968, Cotter 1978, Fuller 1985, Bose and Chakraborty 1994) lend support to the conclusion.

Red Mudstone-Sandstone (F2)

The facies consists of few millimetres to few centimetres thick beds of red mudstone/fine- to coarse-grained sandstone. Mudstones are generally thin discontinuous to fairly continuous sheets and form thin drapes along sandy ripples. Thin parallel laminations, ripple laminations and climbing ripple laminations are common in the sandstones. Fine-grained sandy small ripples are found to float within the mudstone sheets. Grain size at places varies from laminae of coarse sand to wisps of silty mudstone. The fine-grained sand are usually discontinuous and display current ripple laminations.

Thin sheets of mudstones and fine-grained sandstones represent deposits from unidirectional waning currents, probably overbank sheet flows. The mudstone-drapes along the ripples, are suspension deposits formed during the low flow stage. This lithofacies of Egalapenta strongly resembles, in terms of facies type and association, the apron sediments deposited downstream of transverse bars in Platte River Basin (Crowley 1983).

Trough Cross-Bedded, Reddish Brown Coarse Sandstone (F3)

The facies comprises trough-shaped cross-bedding sets with thickness varying from few centimetres to ninety-five centimetres. Both small to large troughs are preserved (Figs.2 & 3). The cross-strata are

Fig.2 : Close-view of a trough from the braided-fluvial deposits of the Egalapenta Member.

Fig.3 : Trough cross-bedded granule-bearing sandstone, sequence 1 of Egalapenta Member. Scale: 10cm.

markedly concave-up and tangentially or concordantly related to the underlying bounding surfaces. The lithologies present, range from medium-grained red sandstones to sandstones with scattered pebbles and granules. Some troughs show infillings of ferricretes. Cosets of cross-beds are separated by laterally extensive erosional surfaces. The units bounded by such erosional surfaces, at places, show a fining upward trend. Scale of trough cross-beds and grain size decreases upwards within cosets. At places, vertical transition has been observed between horizontal stratification and trough cross-stratification.

Trough cross-bedding is generated by the migration of transverse to sinuous-crested, out of phase subaqueous dunes and ripples respectively (Rubin 1987, Harms et al. 1975). They are produced by the downstream migration of sinuous-crested dunes, most commonly in the deeper reaches of fluvial channels (Todd and Went 1991, Harms et al. 1975). The erosive surfaces that bounded packages of medium scale trough cross-strata might be interpreted to reflect initiation, filling, abandonment and re-initiation of subchannels within a braided fluvial tract. Trough cross-bedding is common in modern river sands in association with three-dimensional dunes (Frazier and Osanik 1961, Harms and Fahnestock 1965, McGowen and Garner 1970, Shelton and Noble 1974, Cant and Walker 1978, Schwartz 1978).

The vertical transition from horizontal stratification to medium or small scale troughs records waning from upper to lower flow-regime condition within a single depositional event (McKee et al. 1967, Todd and Went 1991, Harms et al. 1975). Presence of thinly preserved small scale trough cross-beds in the facies reflects reduction in aggradation, low climb angles or variation in scour depth. Arkosic and locally pebbly character of the facies, ubiquitous presence of unimodally directed trough cross-beds and extensive erosion surface bounded sequences, collectively point to a braided fluvial origin of the facies.

Planar Cross-Bedded Coarse-Grained Reddish Brown Sandstone (F4)

The facies comprises medium- to coarse-grained reddish brown sandstone with subordinate mudclasts, quartz granules and dispersed pebbles. It is characterised by large planar cross-beds at an angle of 18°-32° (Fig.4). Tabular cross-beds at low angles (5°-10°) have also been recorded. They have sharp, flat or slightly scoured bases and tops. They are found both in solitary form as well as in groups. Set thickness ranges from a few centimetres to a few metres and descends in a down-palaeoflow direction. Both large and small scale tabular cross-strata are discernible. The large ones range in height from 0.5m to 4.5m and the smaller ones reach heights of 0.05m to 0.1m. The smaller sets are commonly found to occur immediately above a unit of large scale tabular

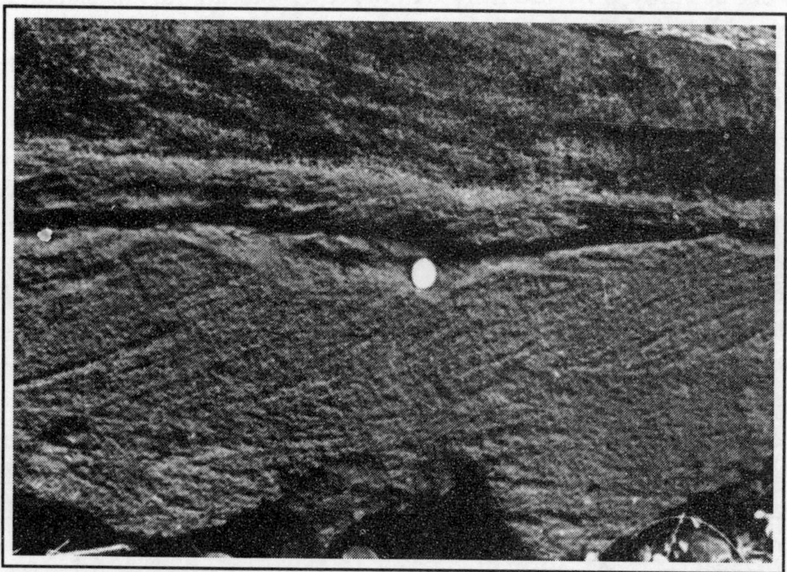

Fig.4 : Planar cross-bedded coarse sandstone in association with thin-bedded rippled sandstone and trough cross-stratified sandstone, sequence 1 of Egalapenta Member. Scale: 2.5cm diameter coin.

cross-strata. Cosets of cross-strata descend in a down-palaeoflow direction, thicken as they overrun a preceding coset, but gradually thin as they reach their downcurrent limits. Individual sets persist laterally and downcurrent for quite a few metres. Compound cross-beds occur at places. Reactivation surfaces are common. The planar, tabular cross-bedded facies are found at places underlain by thinly interbedded sandstone-mudstone and also interbedded and/or interfingered at places with horizontal stratification and trough cross-beds.

Planar tabular cross-beds of the facies were formed by migrating two dimensional bedforms and laterally accreting bars (cf. Allen 1983, Harms et al. 1982, Todd and Went 1991). The height of the cross-beds indicate that the sandstone was deposited in fairly deep channels. Despite their existence as point bars in high-sinuosity streams, the present association of planar cross-bedded sandstone facies with sheet flood deposits also point to their occurrence as point bars in low-sinuosity streams (cf. Fedo and Cooper 1990). The reactivation surfaces might result either due to fluctuating flow stage or due to scouring by the superposed smaller bedforms (McCabe and Jones 1977, Collinson 1970). The occurrence of thinly-bedded sandstone-mudstone beneath the facies precludes their origin from overbank flows. The presence of mudclasts, clasts of ferricrete and scarcity of fines in the sandy fluvial sequence attests to their deposition temporarily in overbank and protected channel environments and subsequent reworking (Long 1978). However, overall short transport, reworking and quick burial were responsible for the presence of clasts of ferricrete and mudclasts. The large scale tabular cross-strata of the Egalapenta sandstones are interpreted as the product of outbuilding of large, probably bank-attached and at least partly slipfaced compound bars. This interpretation is broadly supported by the geometry and directional properties of the structures. Presence of compound cross-beds points to low flow conditions, when smaller bedforms migrate over the lee of slower moving or stationary larger bedforms (Blodgett and Stanley 1980). Lateral persistence of the tabular cross-strata (over tens of metres) attests to the stability of these barforms over a considerable period of time. Variability in their geometry and orientation reflects subtle differences in depositional processes. In contrast, the small scale tabular cross-strata in the study area indicates their generation by small, repetitive sandwaves, which are formed in flows of lower velocities than dunes. In fluvial environments, these conditions are commonly met with in shoaled waters such as bar tops (Todd and Went 1991, Allen 1983). Case histories in support of the conclusion can be cited from literature (Crowley 1983, Cant and Walker 1978, Todd and Went 1991, Shepherd 1987, Ori 1982, Arche 1983).

Disorganised Fine-Grained To Pebbly Sandstone (F5)

The facies includes very poorly sorted fine- to very coarse-grained sandstone grading locally to pebbly sandstone. It is characterised by abrupt vertical and lateral variations of grain size and internal structures. Lower part of the facies is generally conglomeratic and consists of nearly massive beds (a few poorly differentiated foresets) whereas the upper part is characterised by profuse development of shallow trough cross-beds and planar cross-beds. Granule to small pebbles (5mm to 30mm) are dispersed throughout the unit. Cross-beds range in thickness from 15cm to 50cm. Foresets are generally planar. Down-current transition from planar to sigmoidal foresets has been observed locally. Mean palaeoflow direction of this facies measured from particular levels is towards northwest and west. This facies occurs in lower part of some channels and is found intimately associated with trough cross-bedded sandstones, which at places overlie the facies. But in some exposures, the planar cross-bedded facies overlie the fine-grained sandstone and pebbly sandstone facies with an erosional contact. The facies is volumetrically significant in sequence 1.

Generally, very coarse to pebbly nature of the facies and local occurrence of planar cross-beds with sigmoidal foresets in the sandy parts indicate high energy flow conditions (Roe 1987). Massive beds and poorly differentiated foresets indicate rapid sedimentation from highly sediment-charged flow (Smith 1986). The attributes indicate that the facies originated owing to sediment gravity flow during flood stage. High concentration of sediments is known to suppress the normal hydrodynamic sorting, turbulence and the development of bedforms (Pierson and Scott 1985, Turner and Monro 1987). Rapid variations in grain size and structures point to the fluctuating flow conditions. Hence sedimentation from unconfined flows on the flood plains will be significant. Such deposition on the flood plain may be volumetrically greater than in-channel sand accumulation. Existence of volumetrically significant, ill-sorted, structureless massive sand bodies justifies their deposition from unconfined flows during overbank flooding.

Convolute Bedded Reddish Brown Coarse Sandstone With Deformation Structures (F6)

Soft sediment deformation of trough-, planar cross-bedded, and horizontal/planar-stratified sandstones characterises the facies. Continuity of the deformed structures was of limited lateral extent. The intensity of deformation varies from oversteepening or gentle perturbation on the foresets to total disruption and destruction of sedimentary structures such as torn ripples, deformed ripples and slump structures. Planar cross-beds show overturning, wavy undulations and convolutions. Locally beds have been completely homogenised and fragments of disrupted mudstone beds float in the coarse- to fine-grained massive sandstone. These convolute beds and deformed structures are found over- and underlain by undeformed strata. They have been delineated from channels in Egalapenta. The thickness of such facies varies from 1m to 1.6m.

Soft sediment deformational structures in beds, laterally extensive over hundreds of metres are usually of seismic origin (Sims 1975, Bhattacharyya 1998, Owen 1995). However, existence of seismites in beds of limited extent is highly debatable. On the other hand, one of the reasons for soft sediment deformation is due to the expulsion of pore fluid under the influence of overburden stress which is favoured by poorly sorted coarse sediments with frequent impervious mud interlayers (McKee et al. 1962, Allen 1982, Plint 1983). Convolute bedding frequently originates in point bar coarse sands owing to partial liquefaction and deformation of water-saturated sediments under the influence of lateral shearing (McKee et al. 1962, Allen and Banks 1972). The deformed beds were deposited over limited lateral extent and are found associated spatially and vertically with undeformed strata of fluvial origin. Moreover, intergranular space of the sediments was filled by water due to subaqueous deposition. Initially and texturally the facies comprises ill-sorted sands in association with mud interlayers. Deformation, therefore can be attributed to liquefaction of sediments in the cross-stratal set causing restricted movement of the disrupted fabric. Elevated pore pressure is a phenomenon characteristic of water saturated sands. Deformation and destruction of bedforms, homogenisation of strata and disruption of structures in the facies might have been promoted by such raised pressure of pore fluids.

Stratified Clast- and Granule-bearing Sandstone (F7)

Reddish brown sandstones with granules occur in beds reaching in thickness from 0.6m to 1.8m. Individual beds are laterally persistent and can be traced for tens or hundreds of metres. Overall bed geometry varies from sheet-like to lenticular forms. The lower contacts of the beds are sharp, flat, gently undulating or highly irregular. Parallel stratifications are common. These sandstones include profuse clasts of quartz and ferricrete granules dispersed throughout. The size and concentration of the clasts varies considerably from bed to bed (15-20%). Maximum particle size (average of ten large clasts) measured in different beds varies from 25mm to 70mm and size of the largest clast is about 12.5cm. The clasts are moderately-sorted – angular to subangular. Elongate clasts show crude parallelism with the bedding surfaces at places. Clast imbrication, however, is generally not well developed. Low angle stratifications and crude planar stratifications are common. The texture of the facies shows rapid variations. The clast- and granule-rich sandstones, at places, grade into planar and trough cross-bedded sandstones with granules and trough cross-stratified sandstones, both vertically and laterally. Poorly developed planar cross-stratification are common at places.

Erosive basal contacts of the, clast- and granule-rich reddish brown sandstones, sheet-like and lenticular geometry and presence of parallel stratifications indicate that the sediments were laid down by channelised tractive currents (Nemec and Steel 1984, Reading 1996, Miall 1996). This facies bears some similarities with Miall's 'St' and Sp' facies. The sheets of granule-bearing sandstones were probably deposited in wide shallow channels (Allen 1983, Carling 1990, Southard et al. 1984). Low angle stratifications and crude planar stratifications with very coarse sand, granules and dispersed clasts suggest that the facies originated under an upper flow regime condition (e.g. Stear 1985). Variations in the texture and structures of the facies reflect the extreme unsteadiness of flow and transport rates. Capping of the mesoforms by clast- and granule-rich sandstones which interfinger with planar and trough-cross-stratified sandstones were probably the product of changing water velocity and depth over the bar crests during active bar growth (Crowley 1983, Massari 1983, Miall 1996).

Small Trough And Planar Cross-Bedded Coarse To Pebbly Sandstone (F8)

The facies consists of moderate to poorly sorted coarse sandstones and dispersed pebbles. It is dominated by trough cross-beds. The troughs occur as solitary or grouped forms. Planar cross-beds also occur commonly. Cross-beds generally range in thickness from 8cm to 20cm and occur in cosets that commonly vary from 0.5m to 2.5m in thickness. The cosets usually have sheet-like geometry and often show thinning and fining-upward (FU) trends. Cosets of cross-beds occur mostly in sheet-like units separated by laterally extensive erosion surfaces. Long tabular sets with unidirectional current and lenticular sets with variable current direction are present. Thin layers of muddy fine-grained sandstone drape the foresets, occasionally. In places, small scale (height: width=1:12), asymmetrically filled channels and isolated larger planar cross-beds occur within the facies.

Sheet-like cosets of long cross-beds appear to record its depositional history in shallow, wide channels during the flood stage (Williams 1971, Muir and Rust 1982). The planar cross-bedded sands are formed by the migration of two-dimensional dunes (Miall 1996, Reading 1996, Long 1978). Drapes of muddy sandstone point to deposition from rapidly waning flows with high load of fines (Bull 1972,1977, Houseknecht and Ethridge 1978). The cosets with lenticular sets and variable palaeocurrent directions probably represent deposition during the low stage conditions. The small trough cross-beds record migration of three-dimensional dunes within channels (Allen 1983, Miall 1996, Reading 1996, Harms et al. 1975). The erosive surfaces that bound packages of medium scale trough cross-strata are interpreted to reflect initiation, filling, abandonment and re-initiation of subchannels within a braided fluvial tract (Todd and Went 1991, Smith 1971, Cant and Walker 1978). Facies attributes of F8, such as, ubiquitous presence of unimodally directed trough cross-beds, locally pebbly character of the sandstones, extensive erosion surface-bound fining-upward sequences and planar cross-beds point to a braided fluvial origin.

Facies Associations

Vertical and spatial facies variations delineate several facies associations (FA) within sequence 1 of the Egalapenta Member. Interrelationship of different facies elements and their organisation vary significantly in each of such associations. These are five in number and termed as FA-I to FA-V.

FA-I

The association is characterised by a complex arrangement of trough-, tabular cross-beds (F3, F4) and horizontal stratified sandstones along with low angle laminations and ripple laminae at places (F1). Cosets of small troughs pinch out or alternate with horizontal to low angle stratifications (F1) in the downcurrent direction. Cosets of large cross-beds are separated by horizontally stratified beds and small trough cross-bedded sandstones. Majority of these cross-sets thicken downslope and thin out further downstream, resulting in a lenticular set geometry. Cosets of large cross-beds intertongue with and are overlain by a thick unit of fine-grained sandstone comprising ripple and small cross-bedded sandstone. This sequence is overlain by trough- and planar cross-bedded coarse to pebbly sandstone (F8). Palaeoflow directions are towards northwest and west. Moreover, it has also been observed that F1 and F3 are interfingered and/or interbedded with two other facies, completely different from the above lithofacies in their characteristics. These two facies are rhythmically cross-stratified glauconitic sandstone (M1) and mudstone intercalated with glauconitic sandstone (M3).

The interrelations of the different lithofacies of this sequence closely resemble the depositional architecture of many recent and ancient braid bars (Enyon and Walker 1974, Bluck 1976,1979,1980,1981, Haszeldine 1983, Kirk 1983). The cosets of large down-dipping sets denote accretion in front of a depositional 'high' which forms the core of the bar. During the low water stage shallow flows over the 'highs' deposited horizontal stratification of lower flow regime and small trough cross-beds. Fine-grained rippled sandstones were deposited further downstream in the lee of the 'high' and resemble the bar tail deposits of Bluck (1976,1979,1980). The coarse pebbly sandstones represent the winnowed bar top deposits (Bluck 1979,1980, Haszeldine 1983). But the interfingering and interbedding of F1 and F3 with M1 and M3 (see chapter 4 for facies attributes of M1 and M3) reveals the close association of the fluvial facies with the shallow marine sediments. Such fluvial and shallow-marine interaction has been documented in the lower and upper part of the sequence 1 of the Egalapenta Member. The two shallow-marine lithosomes within the fluvial facies have been named as 'LS' (lower) and 'US' (upper). Interbedding of marine and continental facies are suggestive of transgressive-regressive pulses caused by shallow-marine incursions on land. The conclusion is supported by many case-histories (Simpson and Eriksson 1993, Bose and Chakraborty 1994).

FA-II

The facies association comprises granule- and clast-bearing sandstone with trough cross-beds (F7), gradationally overlain by medium-grained units of planar cross-strata (F4) and defines a fining-upward trend. Sandstones with deformation structures (F6), at places, are associated with F4. The fining-upward units are either lenticular having concave-up basal erosional surfaces or sheet-like (0.35m to 1.5m thick) with flat basal surfaces. In each lenticular sand body, the trough cross-bedded units form the dominant component and exhibit a thinning and fining-upward trend within themselves. Planar cross-beds occur locally. Palaeoflow directions are southwesterly and westerly.

The large cross-beds were deposited in fairly deep water, whereas migration of very small sandwaves in shallow flow condition gave rise to the medium-grained planar cross-strata. The sheets of sandstones with granules were probably deposited in wide shallow channels. Association of granule-bearing, sheet-like sandstones with planar and trough cross-beds suggest the unsteadiness of flow, changing water velocity and depth over the bar crest. The lenticular sand bodies are small channel-fill deposits. Presence of such multiple channel-fill sand bodies within FA-II reflect repeated shifting and entrenchment of the braid channels. The thinning and fining-upward facies have been described as 'in-channel' deposit (cf. Cant and Walker 1978) or 'channel-dominated sequence' (cf. Walker and Cant 1984) and have been studied in both ancient (Rust 1981, Rust and Koster 1984) and recent braided fluvial deposits (Williams and Rust 1969, Donjek model of Miall 1977).

FA-III

The association essentially comprises large tabular cross-stratified, coarse sandstone with dispersed pebbles (F4) and minor amount of mudstone-fine sandstone (F2) forming 1m to 4m thick fining- and thinning-upward sequences. The concentration of some pebbles are also found along the alignment of the foresets. Such associations vary in thickness from few centimetres to even a few metres. The association is often overlain by laterally cutting channels with tabular cross-strata, troughs and ripples.

The facies association reflects deposition from fairly large channels which were dominated by moderately large straight-crested bedforms similar to those recorded in the Platte river (Smith 1971, Blodgett and Stanley 1980, Crowley 1983) and the Tana river (Collinson 1970). The association, delineated from the sequence 1 of the Egalapenta Member bears a strong resemblance with the Platte type braided river deposit (Smith 1971, Blodgett and Stanley 1980, Miall 1977).

FA-IV

The association comprises mudstone-fine sandstones (F2), planar cross-stratified medium-grained glauconitic sandstone (M1) and trough cross-stratified medium-grained glauconitic sandstone (M2). Lithosomes of M1 and M2, laterally and vertically, for tens of metres, intertongue with F2. Association of M1, M2 and F2, at places, overlies horizontally stratified sandstones (F1).

Sheets of mudstone-fine sandstones (F2) of Egalapenta are probably overbank sheet flows formed from unidirectional waning currents in a fluvial set-up. Characteristics of the glauconitic sandstones are markedly different from the attributes of fluvial facies. These facies show intercalated lithosome contacts with the fluvial facies. The attributes of the planar cross-stratified medium-grained glauconitic sandstone and trough cross-stratified medium-grained glauconitic sandstones have been ascribed to a shallow-marine origin. This facies association originated owing to transgressive-regressive pulses.

FA-V

The facies association among disorganised fine sandstone-pebbly sandstone (F5) with trough and planar cross-bedded coarse to pebbly sandstone (F8) and trough cross-bedded medium to coarse sandstone (F3) are laterally traceable for tens of metres. The association shows fining-upward trends in pebble size and decreasing frequency in vertical occurrence of few metres thick pebbly beds. The association is well developed in the basal part of many moderate to large channels. Palaeoflow direction is westerly.

The disorganised fine sandstone-pebbly sandstone (F5) with trough and planar cross-bedded coarse to pebbly sandstone (F8) indicate their deposition in the deeper parts of a river channel characterised by high sediment load of coarse grain detritus and high flow velocity. Presence of small scale trough cross-beds indicate deposition in the shallower, marginal parts of the stream. Sheet-like cosets of cross-beds with dispersed pebbles suggest its deposition in shallow, wide channels during the flood stage (Williams 1971, Muir and Rust 1982, Bull 1972,1977).

Reconstruction of Bars From The Braided Fluvial Deposit Of Sequence 1, The Egalapenta Member

The arenaceous reddish brown sandstone bodies in sequence 1 of the Egalapenta Member reveal a hierarchy of bedding features in both strike and dip sections. They are individually widely extensive, sheet-like with slightly irregular, flat basal erosional surfaces but consist of laterally cutting lenticular complexes, which themselves comprise usually a substantial number of sedimentation units, of which many are composed of more than one facies. Amongst these units a range of bar forms are recognizable and most of the complexes seem to record a limited lateral accretion. Braided sand-bed streams of low sinuosity are implicated (cf. Coleman 1969, Allen 1983).

Depending on the different facies, their associations, three types of bars have been identified which include:

i) cross-bedded simple;
ii) plane-bedded simple;
iii) compound bars.

Coarse-textured cross-bedding sets bounded by irregular tops and bases have been identified as characteristics of cross-bedded simple bars. The plane-bedded simple bars are represented by erosively-bound sedimentation units comprising planar/horizontally medium to coarse sandstones with clasts of shale-pebbles, mudstones and granules of quartz at the base. The base is sharp and erosional. Compound bars have also been identified from the existence of numerous erosively bound sedimentation units which include a cross-bedded facies, grading laterally and/or vertically into a commonly thicker unit of planar/horizontally stratified sandstones. The laminae of this bar form can be traced without any break from the cross-bedded lower part of the set-up into the succeeding plane-bedded sandstones. The cross-bedded and plane-bedded strata are therefore genetically related which are represented by foresets and topsets. Very coarse sands, mudstone clasts and granules have been found at the base of such units in some places.

The facies attributes of three types of bars bear resemblance with those of bar forms recognised in the Brownstones, Welsh Borders by Allen (1983). On the basis of various facies associations different types of bars have been identified from several sections of sequence 1.

Section A

A 60m. long, section aligned subparallel to the Krishna river allows the identification of different fluvial facies associations which bear the characteristics of bars (Fig.5). The lower most part of the section shows clusters of units representing plane-bedded simple bars, associated with

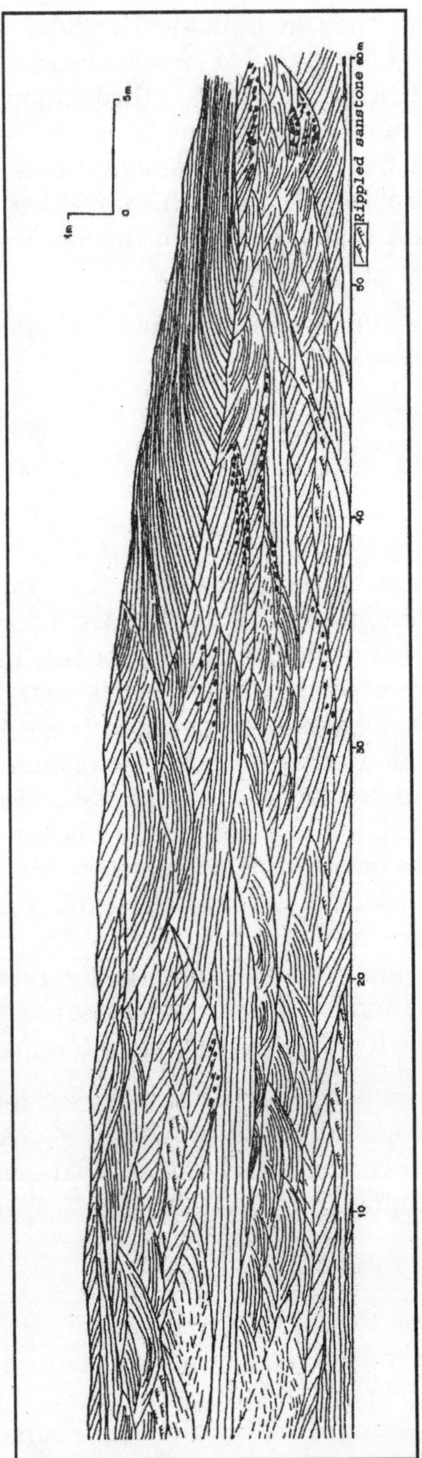

Fig.5 : Schematic summary of sheet sandstone of sequence 1 of the Egalapenta Member. Plane-bedded simple, cross-bedded simple and compound bars discernible.

cross-bedded simple bars and thin lenticular bodies with rippled sandstones in oblique sections. Some dip sections of the cross-bedded simple bars show trough cross-bedded sandstones when traced laterally. The trough cross-bedded sandstones in oblique sections are found laterally associated with cross-bedded bars having granules and clasts dispersed within the unit. Cappings of trough cross-beds by laterally extensive plane/horizontally bedded sands have also been recognised. This is associated with cross-bedded simple bars and small trough cross-bedded sandstones. Downcurrent thickening of the planar cross-beds is also visible. The upper part of this section exhibits significant change in conditions and comprises a series of cross-bedded simple-, plane-bedded simple bars, trough cross-bedded sandstones at places. Compound bars, which have been identified from the cross-bedded facies grading laterally and/or vertically into a thicker development of plane-bedded sandstones, are mainly found in the upper part of the section. Their heights ranged to more than a metre and are laterally traceable for tens of metres.

The different characteristics of the cross-bedded simple bars namely, their great lengths relative to thickness and internal consistency suggest a flat loosely periodic to non-repetitive bars of the broad kind as described from the modern streams as transverse (Smith 1970,1971,1974, Williams 1970), slipface (Cant 1978) or cross-channel bars (Cant and Walker 1978). Tabular planar cross-bedding sets with erosional boundaries within the section are widely associated with such features (cf. Collinson 1970, Smith 1970,1971, Boothroyd and Nummedal 1978, Cant and Walker 1978). Presence of coarse to very coarse sandstone matrix with dispersed clasts and granules indicate the high intensity of the bar-top currents. Downstream fluctuations in grain size imply that either the currents were unsteady or that there was a significant temporal variation in the texture of sediments supplied to the bars, perhaps in response to deposition by channels wandering upstream. The sheets of plane-bedded sandstones characteristic of plane-bedded simple bars are found only on the tops of transverse or longitudinal bars that were mainly cross-bedded internally (Allen 1983, Smith 1970, Williams 1970). Such plane-bedded sands occur widely amongst the deposits of modern low-sinuosity streams (Harms and Fahnestock 1965, Coleman 1969, Smith 1970, Williams 1970, Shelton and Noble 1974, Cant and Walker 1978, Schwartz 1978). Presence of cross-bedded simple bars in association with the plane-bedded sands suggest the existence of transverse bars and their bar tops. Occurrence of the planar sets over down-dipping surfaces and their downcurrent thickening indicate migration of the bedforms down a depressional high into deeper channel waters (Haszeldine 1983). Small scale structures like various types of ripple cross-lamination within thin lenticular sand bodies are usually found on the upper surface of bars (Harms and Fahnestock 1965) and deposited during falling stage (Miall 1985). Their cappings over the larger bedforms or bars within the section produces local fining-upward sequences. Such sequences are almost ubiquitous in fluvial environment (Miall 1985,1996, Harms and Fahnestock 1965). The trough cross-beds within the exposure, distinguishable from those representing bars by their shape, interlocking nature and internal festoon-like bedding were deposited from the shallow channels on the emergent bars at the falling river stage. Concentration of coarser grains in the trough cross-beds implies winnowing of the bartop. Trough cross-bedding is common in modern river sands in association with three-dimensional dunes (Frazier and Osanik 1961, Harms and Fahnestock 1965, Levey 1978, Schwartz 1978, Williams 1968). The dunes, however, characteristically occupy the deeper portions of active channels wherever the bedload is predominantly sand, (Miall 1985,1996). These trough cross-bedded sandstones are capped by laterally extensive units representing plane-bedded simple bars. This facies association is interpreted as the lower part of a plugged channel.

The configurations of laminae observed in the near-strike section point to strongly three-dimensional bars. These structures closely resemble Saunderson and Lockett (1981) experimental humpback dunes produced under flow conditions hovering near the dune-plane bed (upper stage) transition, as well as thin sand units from the Red River, attributed by Schwartz (1978) to transverse bars.

Discussion

Facies (F1 to F8) and their associations corroborate a braided fluvial origin for the sequence 1 of the Egalapenta Member, which overlies the Tapasipenta shallow-marine sandstone-siltstone sequence. The characteristics of different lithofacies, their associations and palaeocurrent patterns, collectively point to a braided fluvial origin for the arenaceous sandstones of sequence 1. The fluvial facies is intertongued with shallow-marine strata at two stratigraphic levels. While one (LS) of the shallow-marine incursions has been documented in the lower part of sequence 1 overlying the Tapasipenta shallow-marine sediments, the other one (US) has been recognised in the upper part of the sequence 1. Fluctuations in sea-level caused intertonguing of the shallow-marine and continental facies. Prograding sequence of the braided fluvial facies of sequence 1 is suggestive of emergence of land conditions owing to regression after the Tapasipenta sedimentation.

SHALLOW-MARINE FACIES

Introduction

Palaeoenvironmental interpretations on the basis of facies combinations and successions for the Precambrian shallow-marine environments rely heavily on modern analogues (Eriksson et al. 1998). Approximately 94m thick sequence 1 of the Egalapenta Member, mainly a continental fluvial facies assemblage is interfered by lithosomes of glauconitic sandstones (lower- LS: c.4m to 9m and upper- US: c.3.5m to 12m), laterally extensive for hundreds of metres at two stratigraphic levels. Both LS and US of sequence 1 show coarsening upward trends initially and fining-upward trends in their upper part.

Facies

An attempt has been made to interpret the depositional milieu of the two lithosomes of glauconitic sandstones from arenaceous succession of sequence 1 of the Egalapenta Member. On the basis of lithology, sedimentary structures, grain size and sand body geometry, the rocks have been divided into three facies. The facies have been numbered as M1, M2, M3 which are written in parenthesis.

Rhythmic Cross-Stratified Sandstone (M1)

The facies comprises planar-tabular (Fig.6) to trough cross-stratifications (60cm to 85cm set thickness) with their patterns changing laterally. Sigmoidal sets are as much as 70cm thick and extend upto 15m laterally. The toesets of the sigmoidal beds have a tangential contact with the underlying beds and thin and fine rapidly in the downdip direction. The upper surface of some sigmoidal foresets is truncated by a planar to undulatory scour surface or at places by a single, superimposed set of tabular cross-laminations (5cm to 8cm).

The large scale cross-stratified glauconitic sandstone of the facies may be generated by migration of sandwaves in subtidal environment with intermittent high energy waves. The rhythmic changes in the cross-stratification pattern in the M1 facies of Egalapenta resemble those reported by Kreisa and Moiola (1986) from a tidal sequence. The alternating thick-thin laminae and lamina packages defined by most conspicuous erosional surfaces clearly resemble diurnal/bidiurnal (cf.

Visser 1980, De Boer et al. 1989, Williams 1991) and spring-neap-spring tidal cycles (cf. Kreisa and Moiola 1986, Terwindt 1981, Kohshiek and Terwindt 1981) respectively. Presence of sigmoidal bundles within the facies corroborate tidal origin (cf. Kreisa and Moiola 1986). The tangential contact of the toesets of the sigmoidal beds with the underlying bed, its thinning and finning in the downdip direction results due to sediment fallout from suspension during the slackening phase of tidal flow (cf. Boersma and Terwindt 1981). These structures as documented in Egalapenta have already been described from ancient shallow-marine subtidal analogues (Boersma and Terwindt 1981, Kreisa and Moiola 1986, Bose and Chakraborty 1994, Bhattacharyya 1991). The attributes of this facies within the study area therefore points to a shallow marine origin.

Fig.6 : Planar-tabular cross-stratified sandstone within rhythmic cross-stratified sandstone (M1), sequence 1 of Egalapenta Member. Scale: 2.5cm diameter coin.

Trough Cross-Stratified Medium-Grained Glauconitic Sandstone (M2)

The facies comprises medium-grained glauconitic sandstone which consists of solitary as well as grouped sets of trough cross-strata with varying trough sizes from 15cm to 35cm in thickness and 0.5m to 1.5m in width. They overlie planar erosive and locally channelled bases which are not at all deeply incised on the underlying sediments. The tops of the beds are planar and display symmetrical wave ripples showing laminations concordant with the external shape, parallel laminations and minor channels in cross-section. Ripple marks, which are locally preserved on bedding surfaces are typically linear and symmetrical with wavelengths of c.8cm and amplitude c.1cm. Discontinuous thin layers of mud are found to drape the sandy ripples. The foresets in cross section are found to fill shallow troughs and are occasionally truncated by reactivation surfaces.

The trough cross beds resulted from the migration of large scale ripples or dunes (Allen 1968), possibly of strongly sinuous crested linguoid variety and partly by filling of erosional scours. The currents were unidirectional and apparently never produced bedforms migrating in an opposite direction. The currents fluctuated in strength with maximum flow powers resulting in the erosion of dune bedforms and the development of laterally extensive planar erosional surfaces (cf. Anderton 1976). During the periods of current inactivity, oscillatory wave processes frequently reworked the sediment surfaces producing symmetrical ripples (cf. Anderton 1976). The symmetrical wave ripples observed in the facies are identical to wave ripples recorded from modern environments (Reineck and Singh 1973). The absence of current ripples is probably due to their lack of preservation, possibly resulting from effective wave reworking during slack water periods. The lack of fine-grained sediment testifies to both the paucity of this type of sediment as a whole and to the efficiency of current, particularly wave action (McCave 1970,1971).

This medium-grained facies was deposited in a subtidal environment following its similarity with other shallow marine sand deposits (Narayan 1971, Michelson and Dott 1973, McCave 1973, Swett and Smitt 1972, Dott 1974, Anderton 1976, Stride 1963, Kenyon and Stride 1970). From the facies attributes it can be concluded that the trough cross-bedded, medium-grained glauconitic sandstone facies of Egalapenta probably originated in a subtidal environment.

Mudstone With Intercalated Sandstone (M3)

Interbedded mudstone-sandstone (Fig. 7) consist of stratal package ranging in thickness from a few centimetres to three metres (Fig.7). The sand to mud ratio within the facies varies laterally and decreases upwards. The sandstones are glauconitic. The sandstones are generally characterised by lenticular bedding. Millimetre-thick sand-mud interlaminations are common. In addition some 10cm to 15cm thick sand bodies have been documented. The facies also includes horizontally laminated mudstone layers alternating with massive to locally cross-stratified, fine-grained sandstone units in the lower part and plane-laminated fine sandstone in its upper part. Wave ripples and wavy bedding occur abundantly (Fig.8). The sand-mud alternations are laterally persistent. The facies is associated with the fluvial channels in sequence 1 of Egalapenta sequence (Fig.9 & 10).

The planar-tabular cross-bedded, fine sandstones present in the lower part of the facies display broad channel-shaped geometries with sharp lower contact. Channel-lag deposits were not observed.

The lower part of the facies which contains sandstones with broad channel-shaped geometry most probably developed as tidal channel sandstone (cf. Fedo and Cooper 1990). The sandy deposits of the tidal channels contain cross-bedding, commonly planar-tabular. Similar tidal channel deposits have been recognised by Diehl (1979) in the lower strata of the Wood Canyon Formation in the Death Valley area.

The intercalated mudstone and horizontally stratified fine-grained sandstone contain traits similar to an intertidal depositional environment. Millimetre-thick interlaminations of sand and mud is reminiscent of tidal bedding (Reineck and Singh 1980). Local occurrences of wavy bedding, symmetrical ripples, although not diagnostic also support a tidal flat origin. Sedimentary structures such as dessication

Fig.7 : Interbedded sandstone-mudstone facies. Horizontally laminated mudstones alternate with plane laminated fine sandstone, shallow-marine deposit, sequence 1 of Egalapenta Member. Scale: Marked portion in scale = 10cm.

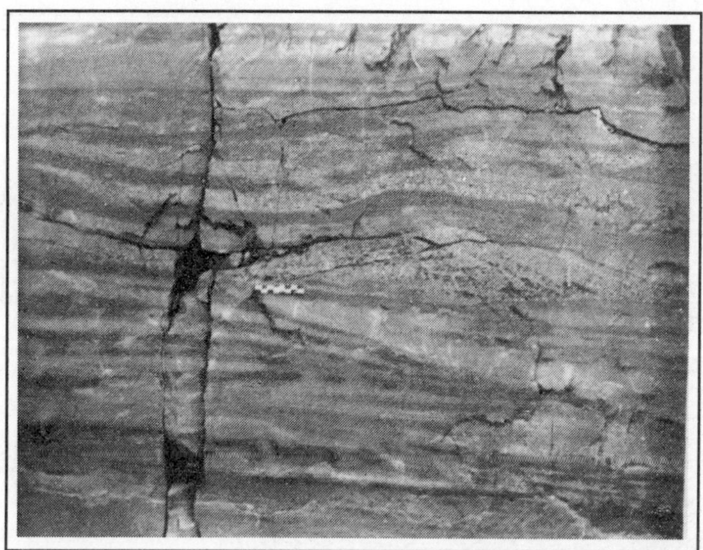

Fig.8 : Wave ripples and wavy bedding occur abundantly within M3 facies, shallow-marine deposits, sequence 1 of Egalapenta Member. Scale: 8cm.

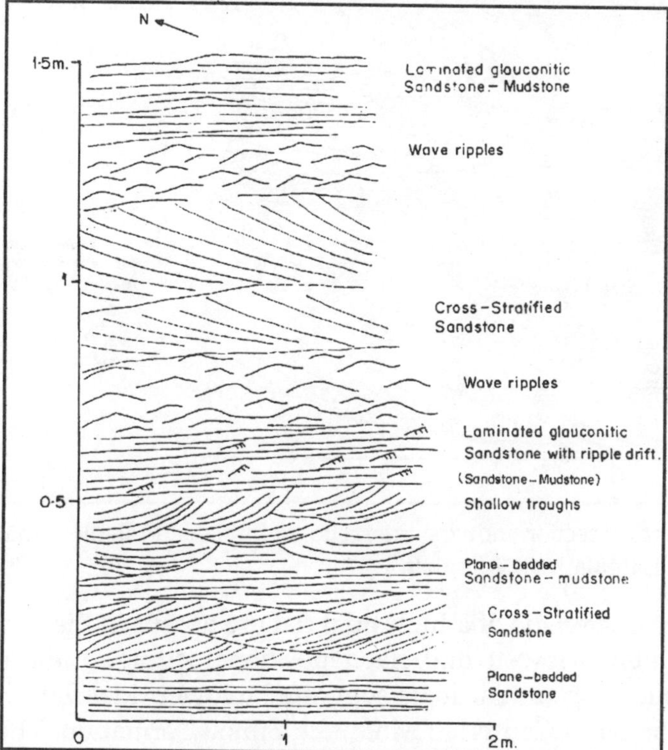

Fig.9 : Field sketch showing facies attributes of shallow-marine sediments(LS), sequence1, Egalapenta Member.

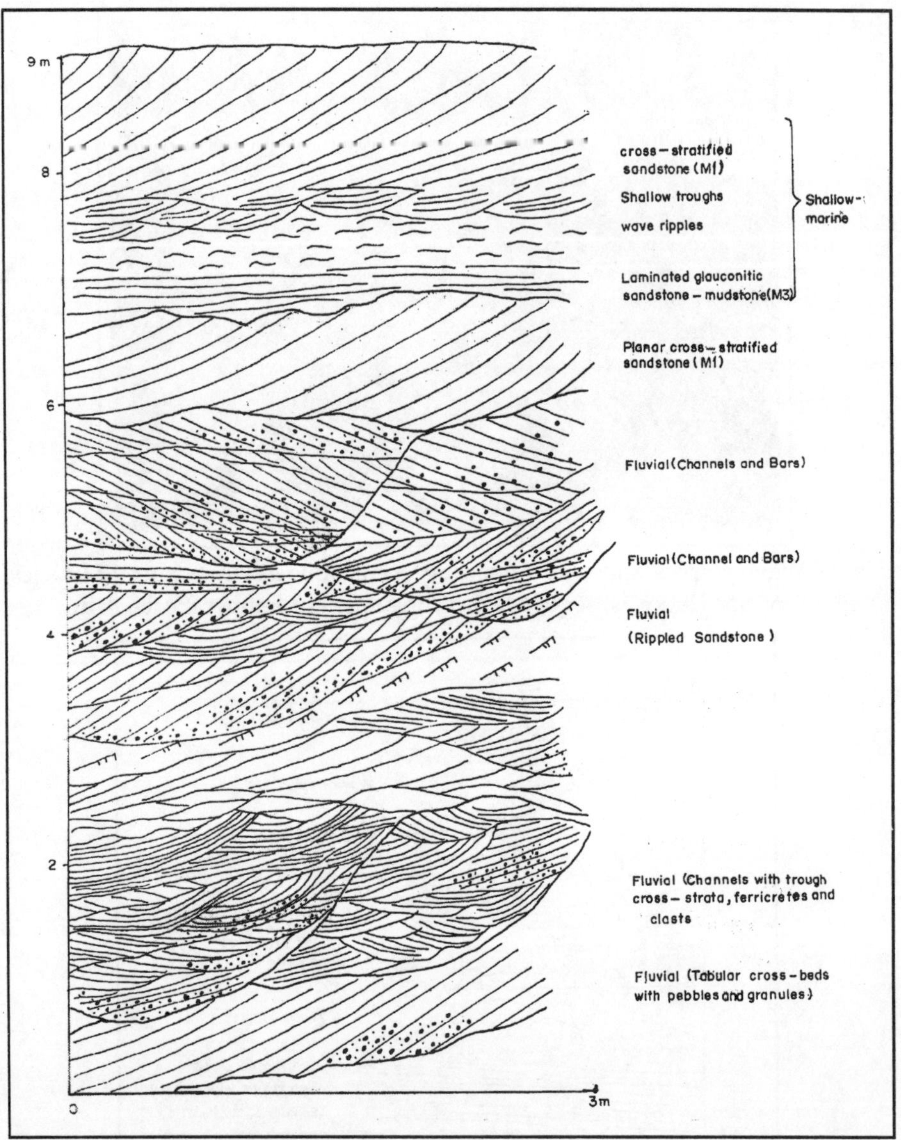

Fig.10 : Vertical section showing interactionof fluvial- and shallow-marine facies, sequence 1, Egalapenta Member.

cracks, characteristic of a high tidal flat are absent in the facies which again points to a low- to mid- flat depositional setting. Driese and others (1981) following similar logic suggested a lower tidal flat origin for a part of the basal Cambrian Mountain Simon Formation. The facies is probably the result of such an environment which prevailed in the Proterozoic Egalapenta Member.

Facies Associations

Section A

Section A (Fig.9) depicts plane-bedded sandstone at the bottom, overlain by cross-stratified sandstones (M1) and trough cross-stratified sandstones (M2) with sharp upper and lower boundaries. The assemblage is overlain by mudstone-sandstone facies (M3) with sand-mud interlaminations and wave ripples. The mixed facies is overlain by planar cross-stratified sandstone comprising superimposed sets of cross-strata with variable orientation and scale. This gives way to another mudstone-sandstone sequence (M3) with sand-mud interlaminations and wave ripples. The plane-bedded sandstone in the section reflects its deposition in an intertidal flat or upper subtidal zone. The cross-stratified sandstones are formed by migrating sand waves and dunes in the upper intertidal zone. Variable orientation and scale of the planar cross-beds might result from changing directions of waves and currents in a tidal flat (Bhattacharyya 1991, Reading 1996). The trough cross-stratified unit on top was generated by migration of large scale ripples or dunes and reflects a subtidal environment. The mudstone-sandstones (M3), suggestive of intertidal environment, overlie the trough cross-stratified sandstone facies (M2). The cross-stratified sandstone finally gives way to another mixed facies of mudstone-sandstone beds (M3).

Section B

The lower 6m of the section B (Fig.10) reveals a dominant fluvial facies assemblage comprising channels and bars. The fluvial sequence is topped by a sharp undulatory erosional surface. This gives way to the deposition of planar cross-stratified glauconitic sandstones (M1), followed up by mudstone-sandstone mixed facies (M3) and again on top by planar cross-stratified beds (M1). The section documents a fluvial-shallow-marine facies association.

Discussion

The sandstones of LS and US units have been ascribed to a shallow-marine origin on the basis of combinations of the following attributes:

- presence of glauconites;
- considerable lateral extent of the different facies (several hundred metres);

- presence of wave ripples, shallow troughs, planar cross-stratifications and millimetre thick mudstone-sandstone interlaminations;
- rhythmically cross-stratified sandstones comprising sigmoidal tidal bundles.

Transgressive-regressive pulses caused shallow-marine incursions on land as suggested from intertonguing of shallow-marine sediments with terrestrial facies at two stratigraphic levels within the sequence 1 of Egalapenta Member.

AEOLIAN AND ASSOCIATED FACIES

Introduction

The reddish brown sandstone of sequence 3, overlying impure carbonate-sandstone-mudstone of arid sea-marginal sabkha facies and interbedded aeolian sand sheets of sequence 2 are laterally extensive over hundreds of metres to few kilometres. The aeolian sand sheets, however, form a significant proportion both in sequence 2 and sequence 3 of the Egalapenta Member. The reddish brown sandstones are supermature quartz arenites mostly with lensoidal and sheet geometry. Although the sandstones are generally medium-grained, sheet-like coarse to very coarse grained variety also forms an important lithic unit. Besides the beds of coarse sand, the sandstone comprises well-sorted medium to fine sands with a large number of well-rounded grains. Uniformly high textural maturity and presence of well-rounded grains are suggestive of aeolian origin, though the features do not rule out their final deposition in shallow-marine environments (Folk 1968, Mukherji and Dasgupta 1972). Ubiquitous presence of several stratification types that form in response to specific aeolian processes confirms the aeolian origin of the sandstones.

Studies of the modern and ancient aeolian deposits have been carried out based on three observational scales. The smallest scale is related to the textural features and process implications of wind-ripple laminae, grainfall laminae, grain-flow wedges, adhesion structures (Reineck 1955, Sharp 1963, Hunter 1977a,b, Kocurek and Dott 1981, Fryberger and Schenk 1981, Porter 1987). A significant observational scale includes the studies of internal stratification, geometry of dunes, sand sheets, interdunes and post-depositional modifications of these deposits (McKee 1966, McKee et al. 1971, Doe and Dott 1980, Horowitz 1982, Ahlbrandt and Fryberger 1981, Kocurek 1981, Rubin and Hunter

1983). The largest observational scale encompasses the spatial organisation, morphology and dynamics of dunes and dune complexes (Wilson 1973, Brookfield 1977, McKee 1979, Kocurek 1981, Ross 1983, Mader and Yardley 1985).

The study of the sheet-like quartz arenites as well as vertically stacked lensoidal reddish brown sandstone of sequence 3 of Egalapenta Member attempts to integrate these observational scales in sedimentological study of Egalapenta aeolian facies. Vertical and lateral lithofacies sequences illustrate the bedform organisation. Precambrian aeolianites are well documented from late Palaeoproterozoic to Neoproterozoic (c.a. 1.8Ga to 0.54Ga) stratigraphic intervals but not from the Archaean and early Proterozoic rock records (Eriksson and Simpson 1998). Sedimentological analysis of aeolian facies of Egalapenta acquires significance in context of growing knowledge on aeolianites of the Proterozoic. Frequent floods and rapid lateral migration of the fluvial systems under the influence of humid climate during the Precambrian probably restricted aeolianite deposition to arid palaeoclimates (Ross 1983).

Aeolian Stratifications in the Egalapenta Sandstone

Grainflow cross-strata

The grainflow cross-strata (Hunter 1977a, Kocurek and Dott 1981) comprises well-sorted medium to very coarse sand having an average thickness of 1cm to 3.5cm. They are inverse-graded with concentration of coarse grains near the toe. They are found as high angle wedge-shaped foresets and intertongue with wind ripple or grainfall strata near the base of the sets. The coarse-grained, inverse-graded layers arise from the oversteepening of dune lee slopes and subsequent avalanching down the slope. Narrow grainflow layers, separated by thin packages of grainfall layers point to the existence of small aeolian dunes. Grainflows also commonly occur on subaqueous dune slipfaces (McKee 1957, Jopling 1966, Allen 1965, Boersma 1967) but in small aeolian dunes, grainflow cross-strata are narrow (width less than length), tongue-shaped, wedge-out upward on the slip face and are most commonly separated by grainfall deposits (Hunter 1976,1977a,1981).

Grainfall strata

The grainfall layers are 3mm to 1.5cm thick, 5m to 10m long, low angle (8°-10°) graded laminae. High angle (18°-22°) grainfall layers have been noted from some exposures in Egalapenta. Each lamina is moderately well-sorted with medium to fine grain size (1.5∅ to 2.5∅). These laminae consist of well-rounded to sub-rounded quartz grains. Individual lamina is tabular and thins down gradually in its down-dip direction. They dominantly occur near the low angle toe region of the foresets. Although laminasets are easily recognisable, it is difficult to recognise the individual strata. The sets are finer grained at the toe region in the down wind direction (cf. Fryberger and Schenk 1981). Grainfall laminae, the most common stratification type in aeolian dunes particularly the larger ones, initiate in the zones of flow separation. The grain-fall deposits appear to be the same as "suspension settling" deposits of McKee and others (1971). The low angle grain-fall laminae indicate that they were formed at the base of dune slip face when the eddy winds or cross winds were not strong enough to form ripples. In contrast to the aqueous bedforms, lack of continuous avalanching in aeolian bedforms ensures preservation of significant proportion of grainfall strata in aeolian deposit.

Wind ripple strata

Two types of wind ripple strata have been identified :

a) thin inverse-graded layers and

b) lenticular discontinuous layers of coarse to medium sand.

Presence of poorly preserved low angle foresets characterise the inverse-graded layers. They are the translatent strata of Hunter (1977b). The translatent strata are low dipping (3°-6°), 5mm to 10mm thick and are 10m to 25m long consisting of windripples, which have small amplitudes and relatively larger wavelengths (Fig.11). Small cycles of thin wind-ripple laminae- to- grainflow and thick wind-ripple laminae followed by truncation surface are identifiable. Coarse grained lag surfaces marked by granules and small pebbles are found. Coarse-grained lenses with high index ripple form, poorly preserved internal foresets and local inverse grading are characterised by poorly sorted bimodal texture. Laterally extensive translatent strata within the reddish brown sandstone are formed by the migration of wind ripples under conditions of net sedimentation. Such thinly laminated aeolian deposits have long been described as 'accretion deposits' (Bagnold 1941), "saltation deposits" (McKee and Tibbitts 1964) or "ripple lamination" (Inman et al. 1966, McKee and Douglas 1971, Goldschmith 1973). The low dip angles of these strata as observed in Egalapenta are mainly due to low angle of ripple climb, less than the critical angle. Climbing translatent strata are also formed by aqueous ripples but aeolian translatent strata are characterised by thin, uniform, inverse-graded laminae with a few visible foresets, ripples of low amplitudes and subcritical climb. They markedly differ from the subaqueous translatent strata which are thicker, normal graded with abundant foresets and with ripples of higher amplitudes (Kocurek and Dott 1981). The wind-ripple strata-grainflow cycles are presumed to represent periodic fluctuations in the wind directions (cf. Crabaugh and Kocurek 1993). Single grain thick lag surfaces characterise wind ripples that form commonly in deflationary wind regime or in an abundant supply of coarse sand (cf. Bagnold 1954, Sharp 1963).

Adhesion Structures

Discontinuous, crinkly layers of fine sand with thickness of few millimetres either occur parallel to the flat depositional surface or as gently dipping cross-laminae (Fig.12). 1mm to 3mm thick crinkly layers of fine to medium sand have been found to be rippled. Adhesion structures are also crinkly layers, common at the dune bases and damp interdune areas (Kocurek 1981, Hummel and Kocurek 1984, Simpson and Loope 1985). These structures have been identified from Entrada sandstone and Padre Island (Kocurek 1981, Hummel and Kocurek 1984).

Fig.11: Cross-laminated, asymmetric ripples with small amplitudes and large wavelengths in reddish brown sandstones, sequence 3 of Egalapenta Member. Scale: 8cm.

Adhesion structures are also reported from beaches and tidal flats (Hunter 1980, Kocurek 1981). The common occurrence of crinkly layers in the interdune deposits and dune bases of Egalapenta strongly suggests their resemblance with adhesion structures. Wet or damp surfaces are usually associated with interdune areas and dune bases where wind blown sands frequently adhere to, forming adhesion structures. Besides existence of adhesion structures in deposits of other environments, they are most common in the interdune and dune bases. Their existence with aeolianites in Egalapenta attests to the prevalence of damp interdune floors.

Facies

The variety of lithology and facies associations in the region has been identified and interpreted. The different facies identified on the basis of stratifications and other sedimentary and lithologic attributes are eight in number (E1 to E8). As the sequence 2 includes E3 and E2 facies in intimate associations, E3 facies, though basically a nonaeolian facies, has been described in this section along with other aeolian and associated facies.

Large scale cross-bedded sandstone (E1)

The facies comprises fine- to medium-grained sandstone with large tabular or wedge-shaped cross-beds within the reddish brown

Fig.12 : Adhesion cross-laminae, sequence 3 of Egalapenta Member. Scale: 10cm.

sandstone. The cross-beds are found as cosets or in isolated sets. The cross-bedded units are laterally persistent for tens of metres. Steeply dipping (25°-32°), medium- to coarse-grained foresets, 15m to 56m long, abutting at a high angle to the lower bounding surfaces are a dominant attribute of the facies. The toe of the foresets are coarse grained. Medium- to fine-grained foresets of grainfall strata which sweep down asymptotically on the lower bounding surface for several metres are common. 0.8cm to 2cm thick translatent strata constitute long, gently dipping cross-beds in a few sections and are laterally traceable for a considerable distance. Superimposition of some cross-beds show a gentle upward curvature over a long distance (Fig.13). Approximately 5.5m thick superimposed sets of large scale cross-beds are identifiable (Fig.13).

In many case histories, the cross-beds are preserved remnants of migrating large dunes, either straight-crested or crescentic (Kocurek and Dott 1981, Hunter 1981). The angular foresets probably represent smaller dunes where grainflows frequently reached the base of lee faces (Kocurek and Dott 1981) and the tangential foresets may represent lower part of large transverse bedforms with abundant grainfall strata in their toe region (Kocurek and Dott 1981, Hunter 1981). The gently convex-up translatent strata may suggest the presence of migrating dome dunes (cf. Fryberger et al. 1983) or may represent a transverse section through a crescentic dune complex. High angle cross-strata, abundant grainflow, wind ripple strata, medium to fine grain size and excellent sorting within the facies in Egalapenta are comparable to those found in modern aeolian dunes with well-developed slip faces (Hunter 1977a, McKee 1966, Inman et al. 1966). Single set of cross-beds observed at places are suggestive of simple dunes without multiple slip faces or superimposed dunes. Wedge-shaped cross-beds of the facies suggest the presence of elongate oblique dunes. Such association is quite common in dune fields (Tsoar 1982, Rubin and Hunter 1983). Multidirectional cross-beds indicate variations in wind direction and multiple sets of such foresets within stacked lensoidal sand bodies represent migration of sand dunes forming dune complex. Superimposed sets of large scale high angle cross-beds associated with grainflow, grainfall and translatent strata may be explained as remnant deposits of downwind migrating dune bedforms (Fig.13) or draas – *sensu* Wilson (1972). The superimposed dune complexes represent the central-erg deposits (cf. Porter 1987) of Egalapenta.

Fig.13 : Overriding dunes or draa represented by superimposed sets of cross-stratifications, sequence 3 of Egalapenta Member. Sharp contact with underlying interdune deposits of mudstone-sandstone beds is discernible. Bar scale = 2m.

Horizontal to low angle cross-stratified sandstone (E2)

The low angle cross-stratified medium- to coarse-grained reddish brown is a major facies in the sequence 3 of the Egalapenta Member. Facies geometry appears tabular and laterally persistent on outcrops (Fig.14). An upsection rapid increase in the bedding thickness of the facies is discernible. Following are the distinctive sedimentary attributes of the facies:

Fig.14 : Section showing aeolian sand sheet deposit, sequence 3 of Egalapenta Member. Low angle stratifications are visible. Ephemeral channel is visible at the top left. Scale: length of tree in the centre of the photograph = 1.2m.

a. individual coarse-grained, low-relief, relatively horizontal ripples are well developed;

b. granule ripples are common;

c. low angle (<10°) to nearly horizontal stratifications of fine- to medium-grained sand alternate with coarse-grained sand lags;

d. few thin layers of terrigenous clastics reveal bimodality of coarse to very coarse (0.71mm to 1.92mm), rounded to well-rounded sands and fine to very fine sand and coarse silt (0.18mm to 0.05mm). Even broken roundness is discernible in populations of coarse to very coarse sand grains (Figs.15 & 15a);

e. gently dipping finely (but discontinuously) laminated or poorly laminated layers are present;

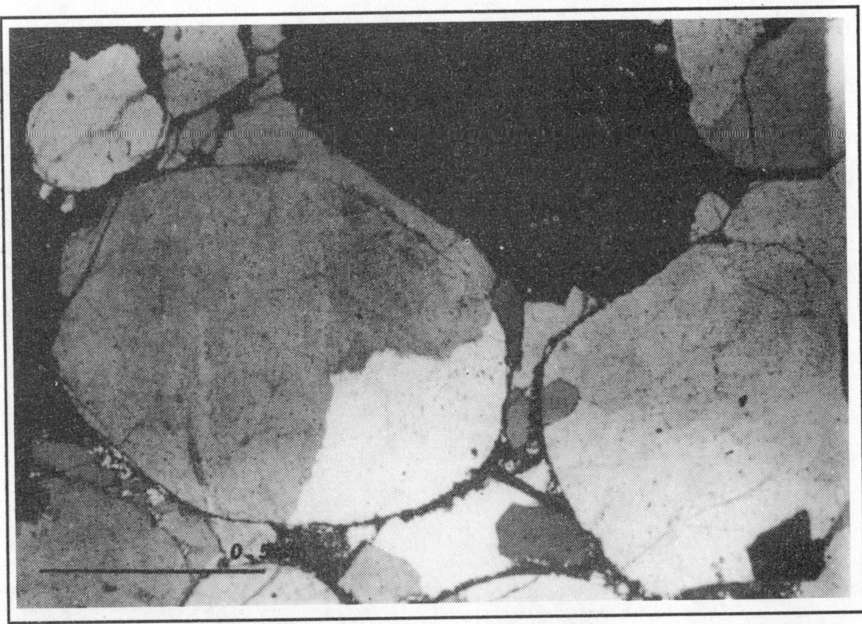

Fig.15: Well rounded, very coarse to small pebbles, sequence 3 of Egalapenta Member. Bar scale = 0.5mm.

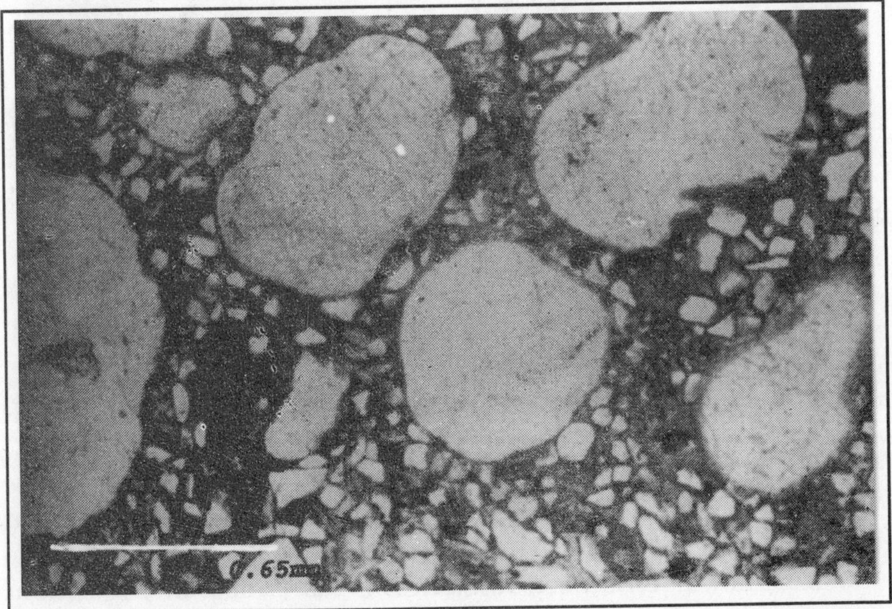

Fig.15a: Aeolian bimodality: coarse sand floating in a pudding of fine sand, sequence 3 of Egalapenta Member. Bar scale = 0.65mm.

f. individual coarse and fine-grained layers (mostly few millimetres thick and 45cm to 55cm long) are discernible;

g. locally, adhesion structures, small lenses of cross-bedded sandstone, isolated shallow troughs are present;

h. Coarse-grained, low amplitude bedforms lacking well-developed slip faces were also identified (Figs.16 & 17). Surface features of such bedforms comprise packages of fine- to medium-grained low angle (6°-10°) flat ripple laminae which are separated from each other by low angle truncation surfaces. The ripple-laminae within each package generally shows a marked parallelism but differ slightly in their orientation from laminae in subsets above or below. These deposits are interbedded with medium- to coarse-grained sandstones with low angle (less than 5°) flat-ripple laminae. In absence of the bedforms without slipfaces, these medium- to coarse-grained sandstones with low angle flat-ripple laminae are amalgamated. Out of twenty four grain-size cumulative frequency plots, ten representative curves, five from each of the groups (A&B) representing the medium- to coarse-grained beds with packages of low angle flat ripple laminae and coarse-grained low-amplitude bedforms without well developed slip faces respectively have been focussed (Fig.18) in order to compare the values of the 20th percentiles. The coarsest 20th percentile of samples from Group A averages 1.6φ (standard deviation : 0.6φ) compared to 1.2φ (standard deviation :0.5 φ) for Group B (Table 1).

Sheet-like geometry, dominance of horizontal to low angle stratifications, presence of wind-ripple laminae, small dune cross-sets (Figs.14, 15 and 16) and the stratigraphic position of the facies are the basis on which it has been interpreted to represent an aeolian sand sheet deposit. Similar aeolian sand sheet deposits were identified from other stratigraphic records (Chan 1989, Fryberger et al. 1979). The bulk of the facies developed through the migration of wind ripples on topographically flat, dry depositional areas of either interdune or sand sheets. Small aeolian dunes developed locally. Wet conditions prevailed in places for short duration, giving rise to isolated troughs and channels with trough- and tabular cross-beds and adhesion structures. The troughs and channels with smaller infrequent discharge might be engulfed by wind blown sands when the aeolian processes are dominant. Wind erosion might also extensively rework or redistribute sand deposited by ephemeral flows. Packages of low angle wind-ripple strata of fine to medium sand are interpreted to represent zibars, the low-amplitude aeolian bedforms that lack slipfaces and generally develop in the marginal parts of dune field or in sand sheets (cf. Nielson

Fig.16 : Coarse-grained low amplitude bedforms lacking well-developed slip faces (zibar) is discernible. Packages of low angle wind-ripple laminae visible. They overlie the aeolian sand sheet deposits, which comprises low-angle wind-ripple stratifications (translatent strata) sequence 3 of Egalapenta Member. Scale: height of the shrub on top = 45 cm.

Fig.17 : Packages of medium- to coarse-grained wind-ripple laminae are interbedded with medium- to fine-grained wind-ripple laminae forming the amalgamated interzibar, zibar deposits. They overlie the aeolian sand sheet deposit (represented by wind ripples in the photograph), sequence 3 of Egalapenta Member. Scale: height of the visible part of shrub to the right = 55cm.

Table 1 : Grain size data of zibar and interzibar facies of Egalapenta Member showing mean grain size (ϕ Mz), average of coarsest 20th percentiles (C_{20}) and their standard deviations (σC_{20}) as compared to those of Algodones (Neilson and Kocurek, 1986)

Egalapenta member (N=24)		Algodones dune field	
Zibar Facies (Group B)	*Interzibar Facies (Group A)*	*Zibar Lee face*	*Interzibar-stoss slope-crest*
Range : Φ Mz: 1.05-1.91 Φ Average C_{20}= 1.2 Φ σC_{20}=0.5 Φ	Range : Φ Mz: 1.48-2.78. Φ Average C_{20} =l-6 Φ σC_{20}=0.6 Φ	Range : Φ Mz: 1.6-2.8 Φ Average C_{20} = 1.2 Φ σC_{20} = 0.4 Φ	Range : Φ Mz: 0.0-2.5 Φ Average C_{20}= 0.8 Φ σC_{20}= 0.5 Φ

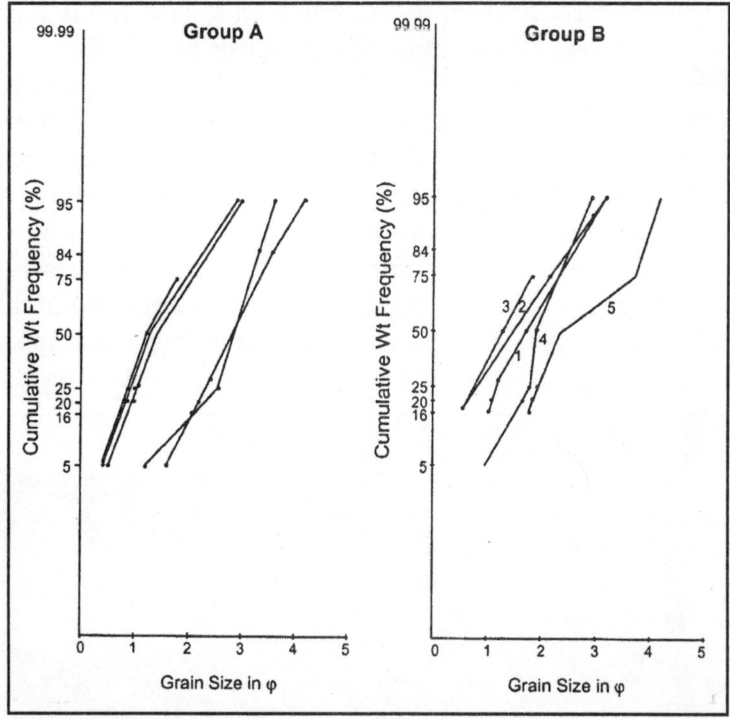

Fig.18 : Ten representative curves showing mean grain size, average of coarsest 20th percentiles (C_{20}) and standard deviations (σC_{20}) of the interzibar and zibar facies of the Egalapenta Member and compared to those of Algodones.

and Kocurek 1986). Wind-ripple laminae of medium to coarse sand represent the amalgamated interzibar deposits (cf. Nielson and Kocurek 1986). The difference in the coarsest 20th percentile is significant by t-test at 95% confidence level. The zibar-interzibar deposits form an important constituent of the Egalapenta aeolian sand sheets. Presence of distinct coarse mode of sand within a fine to very fine mode and absence of medium sand suggest an aeolian bimodality which is a characteristic feature of deflationary flats where the western Saharan term "reg" deposits may be aptly used (Folk 1968). Usually, a grain thick layer of such bimodality are reported in recent analogues (Folk 1968). Presence of few centimetre thick bimodal sand layers in E2 facies might be interpreted as redeposition of grain thick reg deposits owing to occasional torrential downpour (cf. Folk 1968).

Flat-bedded impure carbonate-sandstone-mudstone (E3)

2m to 2.5m above the terrigenous shallow-marine tongue, flat-bedded impure carbonate-sandstone-mudstone, as individual patches, are laterally extensive over hundreds of metres and continue to occur spatially in the outcrop for several kilometers (Fig.19). This lithofacies is repetitive. Five of them, varying in thickness between 0.3m to 2.8m are prominent in a 20m to 32m thick sequence (termed as sequence 2) and are separated by knife-sharp contact from the intervening reddish brown sandstone units (E2) varying in thickness from 1.3m to 2.5m. Besides in sequence 3, the other association of E2 facies is in the sequence 2. Thin sections prepared from samples collected at different stratigraphic intervals of E3 facies were stained by Alizarin Red-S to help in the distinction between calcite and dolomite. As a result of this

Fig.19 : Flat-bedded impure carbonate-sandstone-mudstones (E3 facies), sequence 2, Egalapenta Member. Laminated dolomite with thin gypsum- and anhydrite layers form a dominant attribute of the sabkha facies. Scale : 10 cm.

stain, calcite accepts a pink colouration and dolomite remains colourless (Friedman 1959). The carbonates within the lithofacies are mainly dolomitic. Moreover, quartz silt and medium to fine sands occur as impurities in the carbonates. The lithic units of the carbonate-sandstone-mudstone facies either interfingers or are interstratified within itself. Sedimentary features discernible in the carbonates are laminated dolomite and anhydrites (Fig.19), crudely layered anhydrite units with disturbed fabric, wavy beds, small ripples, ripple-drifts, small troughs and channels with infills of ripples, ripple-drifts and small foresets. Nodular anhydrites termed as chicken-wires are a prominent facies attribute (Fig.20). Such nodules vary in diameter from 0.3cm to 1.8cm and in shape from spherical to strongly flattened. The nodules appear to be squeezed together and are delineated by thin dark seams of clay (the "wires"). Anhydrite veins, birdseye structures (planar and isolated ones: 1mm to 3mm and several millimetres wide; isolated bubble like vugs: 1mm to 3.5mm in diameter) and enterolithic structures are prominent. Medium- to fine-grained sandstones contain laminated units, low angle stratifications, wind ripples, coarse-grained sand lags, adhesion ripples, small lensoidal bodies with foresets, small troughs and channels with infills of horizontal stratifications, trough and tabular cross-beds. Mudstones, calcareous mudstones with thin laminations, ripples and ripple-drifts are also identifiable.

Presence of bird'seye structures within the impure carbonate layers are indicative of a supratidal origin. The planar vugs formed by shrinkage resulting from dessication of exposed sediments (Friedman and Sanders 1978). Laminated dolomite is characteristic of low ridges and natural levees bordering tidal channels. Evaporative processes on exposed tidal flat evidently are involved in the formation of dolomite. Presence of nodular anhydrites, termed chicken-wires is the signature of a sea-marginal sabkha in an arid climate belt (Friedman and Sanders 1978). 'Enterolithic structures' in the carbonates originate due to the mechanical force of crystal growth exerted by the anhydrites growing in a nodule and thereby displacing the surrounding sediment. The host sediment hence gets physically deformed and bent around the nodules. Intimate association of the dolomites with anhydrites indicate its penecontemporaneous formation under special condition of molar ratio of Magnesium to Calcium (measured 15:1) of pore fluids in the carbonate mud of the sea-marginal arid sabkha deposit (Friedman and Sanders 1978, Wood and Wolfe 1969). Disturbed fabric within the anhydrite beds appear to be related to the intrasedimentary growth and dissolution of anhydrites (Hunter 1981, Kocurek 1981). The mixture of impure carbonates and sandstones indicates that there is a strong evidence of wind action and water influence within the carbonates. Wind action is evidenced by the presence of erosively based small dunes, wind-ripple strata and horizontal to low angle stratified medium to fine sandstones. Sporadic growth of aeolianites is suggestive of wind action on the dessicated surface of sea-marginal sabkha. Occasional influence of water is evidenced from the presence of small troughs and channels with horizontal stratifications, trough and tabular cross-bedded sandstones.

The coastal sabkhas are characterised by crusts of evaporites produced by evaporation of saline ground waters. Accumulation of these sediments may result in a rapidly prograding coastline (Fairbridge and Bourgeois 1978). Slightly above sea-level, coastal sabkha is a product of coastal offlap. Beneath the surface of the sabkha is a wedge of intertidal sediments and marine shelf. It is capped by supratidal evaporitic crusts and a thin aeolian layer in some cases (McLane 1995). Interbedded aeolian sand sheets and carbonate successions in sequence 2 are probably the results of repetitive marine transgressions over the coastal dune fields, deposition of thin layers of carbonate and subsequent basinward progradation of aeolian sands onto a marine shelf (cf. Fairbridge and Bourgeois 1978, Driese and Dott 1984). The Helderberg Group (Devonian) of the Appalachians is among the best examples of mixed quartz sand and carbonate sediments (Driese and Dott 1984). Repetitive basinward prograding aeolian sand sheets with interbedded sea-marginal sabkhas originated as coastal offlaps.

Mudstone-sandstone (E4)

The facies includes thinly bedded mudstone to fine-grained muddy sandstone with ripples and ripple-drifts. The thickness of beds varies from a few centimetres to as high as 1.2m. At places, calcareous mudstones with ripples and thin laminations are also identifiable. Wind ripple strata, low angle stratifications and adhesion cross-laminae along with adhesion laminae and adhesion warts are a common feature of the facies. Shale-pebbles and mudcracked surfaces are discernible. The mudstone-sandstone facies are closely associated with the aeolian facies, either laterally intertonguing or interstratified.

The mudstone to fine-grained muddy sandstone with ripples, ripple-drifts and thin laminations suggest their deposition in a quiet subaqueous environment such as shallow pools or lakes within the sand

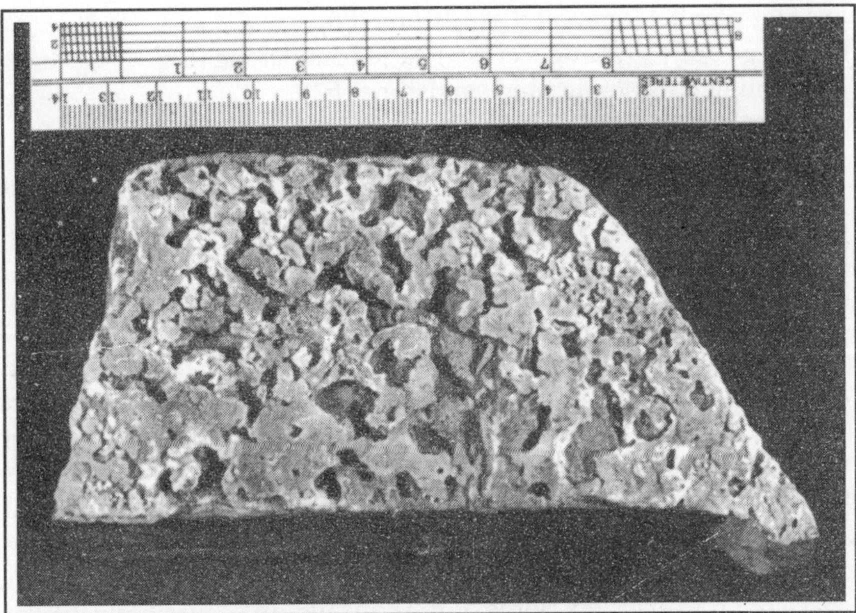

Fig.20 : Sample showing nodular anhydrites (chicken-wires), impure carbonate-sandstone-mudstone beds (E3), sequence 2 of Egalapenta Member. Enterolithic structures are marked by arrow.

sea. Presence of adhesion structures, wind-ripple strata and dessication features like mudcracks, shale-pebbles are indicative of drying up sequences. Adhesion structures form as dry wind-blown sand is trapped by the damp/wet substrate (Kocurek and Fielder 1982). The development of adhesion laminae or cross-laminae is controlled by the moisture content of the substrate, whereas the adhesion warts reflect irregularities, related to the salt crusts or other features of the depositional surface (Kocurek and Fielder 1982). Adhesion structures and wind-ripple strata indicate aeolian origin for a part of the facies. As reported from recent and ancient depositional set-ups, the inland sabkhas or the interdune areas that experienced periodic flooding by rain water are the potential geomorphic sites for the accumulation of such sediments (Kocurek and Fielder 1982, Porter 1987).

Facies E5 to E8 vis-à-vis aeolian sediments

Reddish brown sandstones with sheet geometry in some of the outcrops are poor to moderately-sorted, fine- to coarse-grained. The size range and roundness of the sand grains are comparable to those of the associated aeolian sediments. The sandstones include the following four facies exposed in different proportions in outcrops:

a) Flat parallel laminated sandstone (E5)
b) Large scale low angle cross-bedded sandstone (E6)
c) Trough cross-laminated sandstone (E7)
d) Ripple laminated (E8)

The facies attributes and interpretative aspects related to individual facies have been described separately.

Flat parallel laminated sandstone (E5)

The facies includes horizontal, tabular bodies when traced laterally for tens to hundreds of metres. The frequency of the facies is variable (c.15-60%). Internally the facies comprises thin laminations, 2mm to 6mm thick and forming a coset of 22cm to 142cm thick. Parting lineations trending 105°-285° are identified on bedding planes. Parallel laminated cosets are horizontal to subhorizontal. Small and large channel-like scours are locally discernible within the laminated facies. The concave-up scours have infills of laminae concordant with the lower bounding surface or cross-bedded. Some of the concave-up scours have infills of aeolian translatent strata and wind ripples. The bedsets of the facies at places are separated by thin mudstone-fine sandstone layers and/or erosion surfaces. Mudcrack surfaces, mudclasts and curled-up mudflakes are identifiable.

Laterally extensive, flat parallel laminated tabular sandstones are suggestive of sheet-flood conditions (Picard and High 1973, Harms and Fahnestock 1965, Coleman 1969). Extensively developed parallel laminated sandstones with randomly scattered mudclasts probably represent overbank sheet flood deposits of ephemeral fluvial systems and mark the development of major non-aeolian sequence within an aeolian regime. Absence of persistently developed and well-defined channel complexes attest to the existence of unconfined flow which was produced by sheet-flooding during the Proterozoic when bank stability was meagre due to absence of vegetation. Derivation and intermixing of aeolian sands from the surrounding during the sheet flood produced poor to moderate sorting of the ephemeral sediments. Moreover various intensities of pulsatory flow during a single sandy ephemeral flood produce wide range of facies types and profiles (Smith 1970). Presence of locally developed channel-like scours with occasional shale-pebbles at the bottom indicates incidence of occasional channelised flows which were set onto the previously formed sheet flood deposits. Such isolated channels rapidly dried up leaving behind lag pebbles at their bottom. Infills of aeolian strata within such channels show that those channels must have been quickly filled by aeolian sands under sand saturated subaerial condition dominated by wind action. Despite insignificant bank stability, preservation of ephemeral, isolated channel forms in subaerial aeolian sandy backdrop was made possible. The absence of coarse extra-formational gravels points to their intra-dune field origin. Preserved channel forms suggest that they did not migrate significantly and developed as ephemeral streams as a result of intense localised surface run-off (e.g. Glennie 1987). Locally developed laminated mudstone-fine sandstone patches with mudcracks, curled-up mudflakes and mudclasts over the erosional surfaces lend support to presence of water pools which were dessicated producing dessication features.

Large scale low angle cross-bedded sandstone (E6)

This facies comprises thick tabular, tangential cross-bed sets with low foreset inclination which are laterally extensive for tens of metres (Fig.21). Set thickness range from 15cm to 130cm with an average thickness of 55cm to 80cm. Each set is separated by sharp, planar lower and upper bounding surfaces which are erosional. These erosional surfaces are carpeted by red shale-pebbles. Slope of the foresets vary between 10°-15°. The sets commonly occur as grouped sets (cosets).

Frequent occurrence of tabular tangential cross-bedded sandstone in the outcrops of Egalapenta appears to have resulted from migration of linguoid bars (cf. Smith 1970) and cosets were deposited when such bars were superposed.

Trough Cross-Laminated Sandstones (E7)

This facies comprises small scale trough sets that are closely interlocked. The width of each small trough varies between 10cm and 20cm and the thickness of the cross-laminations is in the range of millimetres. Locally developed nested trough sets (cosets: 30cm to 120cm thick) have cappings of laterally discontinuous, few centimetres thick muddy/silty laminated unit.

Vertically stacked small dunes possibly generated such trough cross-laminated facies. Nested cosets of troughs with caps of few centimetres thick muddy/silty layers probably stand as responses of waning currents.

Ripple-Laminated Facies (E8)

The facies includes slightly sinuous and locally linguoid ripples and ripple-laminated sets in fine sandstone with subordinate silt and mud at different stratigraphic levels in vertical sections. The average trend of the ripple crests varies between 35° and 220° and between 65° and 275°. Their heights vary 1.5mm and 2cm with wave length on an average being 3.5cm

Presence of ripple laminations and absence of reverse grading suggest their formation in subaqueous environment.

Facies association of E5 to E8

Attributes of E5 to E8 facies associations closely resemble characteristics of braided ephemeral rivers. Planar-parallel laminations, climbing ripple laminations and low angle cross-beddings are characteristics of high energy ephemeral flash floods. Aggradation, rather than lateral accretion randomly distributed mudclasts in a backdrop of fine to coarse sand population. Uncommon presence of lateral accretion bedding, tabular geometry of the sandstone bodies, frequent dessication features, mud draping at the top of thin waning flow sequence, erosional surfaces carpeted by red shale-pebbles are common attributes of E5 to E8 ephemeral facies association and ephemeral fluvial deposits of present and past (Rust 1978, Sneh 1983, Steel et al. 1977, Allen 1983, Cant and Walker 1976). Furthermore, the

Fig.21 : Low-angle cross-bedded sandstone as ephemeral channel-fills (E6 facies), sequence 3, Egalapenta Member. Scale : Length of hammer=42cm.

sheet-like geometry of the beds with rare channel incision implies that the sediment depositing flows might not have been as clearly channelised as the most present day flows (cf. Friend 1978) and that much of the sedimentation is a product of sheet flood. Long tangential foresets (E6) in the Egalapenta sandstone are indicative of high flow velocity coupled with sediment fall out from suspension (Roe 1987). Moreover, long tangential foresets in association with sets of parallel laminations (E5) with particles streaming on the bedding surfaces are direct results of the proximity of dune-plane bed transition (Jopling 1965, Roe 1987). Presence of parallel laminations, cross-beds, ripple-laminations, trough cross-beds, all reflect the varying flow stages of the streams. Their upward coarsening sequence further points a strong progradational tendency of the braided plains (Williams 1969, Bluck 1971, Costello and Walker 1972, Steel et al. 1977). Facies associations of E5 to E8 record internal erosion surfaces covered with intraclasts, overall dominance of aggradational tendencies and rapid alternation of upper and lower flow regime bedforms in vertical sections. In addition, the ephemeral fluvial facies (E5 to E8) are identifiable in association with dune and sand sheet deposits.

Facies Association

A further attempt has been made to show variations of environments in space through time in a part of the Egalapenta erg, as revealed from a 360m long traverse parallel to the palaeowind direction (Section A, Fig. 22). Aeolian facies associations have also been studied from various other sections in and around the Krishna gorge. Section A, drawn along the longest traverse, reveals the juxtaposition of different aeolian facies mosaic has been chosen for detailed description.

Section A

Large high angle cross-stratified lensoidal sandstones are separated by sheet-like low angle cross-stratified sandstones. These sheet-like sandstones are occasionally intertongued and interstratified with mudstone-sandstone lenses. In addition, some sheet-like sandstones with flat parallel laminated (E5), large scale low angle cross-bedded (E6), trough cross-laminated (E7) and ripple laminated (E8) facies as well as occasional channel-like scours are identifiable. Cosets of cross-strata were traced over a total length of 360m along the traverse. Variations occur in these features along the length of the traverse and they are interpreted as representing changes in dune and interdune deposition through time. Lensoidal sandstones with cross-stratified beds are suggestive of superimposed dunes or draas (*sensu* Wilson 1972). The draas are separated by intervening flat-bedded interdune deposits. Single set of cross-strata might hardly generate from the migration of dune bodies or draas (cf. Kocurek 1981). On the other hand, multiple sets of cross-strata, as observed in the section, might be a product of migration of many crescentic dunes superimposed on large compound dune bodies (cf. Kocurek 1981). The compound dunes are coherent bodies separated spatially by interdune deposits. Spatial deposition and migration of generations of compound dunes and interdune deposits through time might lead to the development of deposits revealing the vertical repetition of cosets of cross-strata with intervening thin interdune deposits.

The Egalapenta Erg

Regionally, well established evidences of aeolian regime indicative of various aeolian environments (Fig.23) are recorded from the upper part of the Proterozoic Egalapenta Member. This is comparable with a well developed modern erg. Different changes in facies assemblages from NW to SE include:

a) Above the upper shallow-marine tongue, flat-bedded carbonate-sandstone-mudstone facies are interbedded with horizontal to low angle stratified sandstones. The association represents repetitive occurrence of the coastal sabkha with aeolian sand sheets.

b) This is followed by the increase in horizontal to low angle stratified sandstone facies representing the dominance of aeolian sand sheet deposits.

c) The sand sheet deposits gradually decrease with the increase of high angle cross-stratified sandstone facies. This is indicative of overall dominance of superimposed dunes or dune complexes (Fig.24).

Intercalations of the coastal sabkhas with aeolian sand sheets of sequence 2 above the shallow-marine tongue of sequence 1 delineate the coastal erg margin and is suggestive of the northwestern limit of the sand sea. The dominance of aeolian sand sheets characterises the 'fore-erg' deposits. The presence of superimposed dunes or dune complexes marks the 'central-erg' deposits (cf. Porter 1987) in Egalapenta. The major trend in facies variations from NW to SE reflects the transition from the erg margin to the erg interior.

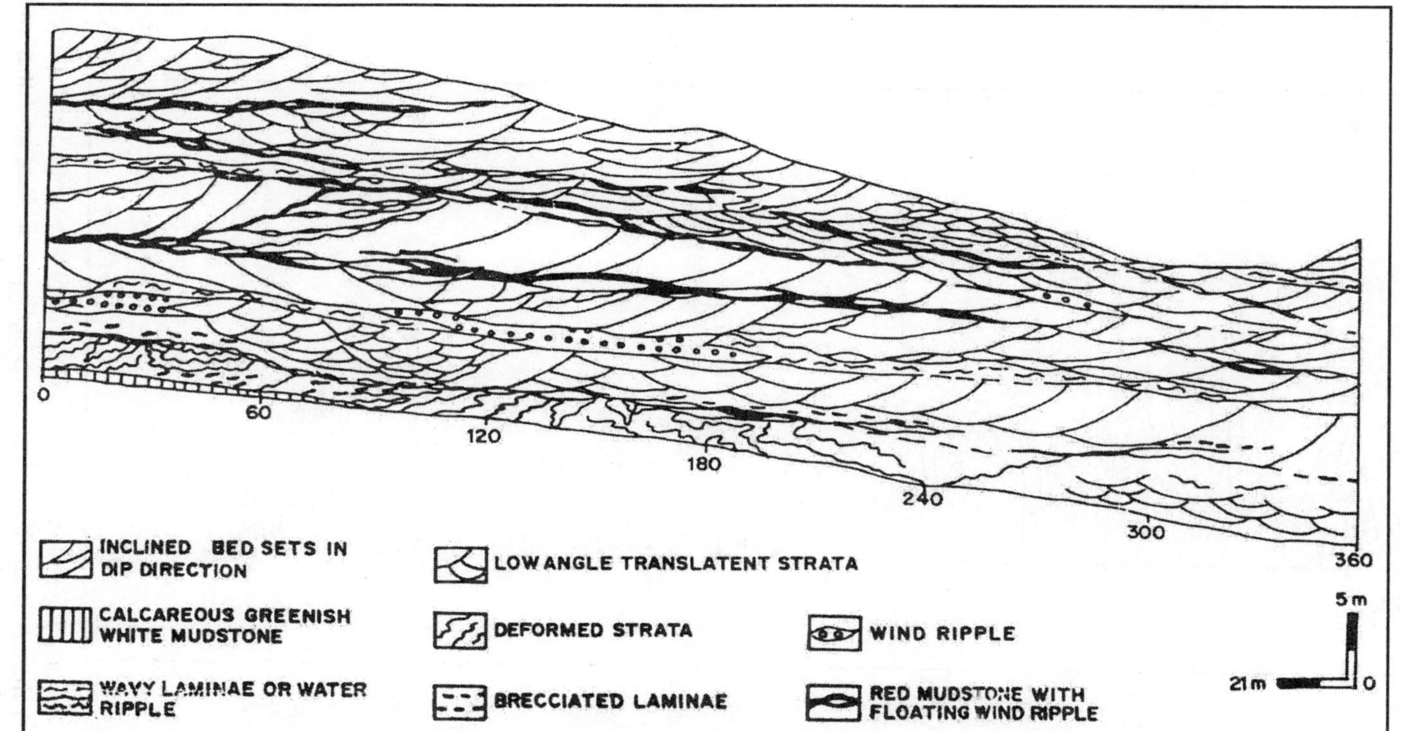

Fig.22 : A 360m panel diagram showing cyclic association of climbing dune (draa) with intervening interdune deposits, forming a part of an erg, Egalapenta Member, Srisailam Formation, Cuddapah basin, India.

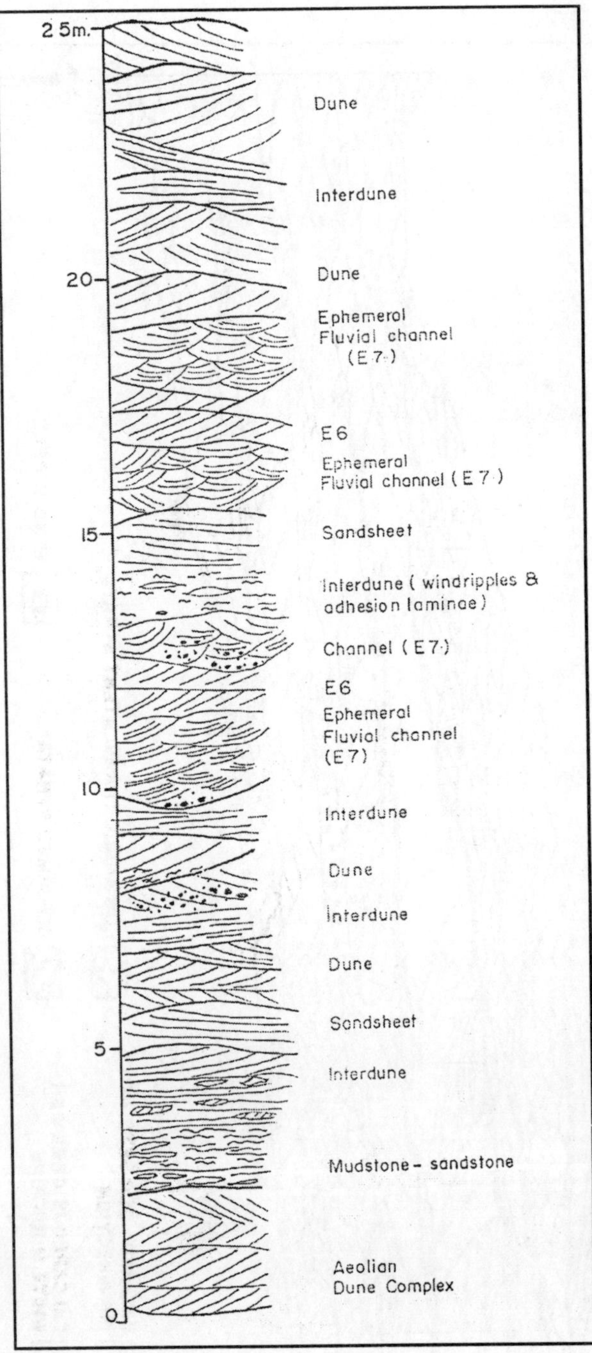

Fig.23 : Vertical log showing the mosaic of different aeolian and associated facies, near the east bank of Krishna river, Egalapenta. Aeolian dune, interdune, ephemeral and interdune deposits are discernible.

Reconstruction of different major erg environments was made from the study of different facies associations in various parts of the Egalapenta Member. Source of aeolian sands evidently appears to be the earlier deposited fluvial sands. Derivation of fluvial sands, on the other hand, as suggested by palaeocurrents was from the lower Cuddapah sediments and granitic gneissic complex. The different erg environments are dune field, sand sheet and coastal erg margin. The erg environments of the Egalapenta scenario are comparable to those of ancient aeolianites (Porter 1987, Kocurek 1981) and modern sand seas (Sweet et al. 1988).

Dune Field

Laterally extensive high angle large aeolian cross-beds characterise the dune field sequences of Egalapenta sand sea. The large aeolian cross-stratified sandstones form 85-90% of the central-erg division. Both solitary as well as superposed cross-beds or draas (*sensu* Wilson 1972) with intervening thin interdune sediments characterise the sequence (Fig.25). Decline of thickness of sabkha sediments within the dune field suggests that water supply gradually decreased. Sand saturated environment under limited water supply led to the development of dunes and dune complex. Vertical change from amalgamated interdune deposits to dunes and draas records dune-field initiation or lateral migration of peripheral to more centralised portion within the dune-field.

Sand sheet

Horizontal to low angle stratified sandstone facies containing abundant horizontal to low angle wind-ripple strata, in 0.5m to 9.5m thick units, characterise the sand sheet environment. The sand sheets are intercalated with sabkha and dune sediments. Abundant wind ripples, sabkhas, small dune complex, isolated small dunes, ephemeral channels, zibars and amalgamated zibar-interzibar deposits occur in the sequence. The sabkhas form an important facies variant in association with the sand sheets in extra-dune field areas or coastal areas as well as in the dune-dominated areas. The sand sheets form a major erg facies comparable to that of a fore-erg and extend over a large area between the erg margin and the well developed dune field forming the central-erg deposit.

Erg Margin

The erg margin sequence, represented by the intercalation of aeolian deposits with coastal sabkhas is stratigraphically slightly above the shallow-marine sediments (US). The coastal sabkhas are represented by flat-bedded carbonate-sandstone-mudstone beds with facies variants such as nodular anhydrites, birdseye dolomite, dessication features, adhesion structures, wind ripples, translatent strata, small channels and dunes. It occurs repetitively with the aeolian sand sheets which are represented by horizontal to low angle stratified sandstones. An upsection rapid increase in the bedding thickness of E2 is discernible. Rapidly deposited significant sediment volume of aeolian sand sheets during the closing phase of sedimentation of sequence 2 were modified by alluvial incursions and wind action in the coastal scenario. The set-up is comparable to an erg-margin.

Prominent Features of The Egalapenta Erg

The Egalapenta erg illustrates some characteristics which are comparable to those of ancient and modern ergs. Facies association in Egalapenta Member suggests that there were two successive marine incursions separated by basinward prograding braided fluvial deposits. The upper shallow-marine tongue, offlapped by basinward prograding fluvial sediments and repetitive aeolian sand sheets with intervening arid sea-marginal sabkha deposits indicates an overall land growth in a sand saturated environment under variable wind conditions.

The Egalapenta sand sea includes a wide sandy plain extended over a large area. Vertical and spatial facies associations suggest that at a point of time, the sandy plain, a mosaic of small pools, sand sheets interspersed with isolated dunes, zibars and interzibars were spatially spread over leading to the sea towards the northwest and a well developed dune field in the southeast.

It is evident that the superimposed dunes of Egalapenta were spatially separated by interdune areas which were occupied by deflationary desert flats and shallow inland sabkhas. Such fields were peripherally studded with coastal sabkhas which were connected to the sea by extreme low angle tidal flats (Fig.26). The dominant assemblage of aeolian sand sheets with zibars and interzibars as recorded in the Egalapenta erg is rather uncommon in recent and ancient analogues. Despite many common aspects in facies assemblages with ancient and modern ergs (Porter 1987, Kocurek 1981, Sweet et al. 1988), the Egalapenta erg includes development of

a) dominant very coarse-grained sand sheets without dunes;
b) dunes without slip faces;
c) dunes with slip faces mainly formed of fine sand population.

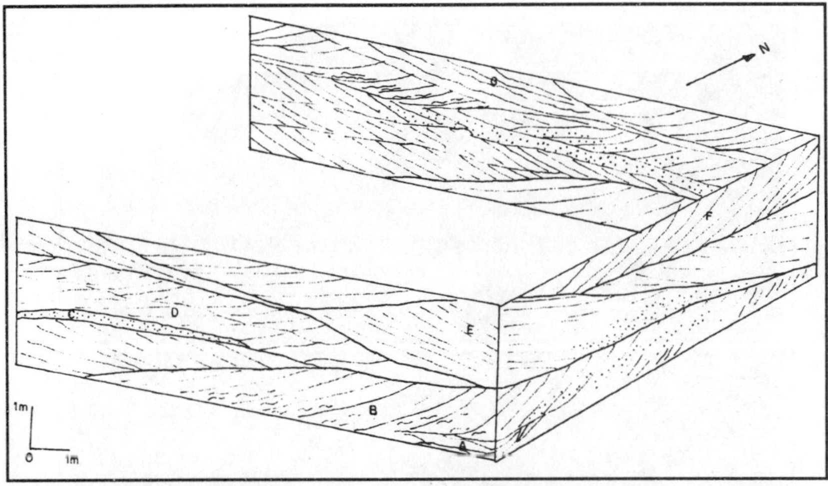

Fig.24 : Field sketch showing : A-wind ripples, B-aeolian dune cross-beds with wind ripples, C- grain flow-grain fall layers, D, E and F- multidirectional dune cross-beds with wedges of grainflow layers at places, G-translatent strata, from sequence 3, Egalapenta Member.

Palaeographic Shifts

Palaeographic shifts occur among the environments such as fluvial, shallow-marine, coastal sabkha, aeolian sand sheets, dunes and interdunes during the Egalapenta sedimentation. Spatial distribution of environments at a point of time is diagrammetically represented (Fig.26).

Gradational basal contact of the Egalapenta fluvial sandstone with the underlying Tapasipenta marine facies corroborates the initiation of basinward progradation. This was, however, interfered later. Onlaps by lithosomal fine- to medium-grained glauconitic terrigenous clastic tongues and offlaps by ferricrete-rich coarse lithosomal sand bodies of sequence 1 indicate shifts between offshore and near-shore environments. In addition, interbedded coastal sabkha layers and repetitive aeolian sand sheets of sequence 2 point towards transgressive-regressive fluctuations. Basinward progradation, however, was reset and landgrowth was substantial as evidenced from the formation of thick erg deposits of sequence 3.

Two shallow-marine incursions and repetitive offlaps by basinward prograding aeolian sand sheets interbedded with coastal sabkhas were caused by relative sea-level changes. Allocyclic controls on redistribution of environments, spatially and temporally are

Fig. 25 : The section shows superimposed sets of dunes or "draa", separated by thin interdune deposits, together representing a part of the central-erg, sequence 3 of Egalapenta Member. Bar scale (to the bottom left) = 1m.

manifested by facies mosaics and palaeogeographic shifts during the Egalapenta sedimentation. Whether such sea-level changes in Mesoproterozoic extensional regime were eustatic appears to be a matter of debate. There are case histories where extensional basins with high subsidence rates (Mann et al. 1983) provided accommodation space necessary for the accumulation of thick successions and sea-level changes. Moreover, gradual change from wet fluvial condition to dry arid climate might also point to mechanisms such as fluctuations in ground water volume on continents or thermal expansion and contraction of sea water (Ulicny 1999) which could have controlled to some extent the eustatic fluctuations. At the same time climatic oscillations may also have other genetic implications. Association of fluvial and repetitive (cyclic) aeolian dune-interdune and sabkha facies of Egalapenta corroborates climatic fluctuations from humid to arid and possible controls of the climatic variations on the dynamics of the desert systems. Cyclicity in similar interbedded aeolian, fluvial and sabkha deposits appears to be the response of climatic fluctuations which might have been orbitally-controlled (Frederiksen et al. 1998). The Egalapenta complex-composite represents a spectrum of aeolian system types in which end members are dry, wet and stabilised systems with records of interference by fluvial-aeolian interaction, sabkha phase and allocyclic shallow-marine milieus.

CONCLUSIONS

The Egalapenta beds constitute the youngest Member of the Mesoproterozoic Srisailam Formation of the Cuddapah basin, India. Three sequences (designated 1, 2 and 3) gradational to one another are distinguishable within the Egalapenta Member. The Egalapenta itself is gradational to the underlying Tapasipenta beds and display complex interplay of fluvial, shallow-marine, arid sea-marginal sabkha and aeolian processes.

Progradational configuration, association of fluvial facies, palaeocurrent trends, gradual increase of braided fluvial signatures in sequence 1 and intercalated lithosomes of glauconitic sandstone with tidal imprints reflect westerly basinward progradation by a terrigenous wedge of fluvial clastics interfered by shallow-marine incursions.

Sequence 2 is marked by repetitive interbedding of anhydrite-bearing carbonate-sandstone-mudstone (coastal sabkha) with aeolian sandsheets. Abrupt facies transitions, however, resulted from the thinly interbedded facies. Rapid increase in aeolian sandsheets, their partial changes by ephemeral stream discharges, wind action and the overall small size of aeolian bedforms later dominated the sedimentation patterns. The facies association attests to further basinward progradation of land alternating with arid sea-marginal sabkha sedimentation and subsequent modifications by alluvial incursions and wind action of rapidly deposited aeolian sands in the coastal scenario. The set-up is comparable to an erg-margin.

Aeolian and associated facies are characteristic of sequence 3 of the Egalapenta Member. Dominant attributes of the facies associations, the vertical arrangement, lithology, geometry record the preservation of dunes without intervening interdune deposits, individual aeolian sandsheets of interdunes, dune-interdune and amalgamated interdune deposits comprising aeolian sandsheets, flash-flood-, ephemeral- and inland sabkha deposits. Vertical change from amalgamated interdune deposits to dunes and draas records dune-field initiation or lateral migration of peripheral to more centralised portion within the dune-field. Sequence 3 shows signatures of aeolian environments which are comparable to those in a modern erg.

Facies associations in the Egalapenta Member indicate development of erg environment in all its three divisions, namely erg-margin, fore-erg and central-erg where detritas was supplied from earlier fluvial deposits.

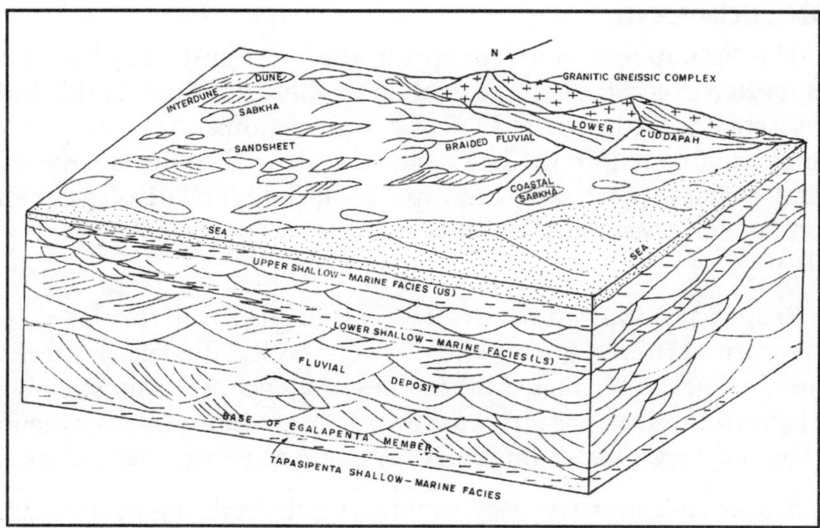

Fig. 26 : Schematic diagram showing palaeogeographic reconstruction during Egalapenta sedimentation, Srisailam Formation. The Egalapenta draa were spatially separated by aeolian sandsheets, braided-fluvial deposits, ephemeral- and coastal sabkha deposits.

Transition from braided fluvial deposits to a repetitive sequence of aeolian sandsheets alternating with coastal sabkha carbonates, followed by vertically arranged dunes, interdunes and ending in draas reflects alternation of humid and arid conditions which might have been orbitally-controlled. Eustatic fluctuations account for repetitive occurrence of environments and palaeogeographic shifts.

REFERENCES

Ahlbrandt,T.S., and Fryberger,S.G.,1980. Eolian deposits in the Nebraska Sand Hills. US Geol.Surv.Prof.Pap., v.1120, pp.1-24.

Allen,J.R.L.,1965. The sedimentation and palaeogeography of the Old Red Sandstone of Anglesey, North Wales. Yorkshire. Geol.Soc.Proc, v.35, pp.139-185.

Allen,J.R.L.,1968. Current ripples. Their relation to patterns of water and sediment motion. Amsterdam North Holland Publ.Comp., 433p.

Allen,J.R.L.,1982. Developments in sedimentology,: Sedimentary structures. Their character and physical basis. Elsevier (Amsterdam). v.30A and B.

Allen,J.R.L.,1983. Studies in fluviatile sedimentation : bars, bar complexes and sandstone sheets (low-sinuosity braided streams) in the Brownstone (L.Devonian), Welsh Borders. Sed.Geol., v.33, pp.237-293.

Allen,J.R.L and Banks,N.L.,1972. An interpretation and analysis of recumbent-folded deformed cross-bedding. Sedimentology, v.19, pp.257-283.

Anderton,R.,1976. Tidal-shelf sedimentation : an example from the Scottish Dalradian. Sedimentology, v.23, pp.429-458.

Arche,A.,1983. Coarse-grain meander lobe deposits in the Jarama River, Madrid, Spain. In: Modern and ancient fluvial systems. (J.D.Collinson and J.Lewin, eds.). Int.Ass.Sediment.Spec.Publ., v.6, pp.313-321

Bagnold,R.A.,1941. The physics of blown sand and desert dunes. London, Meuthuen, pp.265.

Bagnold,R.A.,1954. Experiments on a gravity-free dispersion of large solid spheres in a Newtonian fluid under shear. Proc.R.Soc.Lond.Ser. A, v.225, pp.49-63.

Bhattacharyya,H.N.,1991. A reappraisal of the depositional environment of the Precambrian metasediments around Ghatsila-Galudih, eastern Singhbhum. J.Geol.Soc.Ind., v.37, pp.47-54.

Bhattacharyya, H.N.,1998. Seismites in a Proterozoic tidal sequence, Singhbhum, Bihar, India. Sedimentary Geology. vol.119; pp.239-252.

Blodgett,R.H. and Stanley,K.O.,1980. Stratification, bedforms and discharge relations of the Platte braided river system, Nebraska. J.Sedim.Petrol., v.50, pp.139-148.

Bluck,B.J.,1971. Sedimentation in the meandering river Endrick. Scott.J.Geol., V.7, pp.93-138.

Bluck,B.J.,1976. Sedimentation in some Scottish river of low sinuosity. Trans.R.Soc.Edinb., v.69, pp.425-456.

Bluck,B.J.,1979. Structure of coarse-grained braided stream alluvium. Trans.R.Soc.Edinb, v.70, pp.181-221.

Bluck,B.J.,1980. Structure, generation and preservation of upward fining, braided stream cycles in the Old Red Sandstone of Scotland. Trans.R.Soc.Edinb, v.71, pp.29-46.

Bluck,B.J.,1981. Upper Old Red Sandstone, Firt of Clyde. In: Field Guides to Modern and Ancient Fluvial Systems in Britain and Spain (T.Elliot, ed.). Proc.2nd Cong.Fluvial Seds.Univ.Keele, U.K., Int.Ass.Sediment., 5.14-5.20.

Boersma,J.R.,1967. Internal structures of some tidal megaripples on a shoal in the Westerschelde estuary, the Netherlands. Report of a preliminary investigation: Geologie en Mijnbouw, v.48, pp.409-414.

Boersma,J.R. and Terwindt,J.H.J.,1981. Neap-spring sequences in intertidal shoal deposits in a mesotidal estuary. Sedimentology, v.28, pp.151-170.

Boothroyd,J.C. and Nummedal,D.,1978. Proglacial braided outwash: A model for humid alluvial-fan deposits. In: Fluvial Sedimentology (A.D.Miall, ed.). Can.Soc.Petrol.Geol.Mem., V.5, pp.641-668.

Bose,P.K. and Chakraborty,P.P.,1994. Marine to fluvial transition : Proterozoic Upper Rewa Sandstone, Maihar, India., Sed.Geol., v.89, pp.285-302.

Brookfield,M.E.,1977. The origin of bounding surfaces in ancient aeolian sandstone. Sedimentology, v.24, pp.303-332.

Bull,W.B.,1972. Recognition of alluvial fan deposits in the stratigraphic record. In: Recognition Of Ancient Sedimentary Environments (J.K.Rigby and W.K.Hamblin, eds.). Soc.Econ.Paleontol.Mineral Spec. Paper, v.60, pp.63-83.

Bull,W.B.,1977. The alluvial fan environment. Prog.Phys.Geogr., v.1, pp.222-270.

Cant,D.J.,1978. Development of a facies model for sandy braided river sedimentation: comparison of the south Saskatchewan River and the Battery Point Formation. In: Fluvial Sedimentology (A.D.Miall, ed.). Can.Soc.Petrol.Geol.Mem., v.5, pp.627-640.

Cant,D.J. and Walker,R.G.,1976. Development of a braided-fluvial facies model for the Devonian Battery Point Sandstone, Quebec. Can.J.Earth.Sci., v.13, pp.102-119.

Cant,D.J. and Walker,R.G.,1978. Fluvial processes and facies sequences in the sandy braided South Saskatchewan River, Canada. Sedimentology, v.25, pp.625-648.

Carling,P.A.,1990. Particle over-passing on depth-limited gravel bars. Sedimentology, v.37, pp.345-355.

Chan,M.A.,1989. Erg margin of the Permian White Rim Sandstone, SE Utah. Sedimentology, v.36, pp.235-251.

Chanda,S.K. and Bhattacharyya,A.,1982. Vindhyan sedimentation and paleogeography: Post Auden development. In: Geology of the Vindhyachal (K.S.Valdya, S.B.Bhatia and V.K.Gour, eds.). Hindusthan Publishing Corporation, India., pp.88-101.

Chaudhury,A.K. and Howard,J.D.,1985. Ramgundam Sandstone: A Middle Proterozoic shoal bar sequence. J.Sedim.Petrol., v.55, pp.392-397.

Chaudhury,A.K.; Chanda.S.K. and Dasgupta,S.,1994. Proterozoic glauconitic peloids from south India: their origin and significance. J.Sedimentary Research, v.A64(4), pp.765-770.

Coleman,J.M.,1969. Brahmaputra River : Channel processes and sedimentation. Sed.Geol., v.3, pp.129-239.

Collinson,J.D.,1970. Bedforms of the Tana river, Norway. Geogr.Ann., v.52A, pp.31-55.

Costello,W.R. and Walker,R.G.,1972. Pleistocene sedimentology, Credit River, southern Ontario: A new component of the braided river model. J.Sedim.Petrol., v.42, pp.389-400.

Cotter,E.,1978. The evolution of fluvial style, with special reference to the central Appalachian Paleozoic. In: Fluvial Sedimentology (A.D.Miall, ed.). Can.Soc.Petrol.Geol.Mem., v.5, pp.361-383.

Crabaugh,M and Kocurek,G.,1993. Entrada Sandstone: an example of a wet aeolian system. In: The Dynamics and Environmental Context of Aeolian Sedimentary Systems (K.Pye, ed.). Geol.Soc.Sp.Publ.No.72, pp.103-126.

Crowley,K.D.,1983. Large-scale bed configurations (macroforms), Platte River basin, Colorado and Nebraska: Primary structures and formative processes. Geol.Soc.Am.Bull., v.94, pp.117-133.

Cudzil,M.R. and Driese,S.G,1987. Fluvial, tidal and storm sedimentation in thc Chilhowee Group (Lower Cambrian), northeastern Tennessee, USA. Sedimentology, v.34, pp.861-883.

De Boer,P.L., Oost,A.P. and Visser,M.J.,1989. The diurnal inequality of the tide as a parameter for recognising tidal influences: J.Sedim.Petrol., v.59, pp.912-921.

Diehl,P.E.,1979. The stratigraphy, depositional environments, and quantitative petrography of the Precambrian-Cambrian Wood Canyon Formation, Death Valley. (unpubl.Ph.D. Diss.) University Park, Pennsylvania State University., 365p.

Doe,T.W. and Dott,R.H.Jr.,1980. Genetic significance of deformed cross-bedding – with examples from the Navajo and Webeer Sandstones of Utah : J.Sedim.Petrol., v.50, pp.793-812.

Dott,R.H.Jr.,1974. Cambrian tropical storm waves in Wisconsin. Geol., v.2, pp.243-246.

Driese,S.G. and Dott,R.H.Jr.,1984. Model for sandstone-carbonate "cyclothems" based on upper member of Morgan Formation (Middle Pennsylvanian) of northern Utah and Colorado. Am.Assoc.Petrol.Geol.Bull., v.68(5), pp.574-597.

Driese,S.G., Byers,C.W. and Dott,RH.Jr.,1981. Tidal deposits in the basal Upper Cambrian Mt. Simon Formation in Wisconsin. J.Sedim.Petrol., v.51, pp.367-381.

Enyon,G. and Walker,R.G.,1974. Facies relationship in Pleistocene outwash groups, southern Ontario: A model for bar growth in braided river. Sedimentology, v.21, pp.43-70.

Eriksson,K.A. and Simpson,E.L.,1998. Controls on spatial and temporal distribution of Precambrian eolianites. Sedimentary Geology., v.120, pp.275-294.

Eriksson,P.G.; Condie,K.C.; Tirsgaard,H.; Mueller,W.U; Altermann,W.; Miall,A.D.; Aspler,L.B.; Catuneanu,O. and Chiarenzelli,J.R.,1998. Precambrian clastic sedimentation systems. Sedimentary Geology., v.120, pp.5-53.

Fairbridge,R.W. and Bourgeois,J.,1978. The encyclopedia of sedimentology, encyclopedia of earth sciences. Dowden,Hutchinson and Ross Inc., Stroudsburg. v.6, 901p.

Fedo,C.M. and Cooper,J.D.,1990. Braided fluvial to marine transition : The basal lower Cambrian Wood Canyon Formation, Southern Marble Mountains, Mojave Desert, California. J.Sedim.Petrol., v.60(2), pp.220-234.

Folk,R.L.,1968. Bimodal supermature sandstones : Product of the Desert floor. International. Geol. Cong., v.8, pp.9-32.

Frazier,D.E. and Osanik,A.,1961. Point-bar deposits; Old River Locksite, Louisiana. Trans.Gulf.Coast.Assoc.Geol.Socs., v.11, pp.121-137.

Frederiksen,K.S., Clemmensen,L.B. and Lawaetz,H.S.,1998. Sequential architecture and cyclicity in Permian desert deposits, Brodick Beds, Arran, Scotland. J.Geol.Soc.Lond., v.155(4), pp.677-683.

Friedman,G.M.,1959. Identification of carbonate minerals by staining methods. J.Sedim.Petrol., v.29, pp.87-97.

Friedman,G.M. and Sanders,J.E.,1978. Principles of sedimentology. John Wiley and Sons. 792p.

Friend,P.F.,1978. Distinctive features of some ancient river systems. In: Fluvial sedimentology (A.D.Miall, ed.). Can.Soc.Petrol.Geol.Mem., v.5, pp.531-542.

Fryberger,S.G. and Schenk,C.,1981. Wind sedimentation experiment on the origin of aeolian strata. Sedimentology, v.18, pp.805-821.

Fryberger,S.G., Ahlbrandt,T.S. and Andrews,S.,1979. Origin, sedimentary features and significance of low angle aeolian sand sheet deposits, Great Sand Dunes National Monument and vicinity, Colorado. J.Sedim.Petrol., v.49, pp.733-746.

Fryberger,S.G., Al-Sari,A.M. and Clisham,T.J.,1983. Eolian dune, interdune, sand sheet and siliciclastics sabkha sediments of an offshore prograding sand sea, Dhahran area, Saudi Arabia. Bull.Amer.Petroleum Geologists, v.63, pp.280-312.

Fuller,A.G.,1985. A contribution to the conceptual modelling of Pre-Ordovician fluvial systems. Trans.Geol.Soc.S. Africa, v.88, pp.189-194.

Galloway,W.E.; Hobday,D.K. and Magara,K.,1982. Frio Formation of the Texas Gulf Coast Basin – Depositional systems, structural framework, and hydrocarbon origin, migration distribution and exploration potential. University of Texas at Austin, Bureau of Economic Geology, report of investigations 122.

Glennie,K.W.,1987. Desert sedimentary environment present and past – a summary. Sediment.Geol., v.50, pp.135-165.

Geological Survey of India,1981. Explanatory Brochure on Geology and Mineral Map of the Cuddapah Basin, 1st ed. (May,1981). Published by Geological Survey of India, Calcutta, 22p.

Goldschmith,V.,1973. Internal geometry and origin of vegetated coastal sand dunes. J.Sedim.Petrol., v.43, pp.1128-1142.

Harms,J.C. and Fahnestock,R.K.,1965. Stratification, bedforms and flow phenomena (with an example from the Rio Grande). Soc.Econ. Palaeontologists Mineralogists Spec. Publ., v.12, pp.84-115.

Harms,J.C., Southard,J.B., Spearing,D.R., Walker,R.G.,1975. Depositional environments as interpreted from primary sedimentaryy structures and stratification sequences. Soc.Econ.Palcontol.Mineral., Short Course, v.2, p.161.

Harms,J.C.; Southard,J.B. and Walker,R.G.,1982. Structures and sequences in clastic rocks. Soc.Econ.Paleontol.Mineral., Short Course, 9.

Haszeldine,R.S.,1983. Fluvial bars reconstructed from a deep, straight channel, Upper Carboniferous coalfield of northeast England. J.Sedim.Petrol., v.53, pp.1233-1248.

Havholm,K.G. and Kocurek,G.,1988. A preliminary study of the dynamics of a modern draa, Algodones, southeastern California, USA, Sedimentology, v.35, pp.649-669.

Hein,F.J. and Walker,R.G.,1977. Bar evolution and development of stratification in the gravelly, braided, Kicking Horse River, British Columbia. Can.J.Earth Sci., v.14, pp.562-570.

Horowitz,D.H.,1982. Geometry and origin of large-scale deformation structures in some ancient wind-blown sand deposits. Sedimentology, v.29, pp.155-180.

Houseknecht,D.W. and Ethridge,F.G.,1978. Depositional history of the Lamotte Sandstone of southeastern Missouri. J.Sedim.Petrol., v.48, pp.575-586.

Hummel,G. and Kocurek,G.,1984. Interdune areas of the back-island dune field, north Padre Island, Texas. Sed.Geol., V.39, pp.1-26.

Hunter,R.E.,1976. Comparison of eolian and subaqueous sand-flow cross-strata. Am.Assoc.Pet.Geol.Bull., v.60, pp.683-684 (abstr).

Hunter,R.E.,1977a. Basic types of stratification in small aeolian dunes. Sedimentology, v.24, pp.361-387.

Hunter,R.E.,1977b. Terminology of cross-stratified sedimentary layers and climbing-ripple structures. J.Sedim.Petrol., v.47 (2), pp.697-706.

Hunter,R.E.,1980. Quasi-planar adhesion stratification — an eolian structure formed in wet sand. J.Sedim.Petrol., v.50, pp.263-266.

Hunter,R.E.,1981. Stratification styles in eolian sandstones : some Pennsylvanian to Jurassic examples from the western interior USA. In : Recent and Ancient nonmarine depositional environments : Models for exploration (F.G.Ethridge and R.M.Flores, eds.), Spec.Publ.Soc.Econ.Paleont.Miner. Tulsa, v.31, pp.315-329.

Hughes,A.D. and Whitehead,D.,1987. Glauconitization of detrital silica in the Barton Formation (upper Eocene) of the Hampshire Basin, southern England. Sedimentology, v.34, pp.825-835.

Inman,D.L., Ewing,G.C. and Corliss,J.B.,1966. Coastal sand dunes of Guerrero Negro, Baja California, Mexico.Bull.Geol.Soc.Amer., v.77, pp.787-802.

Johnson,P.R.,1965. Structure and stratigraphy of part of the upper Pranhita-Godavari Valley with special reference to the Pre-Gondwana. Unpublished Ph.D. thesis, University of Calcutta, 144p.

Jopling,A.V.,1965. Laboratory study of the working processes in cross-bedded deposits. In: Primary Sedimentary Structures and Their Hydrodynamic Interpretation (G.V.Middleton, ed.). Soc. Econ. Paleontol. Mineral. Spec. Publ., v.12, pp.53-65.

Jopling,A.V.,1966. Some principles and techniques used in reconstructing the hydraulic parameters of a paleo-flow regime. J.Sedim.Petrol., v.36(1), pp.5-49.

Kenyon,N.H. and Stride,A.H.,1970. The tide-swept continental shelf sediments between the Shetland Isles and France. Sedimentology, v.14, pp.159-173.

Kirk,M.,1983. Bar development in a fluvial sandstone (Westphalian 'A'), Scotland. Sedimentology, v.30, pp.727-742.

Kocurek,G.,1981. Significance of interdune deposits and bounding surfaces in aeolian dune sands, Sedimentology, v.28, pp.753-780.

Kocurek,G. and Dott,R.H.,Jr.,1981. Distinction and uses of stratification types in the interpretation of aeolian sand. J.Sedim.Petrol., v.51, pp.579-595.

Kocurek,G. and Fielder,G.,1982. Adhesion structures. J.Sedim.Petrol., v.52, pp.1229-1241.

Kohshiek,G. and Terwindt,J.H.J.,1981. Characteristics of foreset and topset bedding in megaripples related to hydrodynamic conditions on an intertidal shoal. Int.Ass.Sediment.Publ., v.5, pp.27-37.

Kreisa,R.D. and Moiola,R.J.,1986. Sigmoidal tidal bundles and other tide-generated sedimentary structures of the Curtis Formation, Utah. Geol.Soc.Am.Bull., v.97, pp.381-387.

Levey,R.A.,1978. Bed-form distribution and internal stratification of coarse-grained point bars, Upper Congaree River, S.C. In: Fluvial Sedimentology (A.D.Miall, ed.). Can.Soc.Pet.Geol.Mem., v.5, pp.105-127.

Long,D.G.F.,1978. Proterozoic stream deposits : some problems of recognition and interpretation of ancient sandy fluvial systems. In. Fluvial Sedimentology (A.D.Miall, ed.). Can.Soc.Pet.Geol.Mem., v.5, pp.313-342.

Mader,D. and Yardley,M.J.,1985. Migration, modification and merging in aeolian systems and the significance of the depositional mechanisms in the Permian and Triassic dunes sands of Europe and North America. Sed.Geol., v.43, pp.85-218.

Mann,P.R., Hempton,M.R., Bradley,D.C. and Burke, K.,1983. Development of pull-apart basins. J.Geol., v.91, pp.529-554.

Massari,F.,1983. Tabular cross-bedding in Messinian fluvial channel conglomerates, southern Alps, Italy. In: Modern and Ancient Fluvial Systems (J.D.Collinson and J.Lewin, eds.). Int.Ass.Sediment.Spec.Publ., v.6, pp.287-300.

McCabe,P.J. and Jones,C.M.,1977. Formation of reactivation surfaces within superimposed deltas and bedforms. J.Sedim.Petrol., v.47, pp.707-715.

McCave,I.N.,1970. Deposition of fine-grained suspende sediment from tidal currents. J.Geophys.Res., v.75, pp.4151-4159.

McCave,I.N.,1971. Wave-effectiveness at the sea-bed and its relationship to bedforms and deposition of mud. J.Sedim.Petrol., v.41, pp.89-96.

McCave,I.N.,1973. The sedimentology of a transgression : Portland Point and Cooksburg Members (Middle Devonian), New York State. J.Sedim.Petrol., v.43, pp.484-504.

McGowen,J.H. and Garner,L.E.,1970. Physiographic features and stratification types of coarse-grained point bars: modern and ancient examples. Sedimentology, v.14, pp.77-111.

McKee,E.D.,1957. Primary structures in some recent sediments : Amer.Assoc.Petroleum Geologists Bull., v.41(8), pp.1704-1747.

McKee,E.D.,1966. Structures of dunes at White Sand National Monument, New Mexico (and a comparison with structures of dunes from other selected areas). Sedimentology, v.7, pp.1-69.

McKee,E.D.,1979. A Study of Global Sand Seas (E.D.McKee, ed.). Geological Survey Professional Paper, U.S.Govt.Printing Office, Washington, v.1052, 429p.

McKee,E.D. and Douglas,J.R.,1971. Growth and movement of dunes at White Sands National Monument, New Mexico in Geology Survey Research 1971: U.S. Geol.Surv.Prof.Pap., v.750-D, D108-D114p.

McKee,E.D. and Tibbits,G.C.,1964. Primary structures of a seif dune and associated deposits in Libya. J.Sedim.Petrol., v.34, pp.5-17.

McKee,E.D., Reynolds,M.A. and Baker,C.H.,1962. Laboratory studies on deformation in unconsolidated sediment. U.S.Geol.Surv.Profess. Papers, 450-D, D151-D155.

McKee,E.D., Crosby,E.J. and Berryhill,H.L.Jr.,1967. Flood deposits, Bijou Creek, Colorado, June (1965). J.Sedim.Petrol., v.37, pp.829-851.

McKee,E.D., Douglas.J.R. and Rittenhouse,Suzanne,1971. Deformation of lee-side laminae in eolian dunes: Geol.Soc.Am.Bull., v.82(2), pp.359-378.

McRae,S.G.,1972. Glauconite. Earth Science Review., v.8, pp.367-440.

Miall,A.D.,1977. A review of the Braided-river depositional environment. Earth-Science Reviews, v.13, pp.1-62.

Miall,A.D.,1985. Architectural-Element Analysis: A New Method of Facies Analysis Applied to Fluvial Deposits. Earth Sci.Rev., v.22, pp.261-308.

Miall,A.D.,1996. The Geology of Fluvial Deposits: Sedimentary Facies, Basin Analysis and Petroleum Geology. Springer-Verlag, Berlin, 582p.

McLane,M.,1995. Sedimentology. Oxford University Press, 423p.

Michelson,P.C. and Dott,R.H.,Jr.,1973. Orientation analysis of trough cross-stratification in Upper Cambrian sandstones of Western Wisconsin. J.Sedim.Petrol., v.43, pp.784-794.

Morison,S.R. and Hein,F.J.,1987. Sedimentology of the White Channel gravels, Klondike area, Yukon Territory: fluvial deposits of a confined valley. In: Recent developments in fluvial sedimentology (F.G.Ethridge, R.M.Flores and M.D.Harvey, eds.). Spec.Publ.Soc.Econ.Paleont.Miner.Tulsa, v.39, pp.205-216.

Muir,I.D. and Rust,B.R.,1982. Sedimentology of a Lower Devonian coastal alluvial fan complex: the Snowblind Bery Formation of Cornwallis Island, Northwest Territories, Canada. Bull.Can.Petrol.Geol., v.30, pp.245-263.

Mukherji,B. and Dasgupta,P.K.,1972. Depositional environment of the Upper Member of "Muth Quartzite" from Spiti and Kulti regions, the Central Himalayas, India. Proc.24 Session of Int.Geol.Cong., Montreal, Sec.6, pp.424-431.

Narayan,J.,1971. Sedimentary structures in the Lower Greensand of the Weald, England and the Bas-Boulonnaise, France. Sed.Geol, v.6, pp.73-109.

Nemec,W. and Steel,R.J.,1984. Alluvial and coastal conglomerates: their significant and some comments on gravelly mass flow deposits. In: Sedimentology Of Gravels And Conglomerates (E.H.Koster and R.J.Steel, eds.). Mem.Can.Soc.Petrol.Geol.Calgary, v.10, pp.1-31.

Neilson,J. and Kocurek,G.,1986. Climbing zibars of the Algodones. Sedimentary Geology, v.48, pp.1-15.

Ori,G.G.,1982. Braided to meandering channel patterns in humid-region alluvial fan deposits, River Reno, Po Plain (northern Italy). Sed.Geol., v.31, pp.231-248.

Owen,G.,1995. Soft sediments deformation in Upper Proterozoic Torridonian Sandstones (Applecross Formation) at Torridon, Northwest Scotland. J.Sediment.Res. A65, 495-504.

Picard,M.D. and High,L.R., Jr.,1973. Sedimentary structures of ephemeral streams. Developments in sedimentology, Elsevier, New York, v.17, pp.223.

Pierson,T.C. and Scott,K.M.,1985. Downstream dilution of a lahar: transition from debris flow to hyperconcentrated streamflow. Wat.Resour.Res., v.21, pp.1511-1524.

Plint,A.G.,1983. Sandy fluvial point-bar sediments from the middle Eocene of Dorset, England. In: Modern and Ancient Fluvial Systems (J.D.Collinson and J.Lewin, eds.). Int.Ass.Sediment.Spec.Publ., v.6, pp.355-368.

Porter,M.L.,1987. Sedimentology of an ancient erg margin: the lower Jurassic Aztec Sandstone, southern Nevada and southern California. Sedimentology, v.34, pp.661-680.

Ramos,A., Sopena,A. and Perez-Arlucea,M.,1986. Evolution of Buntsandstein fluvial sedimentation in the northwest Iberian Ranges (central Spain). J.Sedim.Petrol., v.56, pp.862-865.

Reading,H.G.,1996. Sedimentary Environments : Processes, Facies and Stratigraphy. 3rd edition. Blackwell Science Ltd., 688p.

Reineck,H.E.,1955. Haftrippeln und Haftwarzen, Ablagerungsfomen von Flugsand. Senckenbergiana Lethaea, v.36, pp.347-357.

Reineck,H.E. and Singh,I.B.,1973. Depositional environments – with Reference to Terrigenous Clastics. Springer, Berlin, 439p.

Reineck,H.E. and Singh,I.B.,1980. Depositional sedimentary environments – with reference to terrigenous clastics. Springer-Verlag, Berlin, 2nd ed. 549p.

Roe,S.L.,1987. Cross-strata and bedforms of probable transitional dune to upper-stage plane-bed origin from a late Precambrian fluvial sandstone, northern Norway. Sedimentology, v.34, pp.89-101.

Ross,G.M.,1983. Proterozoic aeolian quartz arenites from Hormby Bay Group, northwest territories, Canada: implications for Precambrian aeolian processes. Precambrian Research. v.20, pp.149-160.

Rubin, D.M.,1987. Cross-bedding, bedforms, and paleocurrents. Soc. Econ. Paleontol. Mineral. Concepts in Sedimentology and Paleontology, v.1, 188p.

Rubin,D.M. and Hunter,R.E.,1983. Reconstructing bedform assemblages from compound cross-bedding In: Brookfield,M.E. and Ahlbrandt,T.S. (Eds.) Eolian Sediments and Processes. Elsevier, Amsterdam, pp.407-427.

Rust,B.R.,1972. Structure and process in a braided river. Sedimentology, v.18, pp.221-246.

Rust,B.R.,1978. Depositional models for braided alluvium. In: Fluvial Sedimentology (A.D.Miall, ed.). Can.Soc.Petrol.Geol.Mem., v.5, pp.605-625.

Rust,B.R.,1981. Sedimentation in an arid-zone anastomosing fluvial system: Cooper's Creek, central Australia. J.Sedim.Petrol., v.51, pp.745-755.

Rust,B.R. and Koster,E.H.,1984. Coarse alluvial deposits. In: Facies Models (R.G.Walker,ed.). Geosci.Can., Reprint Series, v.1, pp.53-69.

Saunderson,H.C. and Lockett,F.P.J.,1981. Flume experiments on bedforms and structures at the dune plane bed transition. In: Modern and Ancient Fluvial Systems (J.D.Collinson and J.Lewin, eds.). Int.Ass.Sediment. Spec.Publ., v.6, pp.49-59.

Schumm,S.A.,1968. Speculation concerning paleohydrology controls of terrestrial sedimentation. Geol.Soc.Am.Bull., v.79, pp.1573-1588.

Schwartz,D.E.,1978. Hydrology and current orientation analysis of a braided to meandering transition; the Red River in Oklahoma and Texas, USA. In: Fluvial Sedimentology (A.D.Miall, ed.). Can.Soc.Petrol.Geol.Mem., v.5, pp.231-256.

Sharp,R.A.,1963. Wind ripples : J.Geol., v.71(5), pp.617-636.

Shelton,J.W. and Noble,R.L.,1974. Depositional features of braided-meandering stream. Amer.Assoc. Petroleum Geologists Bull., v.58, pp.742-749.

Shepherd,R.G.,1987. Lateral accretion surfaces in ephemeral stream point bars, Rio Puerco, New Mexico. In: Recent developments in fluvial sedimentology (F.G.Ethridge, R.M.Flores and M.D.Harvey, eds.). Spec.Publ.Soc.Econ.Paleont.Miner.Tulsa, v.39, pp.93-98.

Simpson,E.L. and Eriksson,K.A.,1993. Thin eolianites interbedded within a fluvial and marine succession: early Proterozoic Whitworth Formation, Mt. Isa Inlier, Australia. Sed.Geol., v.87, pp.39-62.

Simpson,E.L. and Loope,D.B.,1985. Amalgamated interdune deposits, White Sands, New Mexico. J.Sedim.Petrol., v.55, pp.361-365.

Sims,J.D.,1975. Determining earthquake recurrence intervals from deformational sructures in young lacustrine sediments. Tectonophysics, v.29; pp.141-152.

Smith,N.D.,1970. The braided stream depositional environment : Comparison of the Platte river with some Silurian clastic rocks, North-Central Appalachians. Geol.Soc.Amer.Bull., v.81, pp.2993-3014.

Smith,N.D.,1971. Transverse bars and braiding in the lower Platte River, Nebraska. Geol.Soc.Amer.Bull., v.82, pp.3407-3420.

Smith,N.D.,1974. Sedimentology and bar formation in the upper Kicking Horse River, a braided outwash stream. J.Geol., v.81, pp.205-223.

Smith,G.A.,1986. Coarse-grained nonmarine volcaniclastic sediment: terminology and depositional process. Geol.Soc.Am.Bull., v.97, pp.1-10.

Sneh,A.,1983. Desert stream sequences in the Sinai Peninsula. J.Sedim.Petrol., v.53, pp1271-1280.

Southard,J.B., Smith,N.D. and Kuhnle,R.A.,1984. Chutes and lobes: newly identified elements of braiding in shallow gravelly streams. In: Sedimentology of Gravels and Conglomerates (E.H.Koster and R.J.Steel, eds.). Can.Soc.Petrol.Geol.Mem., v.10, pp.51-59.

Stablein,N.K. III and Dapples,E.C.,1977. Feldspar of the Tunnel City Group (Cambrian), Western Wisconsin: J.Sedim.Petrol., v.47, pp.1512-1538.

Stear,W.M.,1985. Comparison of the bedform distribution and dynamics of modern and ancient sandy ephemeral flood deposits in the southwestern Karoo region, South Africa. Sed.Geol., v.45, pp.209-230.

Steel,R.J., Maehle,S., Nilsen,H., Roe.S.L. and Spinnanger,A.,1977. Coarsening-upward cycles in the alluvial of Hornelen Basin (Devonian), Norway: sedimentary response to tectonic events. Geol. Soc.Am.Bull., v.88, pp.1124-1134.

Strickler,M.E. and Ferrel,R.E.Jr.,1990. Fe substitution for Al in glauconite with increasing diagenesis in the first Wilcox Sandstone (Lower Eocene), Livingston Parish, Lousiana: Clays and Clay Minerals, v.38, pp.69-76.

Stride,A.H.,1963. Current swept floors of the southern half of Great Britain. Q.J.Geol.Soc.Lond., v.119, pp.175-199.

Sweet,M.L., Nielson,J., Havholm,K. and Farrelley,J.,1988. Algodones dune field of southwestern California: case history of a migrating modern dune field. Sedimentology, v.35, pp.939-952.

Swett,K. and Smitt,D.E.,1972. Palaeogeography and depositional environment of the Cambro-Ordovician shallow-marine facies of the North Atlantic. Bull.Geol.Soc.Amer., v.83, pp.3223-3248.

Terwindt,J.H.J.,1981. Origin and sequences of sedimentary structures in inshore mesotidal deposits of the North Sea. In: Holocene Marine Sedimentation In North Sea Basin (S.D.NiO, R.T.E.Schütterhelm and T.C.E. Van Weering, eds.). Int.Ass.Sediment. Spec.Publ., v.5, pp.4-26.

Todd,S.P. and Went,D.J.,1991. Lateral migration of sand-bed rivers : examples from the Devonian Glashabeg Formation, South West Ireland and the Cambrian Alderney Sandstone Formation Channel Islands. Sedimentology, v.38, pp.997-1020.

Tsoar,H.,1982. Internal structure and surface geometry of longitudinal (seif) dunes. J.Sedim.Petrol., v.52, pp.823-831.

Tunbridge,I.P.,1981. Sandy high-energy flood sedimentation – some criteria for recognition, with an example from the Devonian of S.W.England. Sed.Geol., v.28, pp.79-95.

Turner,B.R. and Monro,M.,1987. Channel formation and migration by mass-flow processes in the Lower Carboniferous fluviatile Fell Sandstone Group, northeast England. Sedimentology, v.34, pp.1107-1122.

Ulicny,D.,1999. Sequence stratigraphy of the Dakota Formation (Cenomanian), southern Utah: interplay of eustasy and tectonics in a foreland basin. Sedimentology, v.46, pp.807-836.

Van Houten,F.B. and Puricker,M.E.,1984. Glauconitic peloids and chamositic ooids – favourable factors, constraints and problems: Earth-Science Reviews, v.20, pp.211-243.

Visser,M.J.,1980. Neap-spring cycles reflected in Holocene subtidal large-scale bedform deposits: a preliminary note. Geology, v.8, pp543-546.

Vos,R.G. and Tankard,A.J.,1981. Braided fluvial sedimentation in the lower Paleozoic Cape Basin, South Africa. Sediment.Geol., v.29, pp.179-193.

Walker,R.G. and Cant,D.J.,1984. Sandy flow systems. In: Facies models, 2nd edition (R.G.Walker, ed.). Geosci.Can., Reprint Series, v.1, pp.71-89.

Williams,G.E.,1968. Formation of large-scale trough cross-stratification in a fluvial environment. J.Sedim.Petrol.,, v.38, pp.136-140.

Williams,G.E.,1969. Characteristics and origin of a Precambrian pediment. J.Geol., v.77, pp.185-207.

Williams,G.E.,1970. Piedmont sedimentation and late Quaternary chronology in the Biskra region of the northern Sahara. Z.Geomorph., v.10, pp.40-63.

Williams,G.E.,1971. Flood deposits of the sand-bed ephemeral streams of Central Australia. Sedimentology, v.17, pp.1-40.

Williams,G.E.,1991. Upper Proterozoic tidal rhythmites, South Australia: sedimentary features, deposition, and implications for earth's paleorotation. In: Clastic Tidal Sedimentology (D.G.Smith, C.E.Reinson, B.A.Zaitlin and R.A.Rahmani, eds.). Can.Soc.Pet.Geol.Mem., v.16, pp161-178.

Williams,P.F. and Rust,B.R.,1969. The sedimentology of a braided river. J.Sedim.Petrol., v.39, pp.649-679.

Wilson,I.G.,1972. Aeolian bedforms: their development and origin. Sedimentology, v.19, pp.173-210.

Wilson,I.G.,1973. Ergs. Sed.Geol., v.10(2), pp.77-106.

Wood,G.V. and Wolfe,M.J.,1969. Sabkha cycles in the Arab/Darb Formation off the Trucial coast of Arabia. Sedimentology, v.12, pp.165-191.

6

Paleoseismicity from the Cuddapah Basin

Introduction

To distinguish seismically-triggered soft sediment deformation structures from similar features produced by other auto- and allo-kinetic processes seems to be a difficult task. The debate concerning the role of palaeoslope in generation of earthquake-induced deformation in unconsolidated sediments prior to, or soon after burial arises out of this situation. There are instances in which earthquake-generated deformational structures seem to be related to regional palaeoslope (Plaziat et.al.,1990; Plaziat & Purser,1998).

Faulting within the basin and along the basin margin may trigger off earthquakes in fault-controlled extensional basins (Mastelerz & Wojewoda,1993; Obermeir,1996; Bhattacharyya & Bandyopadhay,1998). Vibratory ground motions lead to liquefaction and fluidisation in unconsolidated sediments which facilitate penecontemporaneous deformation (Seilacher, 1969; Sims,1973, 1975; Lowe,1975; Tutle & Seeber,1991). The term 'seismites' is applied to sediments with features which are believed to have been induced by earthquake shocks (Seilacher,1969, 1984).

Seismites, seismically-affected (reworked) tempestetic turbidites and tsunamiites have been recorded at different stratigraphic levels, including the Cumbum Formation of Palaeo- to Neoproterozoic Cuddapah Supergroup, India, developed in phases under an overall control of extensional regime. Tectonic controls of volcaniclastic-rich background sedimentation of the Mesoproterozoic Cumbum are intimately associated with such seismic signatures, which have been addressed. Earthquake waves, in their wake, were episodic with the attendant explosive volcanism and offsetting by fractures.

Shallow-marine storm deposits are marked by well preserved laminites, wave ripples, hummocky- and swaley cross-stratified beds arranged in several thickening- and thinning-upward sequences. Severe seismic activity in the syn-rift Cumbum scenario is indicated by seismites and seismo-turbidites developed among the storm-induced sediments. They are deposited in transitional offshore (upper bathyal) setting in association with tsunamiites.

Deformational features, present in widely extensive beds, sandwiched between undisturbed strata, include structures developed under undrained condition such as those produced by undersliding, plastic shear and hydroplastic deformations. Sandstone dykes and chaotic facies on the other hand, are products of deformation under drained condition. Low slope of the Proterozoic epicontinental seas precludes the formation of deformational features due to slope failure and/or artesian pressure. In absence of glacial facies in the stratigraphic record, the concept of development of structures due to ice loading does not merit consideration. Shaking by earthquakes remains the only alternative explanation for generation of liquefaction. Episodic volcanism and faulting at the basin margins were accompanied by ground-tremors, tsunamis that induced the development of seismites, seismoturbidites and tsunamiites.

The chapter first describes the sedimentary features of the sandstone-mudstone alternations and associated pyroclastic layers to explain the (sedimentary) setting of the deformed sediments. However, main purpose of the paper is to give an account of the seismic signatures in the Mesoproterozoic Cumbum Formation of the Cuddapah Basin which may be recognised as the product of syntectonic sedimentation in an extensional regime (Kaila & Tewari, 1985; Kaila et.al.,1987; Bhattacharji & Singh,1984; Dasgupta,1998) as revealed from the study of soft sediment deformation structures. The features are prominently displayed in an area around Hostavaram in the Cuddapah District.

Geologic Setting

Peninsular Gneiss with remnants of greenstone belts marks the western margin of the Cuddapah basin. The eastern margin is marked by the Nellore Schist Belt and the Eastern Ghats Granulite terrain. Lower undeformed cuddapah strata lie unconformably on the Peninsular

Gneiss along the westernmargin. The basal formations of the Cuddapah Supergroup lie unconformably on the Peninsular Gneiss along the western margin. The sedimentary sequences in the eastern part of the basin are steeply dipping, isoclinally folded and slightly metamorphosed. The undeformed lower Group is separated by an unconformity from the deformed upper Group forming the Nallamalai Fold Belt (Murthy, 1979). Along the eastern margin of the basin, the Nellore Schist belt is thrust over the sedimentary sequences. Deposition of the sediments appears to have taken place in a Proterozoic rift basin under a regime of continued crustal stretching (Bhattacharji & Singh,1984; Kaila et.al.,1979; Kaila & Bhatia,1981; Kaila & Tewari,1985; Dasgupta,1996). The protobasinal phases are marked by the influx of subaqueous to terrestrial volcanic materials.

The facies assemblage of Cumbum Formation of the Nallamalai Group includes conglomerate-sand facies (now conglomerate-quartzite association), sand-mud alternation facies (now quartzite-phyllite/schist alternation), mud facies (phyllite/schist) and carbonate-mud facies (metamorphosed into crystalline limestone/dolomite-phyllite/schist). Mafic and/or felsic to silicic volcanics are often associated with the facies assemblage. The sequences were involved in greenschist facies metamorphism and moderate to strong deformation (Saha,1994; Matin & Guha,1996; Bhattacharji, 1986 Venkatakrishnan & Dotiwalla,1987).

Depositional settings and original sedimentary features

Original sedimentary features are well preserved in the Cheyyeru succession of the Cumbum Formation. However, disrupted internal fabric of some of the structures are identifiable in discrete zones between the undisturbed tuffaceous sand-silt-mud alternations.

Sandstone Mudstone Alternations

A sequence of 134m thick intercalated and/or interstratified tuffaceous sandstone-mudstone alternations laterally extensive over hundreds of metres occurs in the Cheyyeru river section of the Hostavaram region (Fig.1). Detailed vertical facies association of the Cheyyeru succession reveal presence of widely extended lower tuffaceous and fine sandstone unit intercalated with mudstone-siltstone (silty unit: SU); pyroclastic-rich volcanic bomb-bearing beds forming the marker horizon (volcaniclastic unit: VU) and upper tuffaceous sandstone unit intercalated and interstratified with diastasis crack-bearing mudstone bands (DU). Laminites of alternating normally graded sand and mud range in thickness from a few millimetres to few centimetres. Chertified tuffaceous sandstone-mudstone layers with ballistics are discernible at several horizons. The stratal succession in association with the pyroclast-rich layer characterises the Cheyyeru succession.

Sedimentary structures showing minute details down to the finest scales are preserved in a steeply dipping and northwest striking band of metasedimentary rocks over a stretch measuring 8km by 1.5km within the Nallamalai Fold Belt and metavolcaniclastics of felsic to silicic composition (King,1872; Neelakantam & Roy,1979; Vijayam,1969). The rocks belong to a suite dating back to 1.49-1.09Ga (Crawford & Compston,1973; Rao et al., 1996). Sand and bluish-grey

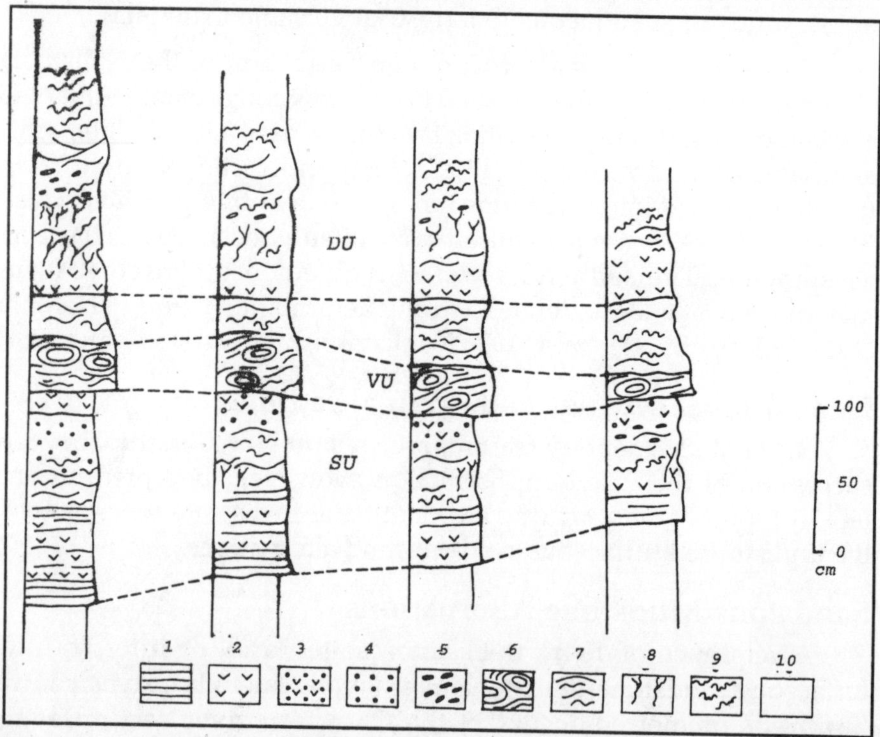

Fig. 1: Detailed columnar sections of the Cheyyeru succession. SU=widely extended lower tuffaceous fine sandstone unit intercalated with mudstone-siltstone; VU=Marker: defined by pyroclast-rich bomb-bearing bed; DU=upper tuffaceous sandstone unit, intercalated and interstratified with diastasis crack-bearing mudstone band; 1=parallel lamination; 2=tuff bed; 3=grey tuffaceous sandstone; 4=pale grey sandstone (slightly tuffaceous); 5=mud clast; 6=bombs with pyroclasts; 7=wavy and contorted lamination; 8=thixotropic deformation and diastasis cracks; 9=contorted lamination; 10=silt and/or clay (slightly tuffaceous).

mud layers are intercalated and/or interstratified in DU unit, a thickening upward sequence. Alternate thin beds of sandstones and mudstones, lenticular beds of ripple-laminated sandstones, graded rhythmites and erosively-based thick channel-fill sandstones with shale clasts, underlain by bluish-grey mudstones are prominent in the DU facies association (Figs. 2, 3 & 4). The sandstones and pyroclast-rich horizons comprise terrigenous and volcaniclastic sands and tuff lenses. Sands are mixed up in different proportions with felsic- to silicic rock fragments, feldspars (few of zoned varieties), volcanic glass shards, pelitic and siliceous rock fragments, biotite and hornblende.

At the lower level of the Cheyyeru succession the sandstones and intercalated mudstones show wavy or contorted laminations, parallel laminations and development of hummocks and swales, while in the upper part wavy bedding, hummocky cross-stratifications, swaley cross-stratifications, low-angle cross-laminations of antidune type, scour surfaces with large slumps in graded sandstones are discernible. Bouma-like divisions are also found immediately above the antiduned sandstones (Figs. 5 & 6). These sedimentary structures are significant as they indicate primary structures of the original sediments which were not disturbed by seismic activity.

Fig. 2: A part of a sandstone-mudstone sequence showing amalgamation. Profusely chertified white tuffaceous layers, erosional top surfaces of mudstone layers underlying amalgamated sandstone bodies with prominent scoured base discernible, Cheyyeru succession.

Fig. 3: Bottom left corner of the Fig.10a blown up: Tsunamogenic sediments as channel-fills at the top right corner of the photograph.

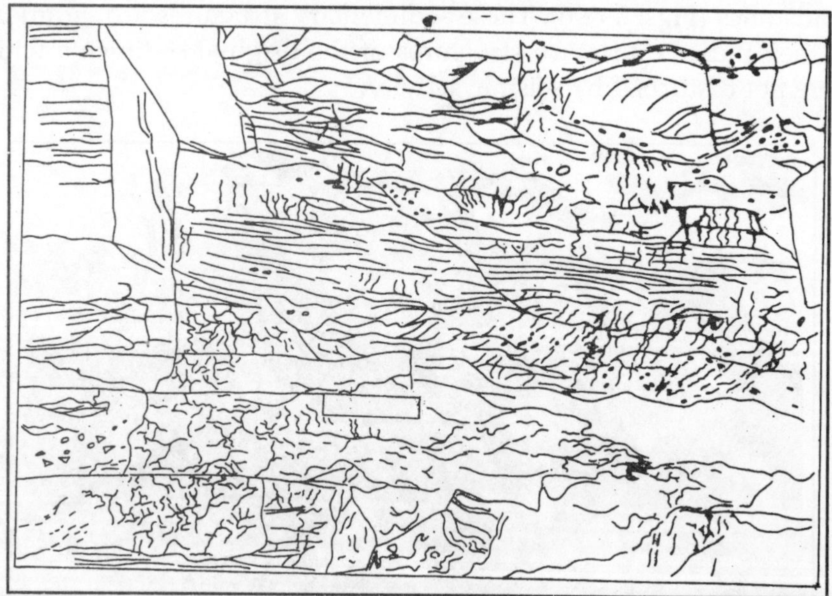

Fig. 4: Partially amalgamated sand rich sandstone-mudstone alternation shows prominent scoured base over erosional top surface of mudstone beds. The mudstone beds show thixotropic deformations restricted within themselves and prominent diastasis cracks both in transverse- and cross-sectional views. Upper right corner shows similar deformations and cracks in mudstone layers which underlie channel-fill sandstone and laterally grade into mud clasts rich pockets. Graben-like downsagging is discernible in amalgamated unit; fine-grained mudstone in the lower part of the amalgamated body is profusely cracked. Bar scale is 10cm.

SCS, HCS and associated structures

The tuffaceous sandstone-mudstone alternations exhibit (Figs. 5 & 6) prominent development of swaley cross-stratification (SCS) which are restricted to relatively thick sand bodies (50cm to 1m) of the Cheyyeru succession and comprise shallow scours ('swales': 45cm-80cm wide, c. 6cm to 20cm deep). Tuffaceous bomb-bearing layers are underlain and overlain by swales. Very few of the swaley layers are convex-upward.

Lenticular bedding and wavy bedding coupled with internal wave ripple laminations are found in sand-mud alternations which range in thickness from a millimetre to a few centimetres (Figs. 7 & 8). The convex-up laminae in the coarse silt and fine sand aggrade with spacing varying from 1 to 4m. The laminae in the hummocks dip (not exceeding 10°-12°) centrifugally forming sets ranging from 2 to 18cm. Sharp boundaries of low-angle lamina may be erosional or nonerosional. They grow either from undulatory beds or from flat substrate which are comparable to the beds studied by Bourgeois (1980), Brenchley et.al. (1979), Ghosh et.al. (1984). Superposed hummocks are occasionally present and 'aggrade' with gutters at higher levels (Figs. 9, 10, 11, 12 & 13e). Associated thicker sand layers show lateral thinning and swelling often with load structures.

Fig. 5: Bomb-bearing pyroclast-rich bed in association with tuffaceous sandstone-mudstone alternations, Cheyyeru succession. Hammer head in the photograph is 15cm.

Fig. 6: Concentrically layered bombs with long axes parallely aligned to the bedding surface of pyroclastic-rich bed, overlain and underlain by hummocks, swales, wavy bedding and inverse grading, cross-lamination rippled phase resembling Bouma-like division near the hammer head, Cheyyeru succession. Hammer head in the photograph is 15cm.

Presence of intercalated (eroded) mudstone streaks with thicker beds suggests that these beds might have amalgamated. Thin sand beds continue as parallel laminations. Plane-laminated fine sands with intermittent low-angle antidune cross-laminations, (Figs. 10 & 12) and incipient HCS grade vertically upward to plane laminations. Sand-silt layers contain soft sediment deformation structures such as load cast of flute marks on the undersurface, flames, low-angle slides, small horsts and grabens and mud-draped symmetrical wave ripples with bi-directional bundles of cross-laminations (Fig.14). Hummocks are overlain by lenticular wavy bedding with internal convex-upward laminations. The wave ripples and hummocks are draped by normally graded sands (Fig.13 c&e). The HCS are often cut across by synsedimentary normal faults. Thickness variations across the faults is conspicuous. The downthrown layers are thicker near the fault. Hummocks also occur in association with thin intraformational breccias and flat-pebble conglomerates, some of them possibly consisting of rip-up microbial mats and gutter casts.

Fig. 7: Sandstone-mudstone alternations showing swales, wavy bedding, ripple marks, laminites, HCS and also casts of rock fragments. Note mild scoured base in many layers, Cheyyeru river section. The height of the shrub to the left of the swale is 42cm.

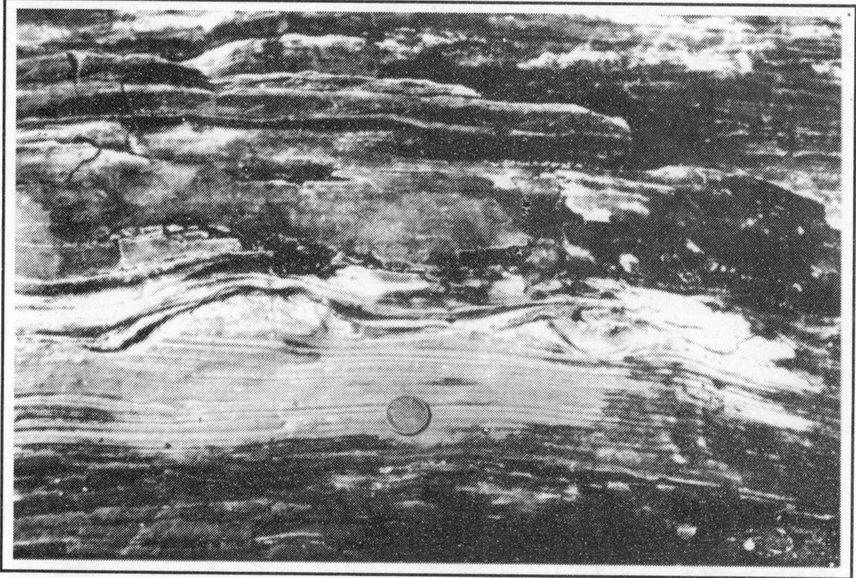

Fig. 8: Hummocks over scoured laminites and ripples in the Cheyyeru succession. Note cross-laminated thin sand layers at the top. Diameter of the coin is 1.7cm.

Fig. 9: Aggraded hummocks showing erosional contact and gutter with sand fills in sand-mud layers, Cheyyeru succession near Hostavaram. Bar scale is 6cm.

Fig. 10: Below the bar scale(6cm), low-angle antidune cross-stratification and above, a part of small hummock (HCS) are visible respectively. Near the top, aggraded hummocks with convolution to the left in sand-mud alternations, Cheyyeru succession, Cheyyeru river section.

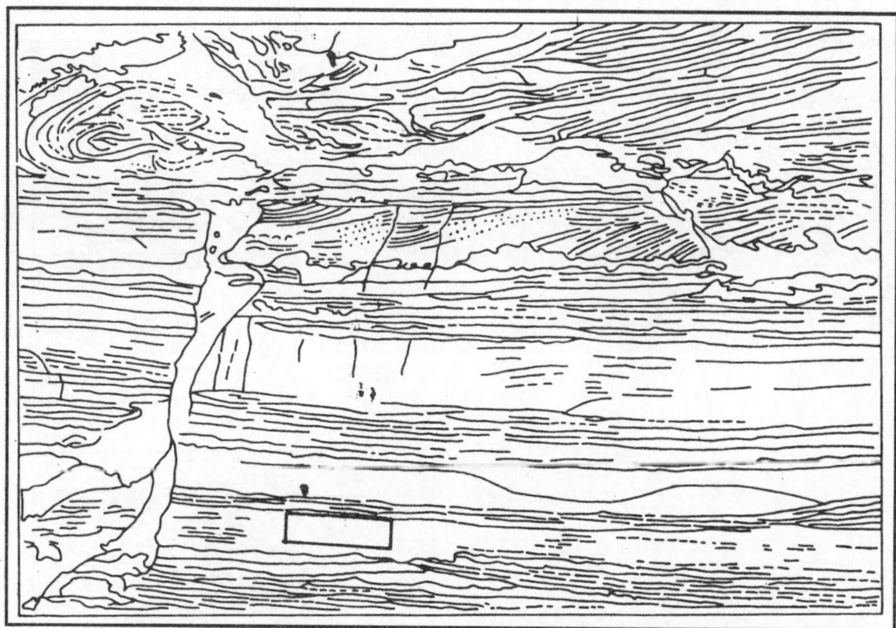

Fig. 11: Below the bar scale(6cm), low-angle antidune cross-stratification and above, a part of small hummock (HCS) are visible respectively. Near the top, aggraded hummocks with convolution to the left in sand-mud alternations, Cheyyeru succession, Cheyyeru river section.

Fig. 12: Disrupted hummocks, erosional contacts and shear plane visible, Cheyyeru succession, Cumbum Formation. Bar scale is 8cm.

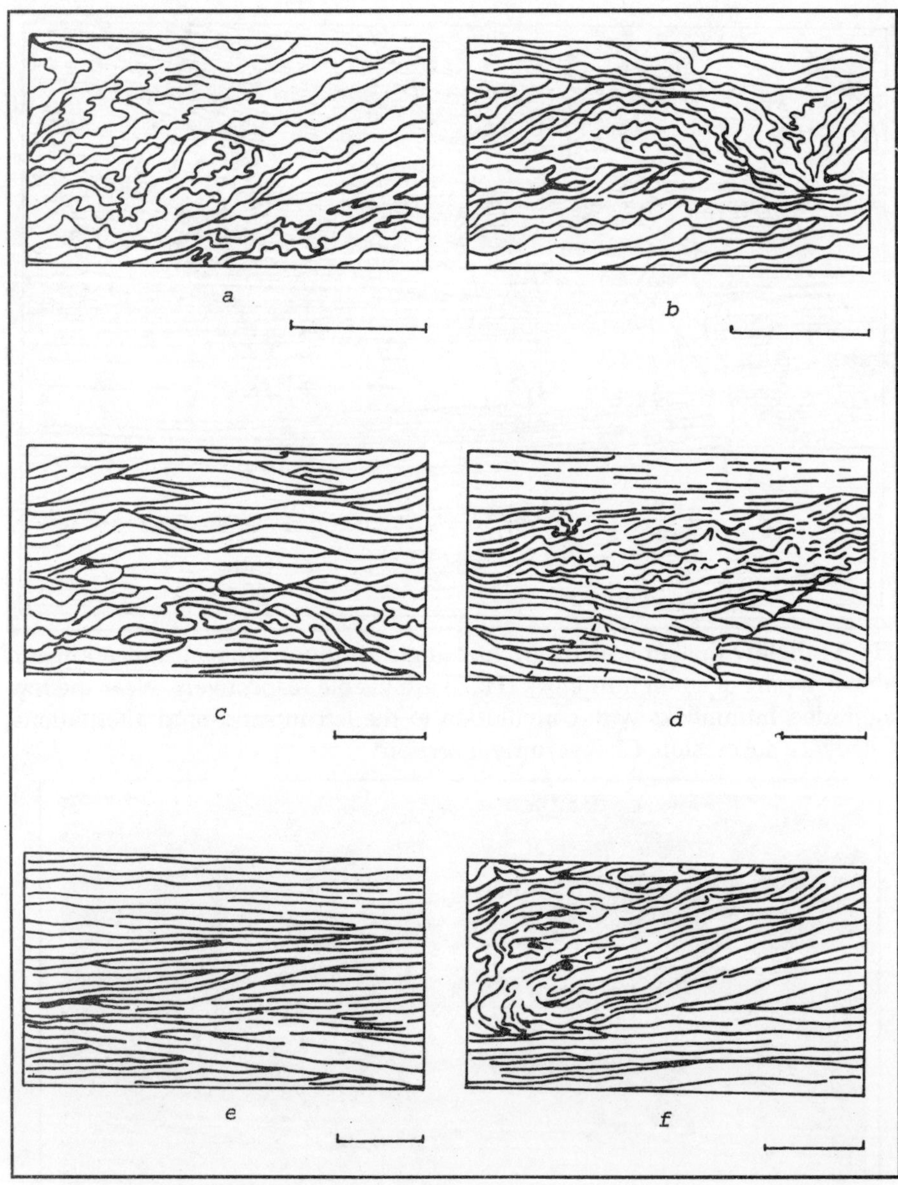

Fig.13a-b: Convolutions in Cheyyeru succession. **c:** Deformed HCS, Cheyyeru succession. **d:** Graben-like downsagging with convoluted laminae on top, Cheyyeru succession. **e:** Eroded, parallel laminae draped by wavy bedding, Cheyyeru succession, Hostavaram region. **f:** Aggraded HCS on flat substrate with large slumps, Cheyyeru succession, Hostavaram region. Bar scale is 25cm.

Cracked mudstone underlying the sandstone layers

Diastasis cracks (cf. Cowan & James,1992) are common in mudstones which are interbedded with normally graded sandstones. These cracks are confined strictly within the erosively-topped mudstone bands of the mudstone-sandstone alternations. Cracked but otherwise undisturbed mudstone layers can usually be traced laterally into rip-up clasts (Figs. 2, 3 & 4). The cracks are filled with sands from overlying sandstones. Thixotropic deformations are also observed in the mudstone layers. Wide shallow channels with sand fillings are preserved in the sandstones.

Fig. 14: Bi-directional bundles of cross-lamination in wave ripple overlying micro horsts and grabens. Low-angle slide and shale clasts visible. The wave ripple is overlain by HCS, capped by normally graded suspension fall out, Cheyyeru succession, Cheyyeru river section. Bar scale is 6cm.

Pyroclast-rich layers and associated sediments

Several pyroclast-rich beds with volcanic bombs serve as markers within the Cheyyeru succession (Figs. 5 & 6). Facies assemblage of such key beds are well-defined (Figs. 5 & 6). Thinning-upward sequences SU with relatively low sand/mud ratio are dominant below and thickening-upward sequences DU with relatively high sand/mud ratio are common above a pyroclast-rich bed VU (Fig.1). Volcanic clasts within the size range of small pebbles to coarse sand show graded disposition. Inversely-graded, non-graded and normally-graded units are also identifiable. Graded-nongraded thin strata over the pyroclast-rich layers are overlain by laterally continuous sandstone-mudstone beds with intercalated tuffaceous material. The mudstone layers show diastasis cracks and give indications of thixotropic deformations (Figs. 2, 3 & 4). Patchy tuffaceous, well-sorted, discontinuous medium- to fine- grained sandstone occasionally overlie the bluish mudstone bands along an erosive contact.

Genetic mechanism of the original sediments

Group A: The alternating sand- and mud- layers in Cheyyeru succession point to processes of deposition with varying intensity. The presence of thickening-upward sequences and thinning-upward beds with HCS and SCS are suggestive of storm wave action (Duke,1985; Dott & Bourgeois,1982; Leckie & Walker,1982; Reading,1996).

Group A1: Stromatolitic carbonates and cross-stratified sandstones represent shore face assemblages as substantiated by presence of gutters, current ripples, troughs and tabular cross-stratifications (Figs. 9, 15 & 16) (Reading,1996; Pettijohn,1975; Walker,1984). Plane-laminated sands, symmetrical wave ripples, low-angle antidune cross-stratified sands (Figs. 10 & 11) with erosional tops, incipient hummocky cross-stratifications, swaley cross-stratifications and graded rhythmites mark further shore-face set-up (cf. Reading,1996; Pettijohn,1975; Walker, 1984).

Group A2: The primary structures of normally graded sand-mud alternations such as hummocks draped with graded laminae clearly suggest deposition of sediments at a depth above the storm wave base by low-density sediment gravity flows which in turn might have been generated by storm action (Walker,1984; Leckie & Walker,1982; Reading,1996). Development of wave ripples through reworking of normally graded sand layers may be related to the activity of high-energy waves.

Group A3: Graded sand/silt-mud alternations with intermittent storm wave-generated structures possibly represent distal transitional offshore setting (Shiki, personal communication,1999).

Group A4: Discrete medium to fine sand/silt alternations include Bouma-like divisions (graded basal unit, low-angle laminites and ripple-like divisions). Variants in the basal division of Bouma sequence include 'a_1': inversely graded, 'a_2': nongraded and 'a_3': normally graded. The 'a_3' division is equivalent to division 'a' for low concentration (density) turbidity currents (cf. Cas & Wright,1987; Allen,1970; Komar,1970; Cas,1979; Walker,1984; Lowe,1982). Such lithounits are intercalated often with some layers showing HCS and SCS and lie above the upper surfaces of the pyroclast-rich layers.

Frequent occurrence of tuffaceous sandstone beds with tempestetic structures like hummocks, swales and wavy bedding along with Bouma-like divisions in Cheyyeru scenario possibly stand for tempestetic turbidites or storm sheets. Case histories in support of this conclusion are ample in literature (Roobol et.al.,1983; Cas et.al.,1990; Reading,1996; Cas & Wright,1987; Fisher & Schmincke,1984, Fenton & Wilson, 1985).

Group B: The Cheyyeru succession includes volumetrically significant volcanic elements. Concentrically layered bombs with their long axes parallely aligned to the bedding (Figs. 5 & 6) indicate that ballistically ejected bombs with spiralling motion adhered layers of ash during their flight in air and subsequently settled in a subaqueous condition (Cas & Wright, 1987; Pettijohn,1975). The thin pink rind (haematitic coating) discernible on the outer surfaces of large blocks and bombs may be ascribed to oxidation in the immediate contact of the atmosphere and water during settling. The bombs whose outer chilled, and wrinkled surfaces cracked owing to expansion of the interior of the clasts during cooling (Hahn et.al.,1979) also corroborate such conclusion.

Seismic sediments

Since the introduction of the notion of 'seismite' by Seilacher,(1969) the definition has been enlarged considerably to include many types of deformation and resedimentation phenomena. The term now covers all sedimentary deposits exhibiting deformation structures or remobilised fabric which can be attributed to earthquake shocks, epicentre of which more or less are distant from the site in question (Cita & Ricci-Lucchi,1984; Plaziat & Purser,1998). However, systematic procedures for reading sedimentary records of palaeoearthquakes and related tsunamis have not been developed sufficiently todate. Discrimination between seismic turbidites and tempest-induced turbidites is not clear in most of the turbidite studies (Shiki & Yamazaki,1996; Cita & Ricci-Lucchi,1984). Constraints faced so far for identifying seismic events in the stratigraphic record is due to the fact that one or more different processes such as gravity pulling, loading and cryoturbation, sometimes acting in combination can produce attributes similar to those of seismic sediments (Ricci-Lucchi,1995).

Despite such overlapping characteristics earthquake-induced structures in lake deposits, young sediments, tsunamogenic landslide and debris flow, liquefaction structures in Holocene sediments of active seismic areas, many sedimentary deformations within syn- and post-rift deposits can be cited with certain degree of certainty as good case histories for seismites (Scott & Price,1988; Von Huene et.al.,1989; Seilacher,1969; Davenport & Ringrose,1987; Kastens & Cita,1981; Karlin & Abella,1992; Plaziat et.al.,1990; Yamazaki et.al.,1989; Shiki & Yamazaki,1996). In recent years good examples of seismic signatures from the Proterozoic rift basins are also reported (Pratt,1992,1993,1994; Bose et.al.,1997; Bhattacharya & Bandyopadhyay, 1998; Dasgupta,1998).

Seismically triggered volcaniclastic turbidites

Pyroclast-rich layers at several horizons in the Cheyyeru succession demonstrate clearly the development of lower cross-bedded division as well as an extra division of horizontal lamination of fine gravels and sands in association with nongraded and normally graded gravels within thin-bedded tuffaceous sandstone-siltstone alternations. Other facies attributes include rhythmically alternating normally graded thin beds of sandstones and mudstones (< few cm thick), chaotic breccia and a few slumped beds. In addition, sand layers, specially those thicker than few millimetres, are normally graded and are gradational to bluish grey mudstones. Similar facies association has been reported in earthquake-induced thin-bedded turbidites from Lake Biwa, Japan (Inouchi et.al.,1996). Volumetrically important fragments of pyroclasts (including unabraded cuspate glass shards and zoned feldspars), terrigenous detritus within medium to fine sand range comprising quartz, feldspar and altered volcaniclastics (probably reworked epiclastics) form thin-bedded layers above the bomb-bearing unit (Figs. 5 & 6). On the other hand, thin-bedded sequence with facies attributes such as eroded and deformed hummocks (Figs. 11, 12 & 13c) and swales with Bouma-like divisions point to the existence of storm-generated volcaniclastic turbidites (cf. Kneller & McCaffrey,2003).

Despite debate in literature (Middleton & Southard,1978; Allen,1982; Lowe,1982), it should be pointed out that slurry flows are clearly transitional between viscous subaqueous debris flows and low viscosity turbidity currents (Hiscott & Middleton,1979; Carter,1975). Subaqueous granular mass-flow processes are justified by virtue of the significance of such deposits in both ancient and modern sedimentary records (Sigurdsson et.al.,1980; Klein & Lee,1984; Wright & Mutti,1981; Cas et.al.,1981; Rickets et.al.,1982). In all of these ancient and modern successions, volumetrically important normal turbidite deposits with volcanic detritus have been identified. In many, the detritus is very juvenile in character (cf. Cas,1979; Wright & Mutti,1981). Presence of unabraded cuspate glass shards and zoned feldspars have been reported from the Cheyyeru succession. This is important from two points of view. Firstly pyroclastically fragmented detritus is being redeposited by epiclastic mass-flow processes and secondly subaqueous mass-flow processes are capable of transporting fragile, pyroclastically fragmented detritus, tens to hundreds of kilometres without significant abrasion. So the modes of fragmentation and final deposition may be quite independent and have to be evaluated independently. It is reinforced that subaqueous mass-flow processes can be nonabrasive and can transport detritus, long distances without significant textural modifications (Cas & Wright,1987). Pyroclastic air fall detritus, terrigenous clastics and volcaniclastics, therefore, can be transported into water and it is then redeposited subaqueously by epiclastic mass-flow processes (cf. Sigurdsson et.al.,1980; Cas,1979,1983; Cas et.al.,1981).

Coupled with seismicity, steep slope of many island volcanoes are obviously unstable and materials readily slump into the ocean basins, generating high concentration debris flows and low concentration turbidity current deposits (Cas & Wright,1987; Reading,1996; Ayres et al,1991). The thin-bedded volcaniclastic turbidites, identified in the Cheyyeru succession, is ascribed a seismically-triggered shallow marine origin. The Qualibone, Dominica & St. Vincent models (Roobol et.al.1983) as recent analogues document similar situations leading to generation of seismically-induced volcaniclastic turbidites. Volcaniclastic sequence of Devonian Bunga beds, Australia also lend support to such conclusion (Cas et.al.,1990; Reading,1996).

Intraformational olistostromes

Association of olistostromes with seismically-triggered thin-bedded volcaniclastic turbidites from the basal parts of the Cheyyeru succession were traced along its strike length. Turbidite systems bear the imprints of volumetrically rich volcaniclastics, impure carbonate and siliciclastics along with olistostromes which are repeated at several intervals in the succession (Figs. 17, 18 & 19). Sandstone-impure carbonate-volcaniclastic rich facies assemblages are clearly demarcated.

Olistoliths are of variable dimensions such as javeline-like, thin disc-shaped, curly, thick needle-shaped, large angular blocks, and canoe-shaped blocks. Size-range is within 22cm/10cm to 3m/1.2m. The shapes of the olistoliths suggest that at the time of slumping the sediment was poorly consolidated. The roundness of the olistoliths in thin-bedded intraformational olistostromes decreases with their size and it is possible that both rounding and fragmentation are related to slumping or creeping movements in a poorly consolidated material. Intraformational olistostromes have also been reported from the Jurassic of the northern Apennines (Passerini,1964; Boccaletti & Bortolotti,1965; Bortolotti & Passerini,1965).

Diagnostic sedimentologic criteria for seismically-triggered intraformational olistostrome at repeated horizons of Cheyyeru succession are the association of seismically generated volcaniclastic turbidites with olistostromes, 'intraclast parabreccia' and significant volcaniclastic elements. This appears to have been caused by earthquake-generated in-situ (autoclastic) brecciation of the early lithified to partially lithified sediments. The frequent seismic shocks were probably related to highly explosive eruptions of felsic to silicic volcanism and syndepositional faulting (McCalpin,1996; Spalletta and Vai,1984; Reading,1996; Holmes,1966; Pettijohn,1975; Doutsos & Piper, 1990). In-situ (autoclastic) brecciation generating intraformational olistostromes was probably followed within seconds to hours by a turbidity current triggered upslope owing to the same earthquake. Such in-situ brecciation in response to intermittent seismic shocks are supported by recent and ancient analogues (Bonilla & Lienkaemper,1991; Asquith,1985; Reynolds et.al.,1980; Spalletta and Vai,1984).

The convulsive aspects of the Cuddapah rift basin attest to such a situation where extensional faulting and intermittent high volcanic pulses were recorded.

Seismites

Despite overlapping by effects of other processes, liquefaction-induced synsedimentary deformation structures, strongly indicative of periodic seismic shaking, are well discussed in literature (Seed,1979; Castro,1987; Dorby,1989; McCalpin and Nishenko,1996; Youd,1973). Such attributes formed during soft to semilithified state are traceable along considerable strike lengths of the Cheyyeru succession of the Cumbum scenario. Liquefaction-induced deformation structures on the basis of different rheological states have been categorised into two major types: (i) as those developed under undrained conditions and (ii) those developed under drained conditions (Lowe,1975; Guiraud & Plaziat,1993; Plaziat & Purser,1998).

Deformation structures related to undrained conditions

Most seismite deformations are the result of a more or less complete liquefaction of a particular horizon near the sedimentary surface. Despite sedimentary loading and localised lithification, sediments retaining some interstitial water are most favourable for this reorganisation (Plaziat & Ahmed,1993; Plaziat & Ahmamau,1997). The increasing water/sediment ratio leads to different rheological states. It evolves into a liquid state which enables flow of a high density fluid even on a very weak slope. Finally the fluidised state is characterised by fluid turbulence enabling an extreme mobility leading to sand and water injection (Lowe,1975; Guiraud & Plaziat,1993; Plaziat & Purser,1998).

Structures owing to undersliding

Graben-like downsagging structures, load structures, graded microfaults, undersliding lystric faults, fissure and wedge openings are discernible.

Fault-bounded graben-like downsagging structures in the range of centimetre to metre are identified in the sand-mud alteration facies in the Cheyyeru succession (Figs. 13d & 20). Such attributes in response to a recent earthquake is cited by Sims (1973). Movements along faults, developed in pairs, were responsible for the graben-like downsagging. The throw along the faults is maximum near the base and decreases upwards. Near the top, the downsagged sandy units can be traced across the graben. Horizontal undeformed sandy mud layers drape the graben structures. Thick muddy sand, in which the grabens are developed, contains isolated load structures of pure interbedded sand detached from overlying layers and these are suggestive of partial liquefaction during graben formation (Collinson,1994; Bhattacharya and Bandyopadhyay,1998). Symmetrical grabens and asymmetrical inclined grabens are some of the variants frequently met with. Slope controls appear to be responsible for generation of asymmetrical grabens.

Horizontal extension of the floor has been claimed to be responsible for the development of such graben-like structures (Ghosh,1993). In the

Fig. 15: Troughs, ripples, foreset laminae are discernible in sand-mud alternation. Thin sandstone layer with scoured base over rippled siltstone-mudstone layers, Cheyyeru succession, Hostavaram region. Scale is 6cm.

Fig. 16: Intricate patterns of laminasets within ripple phase is discernible, Cheyyeru succession, Hostavaram region. Scale is 6cm.

Fig. 17: Broken olistoliths shown detached from lithified units within a thin-bedded intraformational olistotrome, Cheyyeru succession, Hostavaram region. Bar scale is 10cm.

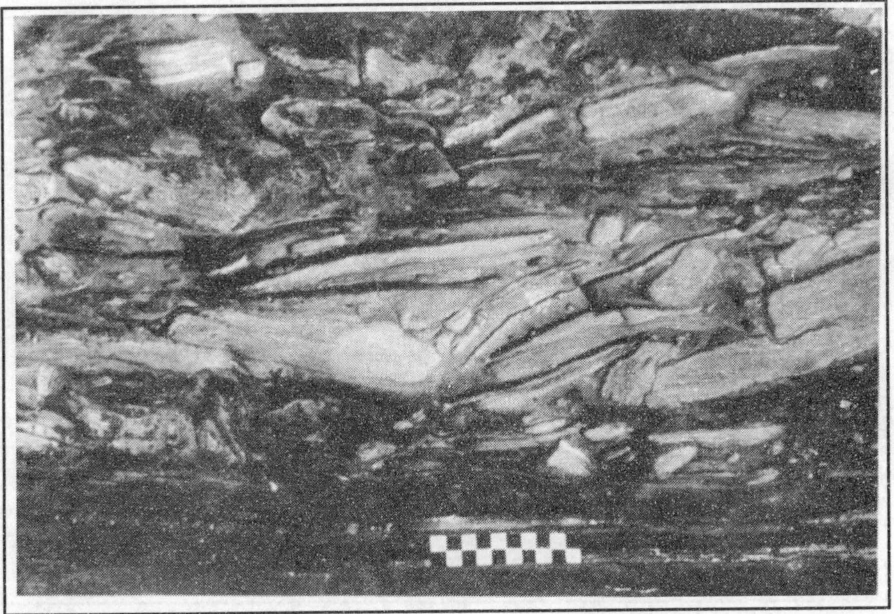

Fig. 18: Javeline-shaped, needle like and spindle-shaped olistoliths, Cheyyeru succession, Hostavaram region. Size of the scale is 10cm.

Fig. 19: Channelised sand bodies with marly matrix show sharply fragmented and few subrounded olistoliths, Cheyyeru succession, Cheyyeru river section. Bar scale is 4cm.

Fig. 20: Fault bounded graben-like downsagging structures in sand-mud alternation facies, Cheyyeru succession, Cheyyeru river section. Scale is 6cm.

present context, lateral flowage of the liquefied substrate under the influence of cyclic loading due to vibratory ground motion might have been responsible for such horizontal extension. Cyclic loading comes about through the oscillatory transgression of seismic waves (Seed & Idriss,1982). Each stress-rise in this repetitive process may well be much less. But if the fluid-pressure response is incompletely dissipated before the next pulse, the incremental accumulation of overpressure can be large and cause liquefaction (Maltman,1994). An upward decrease in the throw along the faults bounding the grabens clearly documents gradual waning of earthquake aftershocks. Similar structures, of probable seismic origin, have been ascribed by Mastalerz & Wojewoda (1993) from alluvial fans developed in an active strike-slip fault setting (Plio-Pleistocene), from southwestern Poland and by Seth et.al. (1990) from the Late Jurassic Katrol Formation of Kutch (India). However, Van Loon & Wiggers (1976) considered storm wave as the trigger mechanism for liquefaction which was responsible for the development of similar structures in the Holocene lagoonal Almere Member (the Netherlands).

Laterally persistent load structures are common in the sand-mud alteration facies of the Cheyyeru succession (Fig.21). In general intensely convoluted or loaded rippled sandy layers overlie such load structure bearing mud layers. The load structures are of variable shape and size (less than a centimeter to 35cm). These are often ropy, conical, cigar-shaped, semicircular to elliptical and kidney-shaped. Parallel orientation of long axes among some markedly elongated load structures are discernible. The internal lamination of the load structures show a form-discordant relationship.

Mudstone beds with load structures continue over tens of meters along their strike lengths. This point to a mechanism that affected a large area. Further, the deep foundering of the load structures well into the mudstone beds, as in this case, is indicative of quasiliquid state generated in the mud layers after successive earthquake shaking (Elliot,1965).

Graded microfaults

Graded microfaults in fine-grained layers are identified (Fig.22). The fault-grading stratigraphy of Seilacher (1969) is discernible. This stratigraphy refers to a four-fold sequence; going from top to bottom shows; (A) completely fluidised sediment, (B) partially coherent sediment layers, (C) faulted layers and (D) undeformed sediment.

Fig. 21: Various types of load casts, dominated by ropy variety on undersurface of tuffaceous sandstone, Cheyyeru succession, Cheyyeru river section. Hammer head in the photograph is 18cm.

Graded microfaults appear to be the prototype of this group (sensu, Seilacher,1969).

The deformation related to graded microfaults was attributable to seismic shocks affecting fine-grained sediments after deposition and compaction. This deformation presents a vertically changing pattern consisting of microfaults (segmented zone) passing upward into a disrupted layer (rubble zone) and finally a homogenous mud (soupy zone). The vertical variation of deformation style should reflect a gradient in the physical state of the sediment (from brittle to plastic and fluid behaviour) reacting to a perturbation coming from below. Both a gravity collapse and a stress applied from above can be excluded by the geometry of this example, which is recalled here not only for historical reasons, but also for its uniqueness: it tells of a case history from which palaeoseismicity can be inferred quite logically and correctly (Ricci-Lucchi,1995).

Structure suggestive of plastic shear

Included in this group are fluidal structures, rotational aspects of internal elements, massive destruction of internal details within the

Fig. 22: Graded microfaults in sand-mud facies indicating liquefaction of sandy mud; Fault bounded graben-like structures en echelon visible in the vicinity of the scale(=10cm). Sandstone dykes (discordant seams of sand) of variable dimensions in sand-mud alternations, Cheyyeru succession, Cheyyeru river section.

bedding and neolayering (Figs. 22, 23 & 24). Plastic shear zones as evidenced from similar attributes are probably of seismic origin. Pore-fluids, initially arrested in multilayered series with impermeable covers building higher fluid pressure owing to repeated seismic shaking, might lead to plastic shearing (Price & Cosgrove,1990; Plaziat & Purser,1998).

Hydroplastic deformation structures

A variety of soft sediment deformation features exist in several widely extended beds disturbed vertically and spatially in the Cheyyeru succession, traceable for kilometres (£ 8km) over the strike length.

Case 1 : A localised fold-fault structure has developed near the junction of a mud-dominated thinly laminated mudstone-sandstone unit A and sand rich thinly-laminated sandstone-mudstone unit B (Fig.23). The bedding laminae within unit A have been doubled up giving rise to a stack of asymmetric synformal folds, the axial plane of which makes a low-angle (8°) with the bedding laminae. The shorter limbs of the asymmetric folds have been truncated against or are nearly asymptotic to the bottom of the unit B, part of which is coincident with a shear plane inclined towards left with angle of inclinaton increasing gradually downwards. The axial plane of the folds is nearly parallel to the shear plane. A triangular space enclosed by the shear plane to the right, a second steep dipping shear plane to the left and bottom of unit B, show structures indicative of fluidisation.

Case 2: An aggregate of diversely oriented, rectangular fragments of thinly-laminated sandstone-mudstone alternations have their longer sides parallel to the bedding laminae (Fig.24). Some of the fragments show portions of highly appressed folds which developed prior to fragmentation. The aggregate defines a band sandwiched between evenly laminated sand and mud layers. Rock layers are frequent; small narrow water-escape tubes, filled up with sands show reddish brown coatings of hydroxydic iron. Smoothness and undisturbed nature of laminae with such structures suggest that sediment was less cohesive during the formation of water-escape pipes. Localised expulsion of pore water from a liquefied layer did take place.

Case 3: Broken-up pieces of laminae with sandy composition have come apart along bedding planes and are of various lengths and thickness (Fig.25). These pieces are mixed up with irregular masses of muddy composition. The broken fragments demonstrate an overall wavy pattern enclosed between the parallel laminae of overlying and underlying rock units (Fig.25).

Case 4: The 5cm thick thinly-laminated sandstone-mudstone layers define a pair of nearly recumbent antiform and synform (Fig.26). Irregular contortion in bedding is recorded in a 2.5cm thick rhythmically alternating layers of sand and mud. Convolute laminations are well developed in the sand-mud alternation facies (Fig.27). Two types of

Fig. 23: Fold fault structure in mud dominated sand-mud alternation, Cheyyeru succession, Cheyyeru river section: bedding laminae doubled up giving rise to stack of asymmetric synformal folds. Triangular area bounded by two shear planes show structures indicative of fluidisation. Water escape pipes seen in cross-section as black spots. Scale is 10cm. Unit A with bar scale shows thinly laminated mudstone-sandstone layers; Unit B, comprising sandstone-mudstone layers above the triangular area in the upper part of the photograph.

Fig. 24: Aggregate of diversely oriented rectangular fragments of thinly laminated sandstone-mudstone alternations. Highly appressed folds in fragments are discernible. Water escape pipes seen in cross-section as black spots, Cheyyeru succession, Cheyyeru river section. Scale is 10cm.

convolutions are recorded: simple and composite (multilobed) ones. Simple convolutions with wide troughs and sharp crests form symmetrical, regular folds. The internal laminations are restricted to certain intervals suggesting that only specific moments presented the required conditions (Figs. 10 & 11, 13a, b, c & f, 27). Crests of folds are pierced locally by fluid-escape channels. Along these channels sand has been carried upward and deposited at the top of the erosional planes which truncate the folds. These simple types are thus metadepositional

Fig. 25: Broken-up pieces of laminae with sandy composition mixed up with irregular masses of muddy composition demonstrates an overall wavy pattern, Cheyyeru succession, Cheyyeru river section.

convolute laminations. The smoothness of the folds, the scarcity of microfaults and the extreme thickness variations shown by individual laminae suggest that the sediment was weak during deformation. Localised expulsion of pore water from a liquefied bed along the crests is regarded as the cause of the formation of convolute laminations (Ten Haaf,1956; Chakraborty,1977; Allen,1977). In some places the geometry of convolution is more complex, so that composite (multi-lobed) convolutions are found. The convolute structures are restricted to certain intervals, suggesting that only specific moments presented the required conditions. Liquefaction and concomitant fluidisation took place during these intervals. The multi-lobed character of the convolutions indicates that the horizon was liquefied several times under the influence of earthquakes.

Fig. 26: Thinly laminated sandstone-mudstone layers define a pair of recumbent antiform and synform, Cheyyeru succession, Hostavaram region. Scale is 5cm.

Fig. 27: Irregular contortions in rhythmically alternating layer of sand and mud, Cheyyeru succession, Hostavaram region. Scale is 5cm.

Extensional structures

These structures are pinch- and- swells, boudinage, dislocation of individual competent beds and brecciation of brittle layers (Figs. 17 & 20).Confining fluid pressure in sediment due to seismicity may exceed that of overburden and cause selective liquefaction. Sedimentary layers with variable rheology under this situation suffer differential stretching and brittle failure with generation of signatures of extensional process (Ramsay & Huber,1983; Price & Cosgrove,1990). Analogues in support of this conclusion are frequent in literature (Price,1973; Middleton,1969; Dobry & Liu,1992; Gerson et.al.,1993).

Deformation structures related to drained conditions

The structures of this category have also been identified. Such attributes on the basis of morphogenetic controls have been grouped separately.

Sandstone dykes

Thick to millimeter-thin discordant seams of sands (sandstone dykes) varying in height from less than one centimeter to one and half metre are developed in sand-mud alternations in Cheyyeru succession (Fig.22). Upward and downward drags are discernible along the contacts of those dykes.

Upward displacement of pore water generates fluidization in vertical paths of least resistance within the overlying sediments. Clastic dykes may develop only when sediments are in a sequence of relatively low permeable and highly cohesive muddy layers, as in the present case. Due to brittle failure of more cohesive sediments, such layers suffer deformation. Most of the fluidised materials follow such fractures for their upward migration (Owen,1995; Plaziat & Purser,1998). For fluid expulsion an external mechanical trigger such as seismic shock is required for liquefaction.

Structure related to disrupted internal fabric

Rafts and broken layers arranged en echelon is suggestive of lateral stretching and limited sliding owing to seismically-induced shocks under an overall control of extensional tectonics (Figs. 17 & 19). Fissures, grabens, back-tilting, step faults are observed on Borah Peak rupture (Crone et.al.,1987). Both the Borah Peak and the 1959 M 7.3 Hebgen Lake, Montana ruptures displayed small thrust faults (dip 25°-30°) on the downthrown block (McCalpin,1996).

It appears that the 1983 Borah Peak ruptures preferentially propagated upward through unconsolidated alluvium along preexisting faults. Such development of faults occurred because of their lower shear stress than the adjacent unfaulted alluvium. Several laboratory sand-box faulting experiments (Horsefield,1977; Lade et.al.,1984) have shown that local dilatancy and strain softening occurred along laboratory-scale normal fault planes and that "such fault planes represent planes of low resistive shear strength on which motion will preferentially occur during subsequent deformation stages" (Vendeville & Cobbold,1988).

This is another example of layer-confined autobrecciation of lithified tempestetic structures. Unlike the olistholiths of intraformational olistostromes, autobrecciated fragments in this context do not show any sign of transport. This is again, a case of brittle failure of cohesive sediments to which seismic origin may be ascribed. Severe pulses of seismic shocks are necessary for such mechanical rupture leading to disruption of lithified layers of tempestetic structures. Presence of earthquake-induced thin-bedded volcaniclastic turbidites in association with storm-sheet sands and disrupted internal fabrics suggest that earthquake shocks were frequent in the syn-rift basinal scenario of Cheyyeru succession. Recent and ancient analogues lend support to the conclusion (Cas et.al.,1990; Reynold et.al.,1990; Roobol et.al.,1983; Abbate et.al.,1970).

Tsunamiites

Faulting at the basin margins may be accompanied with ground tremors owing to explosive volcanic episodes triggering tsunamis which deposit tsunamiites (earthquake-induced wave and/or current related deposits; Shiki & Yamazaki,1996). Grey mudstone bands with thixotropic deformation and diastasis cracks in profusion below the patchy, well-sorted sandstone layers with occasional rip-up clasts bear the signatures of earthquake-generated intermittent tsunami pulses. Since a tsunami, induced by an earthquake has severe pulses, the tsunami-induced ebb currents also act in several pulses. It must also be pointed out that tsunamis, induced very close to the hypocentre of an earthquake, generally have only few pulses (Shiki & Yamazaki,1996; Bourgeois & Reinhart,1993; Plafker & Kachadoorian,1996). The tail of the powerful ebb current of the main tsunami pulse containing a large amount of sorted sand grains came from the shallow-marine set-up and presumably deposited the patchy tuffaceous well-sorted laminated medium- to fine-grained sandstone layers. Analogues in literature lend support to this conclusion (Bourgeois & Reinhart,1993; Plafker & Kachadoorian,1996; Kuenen,1950; Clague et.al.,1994; Shiki & Yamazaki,1996).

The channel-fill tsunamogenic sediments were deposited over a wide area in a laterally correlatable yet discontinuous fashion over the available erosively-topped deformed mudstone wedges and bands. This was possible due to the passage of powerful ebb currents through different channels (Figs. 3 & 4). Tuffaceous sandstones showing similar facies characteristics are found associated with bomb-bearing pyroclast-rich layers at several horizons in the succession. Tsunami-origin is ascribed to such well-sorted tuffaceous sandstones. Giant wave-induced deposition throughout the Cheyyeru succession in a syn-rift basin appears to have close relations with episodic tsunami generation rather than meteorological storms. Proterozoic Chaibasa tsunamiites, India and Miocene tsunami-induced conglomerates from Chita Peninsula, Japan corroborate such inference (Bhattacharya & Bandyopadhyay,1998; Shiki & Yamazaki,1996).

Discussion

It is a well established fact that distinction between shelf and bathyal slope was not clear during most of the geologic history except during post-glacial ages. In this context, the Proterozoic epiric seas were no exception. However, bathymetric relationship among different sequences in the study area, based on structural packages of facies assemblage with the support of modern and ancient analogues, has been attempted. The syn-rift Nallamalai sediments comprise various facies assemblages including shallow-marine facies associations of the Cumbum Formation. The deposits include subaerial and subaqueous basic and felsic to silicic volcanics. The Cheyyeru succession, in general, is a facies mosaic of different shallow-marine milieus. Stromatolitic carbonates and cross-stratified sandstones represent shore face assemblages. The plane-laminated sand, low-angle antiduned cross-stratified sands and hummocky cross-stratified sand-mud alternations mark further distal shore face setting. Graded sand-silt alternations with intermittent storm-waves generated structures possibly represent distal transitional offshore upper bathyal setting.

Emergence of Nallamalai protobasin during a stage of evolution of extensional Cuddapah Basin records abrupt change in tectonic control of sedimentation. The Cumbum turbidite sedimentation, in a basinwide perspective, is an example in that mode which abruptly replaced terrestrial deposition. The stratigraphic sections of Cumbum turbidites in favourable lithofacies, therefore, in all probability, provide the imprints of frequent palaeoseismic events (Dasgupta,1998). Lystric faults, sandstone dykes, graben-like structures, pseudonodules and cyclic olistostromes et ceterra are frequently found together. These facies associations record liquefaction, flowage and fluidization of soft to semilithified and lithified sediments at discrete yet distinct zones which require the impact of sudden repetitive short-lived high energy pulses such as earthquake shocks. It is expected that intermittent highly explosive felsic to silicic eruptions during sedimentation might have produced such seismic shocks. Moreover, possible tsunamiites are developed along with such tempestetic turbidites/storm-sheet-sands in the succession. Association of tsunamogenic sediments with storm-generated resedimented facies in the Cheyyeru scenario is clearly indicative of a major role of earthquakes in depositing and generating deformation features.

However, apart from ground shaking owing to passage of seismic waves, liquefaction is possible near regions of appreciable slopes in confined aquifers with significant artesian pressure and in zones where freshly deposited sands undergo primary compaction through dewatering (Thorson et.al.,1986; Mohindra & Thakur,1998). Other processes, acting individually or in combination, can also produce similar features. In the Cheyyeru succession hydroplastic and other deformation features are associated with shallow-marine sediments deposited on slopes <1°. This eliminates the influence of slope failure and artesian pressure in liquefaction. The possibility of deformation features being generated by glacial or associated processes like ice loading and movement is ruled out as there is no such evidence in the geologic record. The widespread distribution of deformation features further eliminates the possibility of landslide or dropstone events. In addition, fine-grained terrigenous and volcaniclastic facies in the syn-rift Cumbum scenario bear the imprints of extensional tectonism and highly explosive volcanism. Therefore, none of the alternative causes of liquefaction other than shaking by earthquakes is confirmed on the basis of available data.

In conclusion, earthquakes were common in the rift basins which evolved under conditions of crustal stretching. Faulting at the basin margins might have been accompanied by ground tremors, tsunamis that induced the development of seismites, seismoturbidites and tsunamiites. Occurrence of seismites is erratic and scattered throughout the Cheyyeru succession near Hostavaram region. Haphazardly arranged seismic signatures throughout the succession has relevance to cyclic development of facies which is very common during Proterozoic sedimentation.

REFERENCES

K. Aki and W.H.K. Lee, Determination of the three-dimensional velocity anomalies under a seismic array using first P arrival times from local earthquakes: Part 1. A homogeneous initial model, J. Geophys. Res. 81 (1976), pp. 4381–4399.

J.A. Dunn, J.B. Auden, A.M.N. Ghosh, S.C. Roy and D.N. Wadia, The Bihar-Nepal earthquake of 1934, Mem. Geol. Surv. India 73 (1939), pp. 183–268 (391 pp., reprinted 1981, Calcutta).

E. Abbate, V. Bortolotti, and P. Passerini, Olistotromes and olistoliths. Sediment.Geol., 4 (1970), pp.521-557.

J.R.L. Allen, The sequence of sedimentary structures in turbidites, with special reference to dunes. Scot.Jour.Geol., 6 (1970), pp.146-161.

J.R.L. Allen, The possible mechanics of convolute laminations in graded beds. Jour.Geol.Soc.London. 134, (1977), pp.19-31.

J.R.L. Allen, Sedimentary structures: Their character and physical basis. Developments in sedimentology, Elsevier (Amsterdam). 30A & B (1982)

D.O Asquith, Characteristics and alternative mechanisms of late Quaternary faulting, proposed LNG terminal site, Point conception, California. Bull.Assoc.Eng.Geol., 22 (1985) pp.171-192.

L.D. Ayres, N.A. Van Wagoner and W.S. Ferreira, Voluminous shallow-water to emergent phreatomagmatic basaltic volcaniclastic rock, Proterozoic (~1886Ma) Amisk Lake composite volcano, Flin Flon Greenstone Belt, Canada. In: R. V. Fisher and G.A. Smith, Editors, Sedimentation in volcanic settings, SEPM, Tulsa, USA (1991), pp. 175-187

H.N. Bhattacharya and S. Bandyopadhyay, Seismites in a Proterozoic tidal succession, Singhbhum, Bihar, India. Sedimentary Geology. 119 (1998), pp.239-252.

S. Bhattacharji and R.N. Singh, Thermo-mechanical structure of the southern part of the Indian shield and its relevance to Precambrian Basin Evolution, Tectonophysics, 105(1984), pp.103-120.

B.C. Kneller and W.D. McCaffrey, The interpretation of vertical sequences in turbidite beds: the influence of longitudinal flow structure, Jour. Sed. Res. 73 (2003), pp. 706-713.

T. Doutsos and D.J.W. Piper, Listric faulting, sedimentation and morphological evolution of the Quaternary eastern Corinth Rift, Greece: first stages of continental rifting, Geol. Soc. Am. Bull. 102 (1990), pp. 812-829.

M. Boccaletti and V. Bortolotti, Lacune della serie toscana, 1.Serie stratigrafiche giurassico-eoceniche nelle zone di Roggio e Trassilico sul versante orientale delle Alpi Apuane. Boll.Soc.Geol.Ital., 84 (1965), pp. 271-356.

M.G. Bonilla, & J.J. Lienkaemper, Factors affecting the recognition of faults exposed in exploratory trenches. U.S.Geol.Surv.Bull. 54, (1991), pp. 19-47

Bortolotti,V. & Passerini,P.,1965: Segnalazione della presenza delle formazioni dal Calcare Selcifero ai Calcari Variegati ad Aptici alla base della Serie Toscana a Cintoia (Chianti settentrionale), Boll.Soc.Geol.Ital., 84(6), 37-40.

Bose,P.K.; Banerjee,S. & Sarkar,S.,1997: Slope-controlled seismic deformation and tectonic framework of deposition: Koldaha Shale, India. Tectonophysics, v.269; pp.151-169.

Bourgeois,J.,1980: A transgressive shelf sequence exhibiting hummocky cross stratification: The Cape Sebastian Sandstone (Upper Cretaceous), southwestern Oregon. Jour.Sedim.Petrol., v.50, pp.681-702, 6.7.7, Fig.6.97.

Bourgeois,J. & Reinhart,M.A.,1993: Tsunami deposits from 1992 event; Implications for interpretation for palaeotsunami deposits, Cascadia subduction zone. EOS Trans. Am. Geophys. Union 74, 350.

Brenchley,P.J.,Newall,G. & Stanistreet,J.G.,1979: A storm surge origin for sandstone beds in an epicontinental platform sequence, Ordovician, Norway. Sediment.Geol., v.22, pp.185-217, 7.7.2, 7.8.1.

Carter,R.M.,1975:A discussion and classification of subaqueous mass transport with particular application to grain-flow, slurry-flow and fluxoturbidites. Earth Sci.Rev., v.11, pp.145-177.

Cas,R.A.F.,1979: Mass-flow arenites from a Palaeozoic interarc basin, New South Wales, Australia: mode and environment of emplacement. Jour.Sed.Petrol., v.49, pp.29-44.

Cas,R.A.F.,1983: Submarine 'crystal-tuffs', their origin using a Lower Devonian example from southeastern Australia. Geol.Mag., v.120, pp.471-486.

Cas,R.A.F. & Wright,J.V.,1987: Volcanic successions, Modern and Ancient: A geological approach to processes, products and successions. Allen & Unwin (London), 528p.

Cas,R.A.F.; Allen,R.L., Bull,S.W., Clifford,B.A. & Wright,J.V.,1990: Subaqueous, rhyolitic dome-top tuff cones: a model based on the Devonian Bunga Beds, southeastern Australia and a modern analogue. Bull.Volcanol., v.52, pp.159-174. 12.10, Fig.12.54.

Cas,R.A.F.; Powell,C.M.; Fergusson,C.L., Jones,J.G., Roots,W.D. & Fergusson,J.,1981: The Kowmung volcaniclastics: a deep-water sequence of mass-flow origin. Jour.Geol.Soc.Aust., v.28, pp.271-288.

Castro,G.,1987: On the behaviour of soils during earthquakes. In: Developments in Geotechnical Engineering (A. S. Cakmak, ed.). 42, pp.169-204. Elsevier, Amsterdam.

Chakraborty,A.,1977: Upward flow and convolute lamination. Senckenbergiana Marit., v.9, pp.285-305.

Cita,M.B. & Ricci-Lucchi,F., (eds.),1984: Seismicity and Sedimentation. Spec. Issue. Mar. Geol. 55 (1/2).

Clague,J.J., Bobrowsky,P.T. and Hamilton,T.S.,1994: A sandsheet deposited by the 1964 Alaska tsunami at Port Alberni, British Columbia. Estuarine, Coastal Shelf Sci. 38, 413-421.

Collinson,J.,1994: Sedimentary deformation structures. In: Maltman,A. (Ed.), The Geological Deformation of Sediments. Chapman & Hall, London, pp.95-125.

Coulson,A.L.,1934: Asbestos in ceded districts of Madras Presidency with notes on its occurrence in other parts of India. Mem.Geo.Surv.Ind., v.64(2)

Cowan,C.A. & James,N.P.,1992: Diastasis cracks: mechanically generated synaeresis-like cracks in Upper Cambrian shalllow water oolite and ribbon carbonates. Sedimentology, v.39, pp.1101-1118.

Crawford,A.R. & Compston,W.,1973: The age of the Cuddapah and Kurnool Systems, South India. Jour.Geol.Soc.Aust., v.19, pp.453-464.

Crone,A.J.,1987: Introduction. In: Directions in paleoseismology (A.J.Crone and E.M.Omdahl, eds.). U.S.Geol.Surv.Open File Rept., 87-673, 1-6.

Dasgupta,P.K.,1996: Cyclicity in the Proterozoic Cuddapah Supergroup and its bearing on basin evolution, India. Extended abstract, Abstract vol. 2 of 3, p.198, 30th. International Geological Congress, Beijing session.

Dasgupta,P.K,1998: Seismic signatures from thin bedded olistostromes and associated turbidites : Cumbum Shale, Cuddapah Basin, India. Extended abstract, abstract volume, 15th International Sedimentological Congress, Alicante, Spain. pp.210-212.

Davenport,C.A. & Ringrose,P.S.,1987: Deformation of Scottish Quaternary sediment sequences by strong earthquake motions. In: Deformation of sediments and sedimentary rocks (M.E.Jones and R.M.F.Preston, eds.). Geol.Soc.Special Publication. No.29, pp.299-314.

Dobry,R.,1989: Some basic aspects of soil liquefaction during earthquakes. Ann. N.Y. Acad. Sci. 558, 31-38.

Dobry,R. & Liu,1992: Centrifuge modeling of soil liquefaction. Proc.World Conf. Earthquake Eng., 10th, Madrid, Spain, 11, 6801-6809.

Dott,R.H. & Bourgeois,J.,1982: Hummocky stratification: Significance of its variable bedding sequences. Geol.Soc.Amer.Bull., v.93, pp.663-680.

Duke,W.L.,1985: Hummocky cross-stratification, tropical hurricanes, and intense winter storms. Sedimentology, Vol. 32, No. 2. p.167-193.

Elliot, R.E., 1965. A classification of subaqueous sedimentary structures based on rheological and kinematical parameters. Sedimentology, v.5, pp.193-209.

Fenton, M.W. & Wilson, C.J.L., 1985 : Shallow-water turbidites: An example from the Mallacoota Beds, Australia, Sedimentary Geology, v.45, pp.231-260.

Fisher,R.V. & Schmincke,H.U.,1984: Pyroclastic rocks. Springer-Verlag (Berlin).

Gerson,R.; Grossman,S., Amit,R. & Grenbaum,N.,1993: Indicators of faulting events and periods of quiescence in desert alluvial fans. Earth Surf.Processes Landforms, v.18, pp.181-202.

Ghosh,G.; Bardhan,S.; Shome,S. & Bose,P.K.,1984: Storm-affected shoal-lagoon complex, Ghuneri Member (western Kutch) – A stratigraphic model (Abstr), XI Indian Colloquium on Micropaleontology and Stratigraphy. Geol.Min.Met Soc.India Bull., v.51, pp.68-69.

Ghosh, S.K.,1993: Structural Geology: Fundamentals and Modern Developments. Pergamon, Oxford, 598pp.

Guiraud, M. & Plaziat, J.C.,1993: Seismites in the fluviatile Bima Sandstones: identification of paleoseisms and discussion of their magnitude in a Cretaceous Synsedimentary strike-slip basin (Upper Benue, Nigeria). Tectonophysics, 225, 493-522.

Hahn,G.A.; Rose.W.I.Jr. & Meyers,T.,1979: Geochemical correlation of genetically related ryolitic ash-flow and air-fall ashes, central and western Guatemala and equatorial Pacific. In: Ash-flow tuffs (Chaplin,C.E. and Elston,W.E., eds.), Geol.Soc.Amer. Special Paper 180.

Hiscott,R.N. & Middleton,G.V.,1979: Depositional mechanics of thick-bedded sandstones at the base of a submarine slope, Tourelle Formation (Lower Ordovician), Quebec, Canada. In: Doyle & Pilkey (1979), pp.307-326.

Holmes,A.,1966: Principles of physical geology. ELBS edition 1966, 1288p.

Horsefield,W.T.,1977: An experimental approach to basement-controlled faulting. Geol.Mijnbouw, v.56, pp.363-370.

Inouchi,Y; Kinugasa,Y.; Kumon,F.; Nakano,S.; Yasumatsu,S. & Shiki,T.,1996: Turbidites as records of intense palaeoearthquakes in Lake Biwa, Japan. Sed.Geol., v.104, pp.117-125.

Kaila,K.L. & Bhatia,S.C.,1981: Gravity study along Kavali-Udipi deep seismic sounding profile in the Indian Peninsular Shield : Some inferences about origin of anorthosites and Eastern Ghat orogeny. Tectonophysics,79; pp.129-143.

Kaila, K.L. & Tewari,H.C.,1985: Structural trends in the Cuddapah Basin from deep seismic soundings (DSS) and their tectonic implications. Tectonophysics, v.115, pp.69-86.

Kaila,K.L., Tewari,H.C., Roy Chowdhury,K., Rao,V.K., Sridhar,A.R. & Mall,D.M.,1987: Crustal structure of the Proterozoic Cuddapah Basin of India from deep seismi soundings and gravity data. In : S. Asano and W.D.Mooney (editors), Seismic studies of the continental lithosphere. Tectonophysics, v.140, pp.1-12.

Kaila,K.L.; Roy Chowdhury,K.; Reddy,P.R.; Krishna,V.G.; Narain,H.; Subbotin,S.I.; Sallogub,B.B.; Chekunov,A.V.; Kharetchko,G.E.; Lazarenko,M.A. & Ilchenko,T.V.,1979: Crustal structure along the Kavali-Udipi Profile in the Indian Peninsular Shield from Deep Seismic Sounding : Jour.Geol.Soc.India, v.20, pp.307-333.

Karlin,R.E. & Abella,S.E.B.,1992: Palaeoearthquakes in the Puget Sound region recorded in sediments from Lake Washington, USA. Science, v.258, pp.1617-1620.

Kastens,K.A. & Cita,M.B.,1981: Tsunami-induced sediment transport in the abyssal Mediterranean Sea. Geol.Soc.Amer.Bull., v.92; pp.845-857.

King,W.,1872: The Kadapah and Karnul Formations in the Madras Presidency. Mem.Geol.Serv.India, v.8, Pt.1, 346p.

Klein,G. de V. & Lee,Y.,1984: A preliminary assessment of geodynamic controls on depositional systems and sandstone diagenesis in back-arc basins, western Pacific Ocean. Teconophysics, v.102, pp.119-152.

Komar,P.D.,1970: The competence of turbidity currents. Geol.Soc.Am.Bull., v.81, pp.1555-1562.

Krishnan,M.S & Venkataram,M.S.,1942: Report on asbestos and baryte in Pulivendla Cuddapah district, Madras. G.O. No.3408.

Kuenen,P.H.,1950: Marine Geology. John Wiley & Sons. New York, 568p.

Lade,P.V., Cole,D.A.Jr. & Cummings,D.,1984: Multiple failure surfaces over dip-slip faults. Jour.Geotech.Eng., 100(GTS), pp.616-636.

Lake,P.,1890: Basic eruptive rocks of the Cudapah area. Rec.Geol.Surv.Ind., v.23(4), pp.259-261.

Leckie,D.A. & Walker,R.G.,1982: Storm-and Tide- dominated shorelines in Cretaceous Moosebar-Lower Gates Interval — Outcrops Equivalents of Deep Basin Gas Trap in Western Canada. Amer.Assoc.Petroleum Geologists Bull., v.66(2), pp.138-157.

Lowe,D.R.,1975: Water escape structure in coarse-grained sediments. Sedimentology. v.22; pp.157-204.

Lowe,D.R.,1982: Sediment gravity flows: II, depositional models with special reference to the deposits of high density turbidity currents. Jour.Sed.Petrol., v.52, pp.279-297.

Maltman,A.,1994: Introduction and overview. In: A.Maltman (Editor). The geological deformation of sediments. Chapman & Hall, London. pp.1-35.

Mastalerz,K. & Wojewoda,J.,1993: Alluvial-fan sedimentation along an active strike-slip fault : Plio-Pleistocene Pre-Kaczawa fan, SW Poland. Int. Assoc. Sedimentol. Spec. Publ. v.17, pp.293-304.

Matin,A. & Guha,J.,1996: Structural geometry of the rocks of the southern part of the Nallamalai fold belt, Cuddapah Basin, Andhra Pradesh. J. Geol. Soc. India, v.47,pp.535-545.

McCalpin,J.P.,1996: Tectonic geomorphology and Holocene paleoseismicity of the Molesworth section of the Awatere Fault, South Islande, New Zealand. N.Z. J. Geol. Geophys., v. 39, pp.33-50.

McCalpin,J.P. & Nishenko,S.P.,1996: Holocene paleoseismicity, temporal clustering and probabilities of future large (M>7) earthquakes on the Wasatch fault zone, Utah. J. Geophys. Res. 101 (B3), 6233-6253.

Mehdi,S.H.; Shukla,R S. & Babu,B.R.,1978: Detailed geological and geochemical studies of Nallamalai sediments along Mailour-Badval reverse, southern part of Zangamrajupalle Varikunta Mineralised belt, Cuddapah district, Andhra Pradesh. Progress Report for the F.S. 1976-78 — unpublished.

Middleton,G.V.,1969: Turbidity currents and grain flows and other mass movements down slopes, In: The New Concepts of Continental Margin Sedimentation (D.J. Stanley, ed.). American Geological Institute Short Course Lecture Notes. pp.GM-A-1 to GM-B-14.

Middleton,G.V. & Southard,J.B.,1978: Mechanics of sediment movement. SEPM (Eastern Section)short course, no.3.

Mohindra,R. & Thakur,V.C.,1998: Historic large earthquake-induced soft sediment deformation features in the Sub-Himalayan Doon valley. Geol.Mag., v.135(2), pp.269-281.

Murthy,Y.G.K.,1981: The Cuddapah Basin — A review of basin development and basement framework relations: 4th workshop on status problem in the Cuddapah Basin. Institute of Indian Peninsular Geology, Hyderabad, pp.51-72.

Nagaraja Rao,B.K., Rajurkar,S.T., Ramalingaswamy,G. & Ravindra Babu,B.,1987: Stratigraphy, structure and evolution of Cuddapah Basin. In: Purana Basins of Peninsular India (ed. by B.P. Radhakrishna) Bull.Geol.Soc.Ind., v.6, pp.33-86.

Neelakantam,S. & Roy,S.S.,1979: Baryte deposits of Cuddapah Basin. Rec.Geol.Surv.Ind. 112(5), pp.51-64.

Obermeier,S.F.,1996: Using liquefaction-induced features for paleoseismic analysiss. In: Paleoseismology (J. McCalpin, ed.). Academic Press. 588p.

Owen,G.,1995: Soft sediments deformation in Upper Proterozoic Torridonian Sandstones (Applecross Formation) at Torridon, Northwest Scotland. J.Sediment.Res. A65, 495-504.

Pettijohn,F.J.,1975: Sedimentary Rocks. 3rd ed. Harper & Row. 628p.

Passerini,P.,1964: Il Monte Cetona (provincia di Siena), Boll.Soc.Geol.Ital., 83(4), 219-338.

Plafker,G. & Kachadoorian,R.,1996: Geologic effects of the March 1964 earthquake and associated seismic sea waves on Kodiak and nearby islands, Alaska. U. S. Geol. Serv. Prof. Pap. 543-D, D1-D46.

Plaziat,J.C. & Ahmed,E.A.,1993: Diversity of sedimentary expressions of major earthquakes: an example from the 'Pliocene' sandy seismites of the Egyptian Red Sea Coast. Geological Society of Egypt, Special Publication 1, pp.277-294.

Plaziat,J.C. & Ahmamou,M.,1997: Les mecanismes a l'origine des seismites et leur signification tectonique dans le Pliocene du Saiss de Fes et de Meknes (Maroc). Geodynamica Acta, in press.

Plaziat,J.C. & Purser,B.H.,1998: The tectonic significance of seismic sedimentary deformations within the syn- and post-rift deposits of the north-western (Egyptian) Red Sea coast and Gulf of Suez. In : Sedimentation and Tectonics of Rift Basins : Red Sea Gulf of Aden. (B.H.Purser and D.W.Bosence, eds.), Chapman & Hall, London. pp.347-652.

Plaziat,J.C.; Purser,B.H. & Philobbos,E.,1990: Seismic deformation structures (seismites) in the syn-rift sediments of the NW Red Sea (Egypt). Bulletin de la Societe Geologique de France 8(6), pp.419-434.

Pratt,B.R.,1992: Shrinkage features ("molartooth" structure) in Proterozoic limestones – New model for their origin through synsedimentary earthquake-induced dewatering: Geol.Soc.Amer.Abstracts with Programmes, v.24(7), 53p.

Pratt,B.R.,1993: Proterozoic synaeresis cracks and molar-tooth structure: Formation by synsedimentary seismicity: Geol.Assoc. Canada. Annual Meeting Programme with abstracts. 85p.

Pratt,B.R.,1994: Seismites in the Mesoproterozoic Altyn Formation (Belt Supergroup), Montana: A test for tectonic control of peritidal carbonate cyclicity. Geology, v.22, pp.1091-1094.

Price,R.A.,1973: Large-scale gravitational flow of supra-crustal rocks, Southern Canadian Rockies. In : Gravity and Tectonics (ed. by K.de Jong & R.Scholten), Wiley, New York. pp. 491-502.

Price,N.J. & Cosgrove,J.W.,1990: Analysis of geological structures. Cambridge University Press. pp.502.

Ramsay,J.G. & Huber,M.I.,1983: The Techniques of Modern Structural Geology, vol. 1; Strain Analysis. Academic Press, London, p.307.

Rao,N.V.C.; Miller,J.A.; Pyle,D.M. & Madhavan,V.,1996: New Proterozoic K-Ar ages for some kimberlites and lamproides from the Cuddapah Basin and Dharwar Craton, South India : evidence for non-contemporaneous emplacement. Precambrian Research. v.79; pp.363-369.

Reading,H.G.,1996: Sedimentary Environments : Processes, Facies and Stratigraphy. 3rd edition. Blackwell Science Ltd., 688p.

Reynolds,M.A.; Best,J.G. & Johnson,R.W.,1980: 1953-1957 eruption of Tuluman Volcano: rhyolitic volcanic activity in the northern Bismarck Sea. Mem.Geol.Surv.Papua New Guinea, 7, 44, 12.10.

Ricci-Lucchi,F.,1995: Sedimentological indicators of paleoseismicity. In: Perspectives in Paleoseismology (L.Serva and D.B.Slemmons, eds.). Assoc.Eng.Geol.Spec.Publ., v.6, pp.7-18.

Rickets,B.D., Ware,M.J. & Donaldson,A.,1982: Volcaniclastic rocks and volcaniclastic facies in the Middle Precambrian (Alphebian) Belcher Group, Northwest Territories. Can.Jour.Earth Sci., v.19, pp.1275-94.

Roobol,M.J.; Wright,J.V. & Smith,A.L.,1983: Gravity slides or calderas in the Lesser Antilles island arc ? Jour.Volcanol.Geotherm.Res., v.14, pp.121-134.

Saha,D.,1994: Fold-fault structures of the Nallamalai Range, Diguvametta-Nandikanama Pass, Prakashan District, A.P., South India. Ind.J.Geol., v.66(3), pp.203-213.

Scott,B., & Price,S.,1988: Earthquake-induced structures in young sediments. Tectonophysics, v.147, pp.165-170.

Seed,H.B.,1979: Soil liquefaction and cyclic mobility for level ground during earthquakes. J. Geotech. Eng., Am. Soc. Civ. Eng. 105, 201-255.

Seed,H.B. & Idriss,I.M.,1982: Ground motions and soil liquefaction during earthquakes Earthquake Engineering Research Institute, Berkely, CA, 134p.

Seilacher,A.,1969: Fault graded beds interpreted as seismites. Sedimentology, v.13, pp.155-159.

Seilacher,A.,1984: Sedimentary structures tentatively attributed to seismic events. Marine Geology, v.55, pp.1-12.

Sen,S.N. & Narasima Rao,CH.,1967: Igneous activity in Cuddapah Basin and adjacent areas and suggestion on the palaeogeography of the basin. In: Proc.Symp.UMP GRB and NGRI.Publ. Hyderabad, A.P., India, 8-261-285.

Seth,A.; Sarkar,S. & Bose,P.K.,1990: Synsedimentary seismic activity in an immature passive margin basin (lower member of the Katrol Formation, Upper Jurassic, Kutch, India). Sedimentary Geology, v.68; pp.279-291.

Shiki,T. & Yamazaki,T.,1996: Tsunami-induced conglomerates in Miocene upper bathyal deposits, Chita Peninsula, Central Japan. Sedimentary Geology, v.104, pp.175-188.

Sigurdsson,H.; Sparks,R.S.G.; Carey,S. & Huang,T.C.,1980: Volcanogenic sedimentation in the Lesser Antilles arc. Jour.Geol., v.88, pp.523-540.

Sims,J.D.,1973: Earthquake-induced structures in sediments of Van Norman Lake, San Fernando, California. Sci. v.182, pp.161-163.

Sims,J.D.,1975: Determining earthquake recurrence intervals from deformational sructures in young lacustrine sediments. Tectonophysics, v.29; pp.141-152.

Spalletta,C. & Vai,G.B.,1984: Upper Devonian intraclast parabreccias interpreted as seismites. Marine Geolology, v.55, pp.133-144.

Ten Haaf,E.,1956: The significance of convolute lamination. Geol.Mijnbouw, v.18, pp.188-194.

Thorson,R.M., Clayton,W.S. & Seeber,L.,1986: Geological evidence for a large prehistoric earthquake in Curten, Connecticut. Geology, v.14, pp.463-471.

Tutle,M.P. & Seeber,L.,1991: Historic and prehistoric earthquake-induced liquefaction in Newbury, Massachusetts. Geology, v.19, pp.594-597.

Van Loon,A.J., Wiggers, A.J.,1976: Metasedimentary 'graben' and associated structures in the lagoonal Almere Member (Groningen Formation, the Netherlands). Sediment. Geol., v.16, pp.237-254.

Vendeville,B. & Cobbold,P.R.,1988: How normal faulting and sedimentation interact to produce listric fault profiles and stratigraphic wedges. Jour.Struct.Geol., v.10, pp.649-659.

Venkatakrishnan,R. & Dotiwalla,1987: The Çuddapah Salient: a tectonic model for the Cuddapah Basin, India, basd on Landsat image interpretation. Tectonophysics, v.136, pp.237-253.

Vemban,N.A.,1946: A chemical and petrological study of dyke rocks in the precambrian (Cuddapah traps). Proc.Ind.Acad.Sci., v.23(6) Sec A, pp.347-3781.

Vijayam,B.E.,1969: Plutonic rocks in the Cuddapah Basin. Jour.Ind.Geol.Sci.Ass., v.11, pp.75-78.

Von Huene,R.; Bourgois,J.; Miler,J. & Pautot,G.,1989: A large tsunamogenic landslide and debris along the Peru Trench: Jour.Geophysical Research, v.94(B2), pp.1703-1714.

Walker,G.P.L.,1984: Downsag calderas, ring faults, caldera sizes and incremental growth. Jour.Geophys.Res., v.89, pp.8407-8416.

Wright,J.V. & Mutti,E.,1981: The Dali Ash Island of Rhodes, Greece: a problem in interpreting submarine volcanogenic sediments. Bull.Volcanol., v.44, pp.153-167.

Yamazaki,T.; Yamaoka,M. & Shiki,T.,1989: Miocene offshore tractive current-worked conglomerates—Tsubutegaura, Chita Peninsula, Central Japan. Sedimentary Facies in the Active Plate Margin, ed. A. Taira and F. Masuda; pp.483-494. Terrapub, Tokyo.

Youd,T.L.,1973: Liquefaction, flow, and associated ground failure. U.S.Geol.Surv.Circ. 688, 1-12.

7

Cyclic Sedimentation in Palaeo-To Neoproterozoic Cuddapah Supergroup and its Significance in Basinal Evolution

Introduction

Relationships between sedimentary tectonics and their sedimentary responses, long recognized in general terms by stratigraphers, are more explicitly focused by Sloss, Krumbein and Dapples, (1948). The response patterns are now so widely accepted that tectonic response models and cyclicity in sedimentation are highlighted in sedimentology and stratigraphy.

Sedimentary basins are formed owing to subsidence of upper surface of the earth's crust, as a consequence of number of processes like crustal thinning by extensional stretching, mantle to lithospheric thickening due to cooling of lithosphere, sedimentary-, volcanic-, tectonic- and subcrustal loading, asthenospheric flow and crustal densification (Costa et.,al, 1992; Ingersoll and Busby, 1995). All settings of basins formation involve a complex combination of processes, which may suffer modifying influences by additional effects. Cyclicity owing to tectonic events and their consequent sedimentary basin formation clearly points to every pair of tectono-sedimentary submergent and oscillatory emergent episodes leading to generation of a number of basins. The subsidence of the area involved in submergent episodes switches on the formation of a depression, which becomes a sedimentary basin when enough difference in relief is attended. When the tectonic activity during submergence is quite strong, rift formation prevails. To the contrary, during less intense tectonic activity, the basins are generally centroclines or syneclises, superposed often upon an earlier rifted basement. Rifting and accompanying basin formation are more at the end of the submergent episodes and continues during the beginning of the oscillatory emergent episodes that follows which are of course in a small scale as intracontinental small sized rift basins. Character of the lithic infills of these sedimentary basins reflects their supposed origin.

It is known that certain sediment types are preferably deposited during definite periods of the earth's history such as accumulation of BIF, manganese-bearing sediments and extensive deposits of stromatolitic dolostones during the Precambrian. Certain geological periods are therefore characterized by a dominant deposition of sediment types, in which, more intense formation of this types is periodically repeated during the course of earth's history. Periodicity in the formation of sedimentary rock types is determined by the repetition of geological phenomenon of various types during the span of geological time. In this sense, cyclicity of these phenomena is the obvious cause of the periodicity of a particular type of sediment accumulation.

Attention has recently been focussed in Indian subcontinent on development of cyclicities in Proterozoic basins. Various aspects of cyclic sedimentation such as their tectonic controls, role of source area and different types of autocyclic and allocyclic processes through interpretation of facies analysis have put forward enormous work load that is to be addressed (Dasgupta & Sengupta, 1997; Chakraborty & Bhattacharyya, 1996). Moreover, dating the different scales of cyclicities through geochronological approach is a feedback which has just started (Crawford & Compston, 1973; Bhattacharji & Singh, 1984; Jachariah et al, 1999).

Sedimentary cycles and cyclic sequences in basins are composed of atleast three different beds and interbeds which form a repetitive succession. The term 'cyclothem' is widely used in North America in a purely descriptive way specially for coal bearing cyclic sequence (Heckel, 1986; Klein, 1989; Riegel, 1991). In seismic stratigraphy, the term 'cyclothem' is replaced by depositional sequence which is defined by its lower and upper boundaries as well as by low stand, transgressive, high stand deposits and parasequences (Van Wagoner et al, 1987; Vail, 1987; Vail et al 1991; Coe, 2003). In the present chapter primarily the terms 'depositional sequence' or 'depositional cycle' and for a succession of cycles, 'cyclic sequences' have been used.

Evolutionary Trends of the Proterozoic Basins

In India, Proterozoic sedimentary sequences are available in the mobile belts of Singhbhum, Rajasthan and Eastern Ghats as well as in epicontinental/rift basins in the northern fringe of Peninsular India bordering the Indo-Gangetic alluvial trough (Vindhyan Supergroup, Bijawar Group and Chattisgarh basins), along the Pranhita-Godavari valley near the east coast, and further south in the Bhima, Kaladgi and Cuddapah basins.

While addressing cyclicities in Proterozoic basins in India, the evolution of such basins deserve mention. The geometry of the Proterozoic basins in India should have been essentially shaped either by their deposition around the Archaean accretions or nucleii, and the related crustal upwarps and downwarps, or, curving of the 'highs' and 'lows', or, depressions along lineament, or, combination of the two processes. Distribution of such major Proterozoic basins is documented. Ideas are of course are developing but specific models are far from being universally acceptable (Dhoundial, 1987; Narain, 1987; Bhattacharji & Singh, 1984; Radhakrishnan, 1987).

To what extent the present geometry of the exposed Proterozoic basins is a reflection of original depositional geometry is by itself a subject of enquiry. The role of boundary faults and the attendant thickening of sedimentary prisms and structural complications along such selected zones have received considerable attention in deciphering the geometry of the basins (Narain, 1987; Bhattacharji & Singh, 1984). Foreland basins, continental or epicontinental basin-like, graben or half-graben like characteristics have been attributed to some of the earlier stages of the large complex Proterozoic basins. These, underwent strong phases of deformational history before the simpler platform cover deposits were laid down in wider or narrower basins (Dhoundial, 1987; Kaila et al, 1979,). Out of the major thirty two Proterozoic basins, the 44,400 sq.km. Palaeo- to Neoproterozoic Cuddapah basin in Peninsular India housing sediments shed over approximately 1 b.y. bears unique situation of autocyclic as well as allocyclic sequences in its stratal assemblages. Owing to such uniqueness in cyclicities, studied so far, the stratal assemblages of the Cuddapah Supergroup provide case history for genetic interpretation for the development of rhythmic and cyclic sequences.

Geologic Setting of the Cuddapah Basin

The Cuddapah basin is bordered on the west by the Peninsular Gneiss showing remnants of greenstone belts. The Nellore Schist Belt and the Eastern Ghats Granulite terrain appear along the eastern margin. The basal formations of the Cuddapah Supergroup lie unconformably on the Peninsular Gneiss along the western margin. The flat lying strata of the basal units have low easterly dips. In contrast, the sedimentary sequence in the eastern part of the basin are steeply dipping and isoclinally folded. The eastern margin of the Cuddapah basin is an easterly dipping thrust along which the sedimentary sequence is tectonically overlain by the Nellore Schist Belt (3.0Ga to 2.6Ga). Deep seismic soundings indicate that the thrust and the related imbrications are steeply inclined and penetrate the base of the crust (Kaila and Tewary 1985, Kaila et al. 1987, Drury and Holt 1983).

Active rifting, in which impingement of a thermal plume on the base of the lithosphere causes convective thining (dilation), domal uplift, crustal extension, sagging and extensive episodes of igneous activities, have been considered to be responsible for origin of protobasinal phases leading to evolution of Cuddapah basin through time (Sen and Rao, 1967; Bhattacharji and Singh, 1984; Bhattacharji, 1986). But passive rifting in which tensional stresses in the continental lithosphere cause thinning and passive upwelling of hot asthenosphere appears to be widely acceptable model for the origin of sedimentary basins. Crustal doming and volcanic activity are only secondary processes (Turcotte and Oxburgh, 1973; McKenzie, 1978). Cyclic sedimentation in intracratonic basins with its tectonic controls is of relevance to test the role of passive rifting in such basin modeling. Cyclicity in the Cuddapah Supergroup in this context is an important case in hand.

Stratigraphy

Four major subdivisions of the Cuddapah sequence, namely the Papaghni, Cheyair, Nallamalai and Krishna, arranged in ascending order, were recognised by King (1872). Each subdivision was taken to be laterally continuous throughout the basin. Nearly a century later, an attempt was made to develop a scheme of chronostratigraphic classification for the Cuddapah succession (Sen and Rao 1967). This was immediately replaced by a lithostratigraphic classification which appears to be an improvement on the scheme given by King (Nagaraja Rao and Ramalingaswamy 1976, Rajurkar and Ramalingaswamy 1975,1978). Extensive mapping coupled with lateral tracing of the lithostratigraphic units for more than two decades resulted in a lithostratigraphic classification which accounts for most of the observed stratigraphic relations (Murthy 1979).

Lithology

The Cuddapah Supergroup is dominantly arenaceous and argillaceous. Calcareous sediments constitute a small proportion of the total thickness. Volcaniclastics and volcanics make up a considerable proportion of the Cuddapah sequence at different stratigraphic levels, the thickness of the individual units ranging upto a maximum of 3.5km. The Cuddapah sediments mostly belong to fluvial, aeolian and lagoonal facies, though intercalations of marine sediments are rather common (King 1872, Banerjee 1974, Singh 1980, Rajurkar 1979, Dasgupta 1996).

Depositional Sub-basins, facies models and cyclicity from rock-record sequences

Earlier workers recorded progressive shifting of sub-basins with their sub-parallel axes trending NE-SW (Sen and Rao 1967, Murthy 1981). These subbasins (from oldest to youngest) are Papaghni, Chitravati Nallamalai and Srisailam.

Papaghni Sub-basin

Sediments in Papaghni subbasin consist of a clastic-carbonate couple represented by basal formation of Gulcheru and carbonate dominated Vempallae Formation. The couple is separated by a 72 m thick Transition Facies (TF). The impress of cyclicities in its best developed stratigraphic sections near Chitravati Gorge is complex. The stratal assemblages of the section have been subjected to facies analysis and interpretation in order to decipher genetic controls of cycle hierarchy (cf. Einsele, 1992; Goodwin and Anderson, 1985; Schwarzacher, and Fischer, 1982).

Cyclicity in Gulcheru Formation

The Gulcheru Formation comprising mainly coarse clastics develops on broadly undulatory erosion surfaces on Precambrian gneissic basement. The sequence is characterised by dominance of boulder-conglomerate beds around Chitravati Gorge and extends along strike length over a distance of 48 km (Fig.1). Large incised channels with widths of 0.4-0.85 km having channel walls as high as 22 m consist of channel fills (Fig.2). Channel facies mainly comprises thick beds of boulder deposits with or without interbeds. The Lower Gulcheru coarse clastics interfinger with and onlaps over the incised channel facies and shows development of cyclicity in both large and small scales. Cycles

(6-30m) consist of stratified to well stratified discontinuous matrix-poor upward coarsening boulder/conglomerate facies (Fig.3). In addition, their vertical stacking patterns indicate an overall coarsening and thickening upward trend. Intermittent minor cycles which are 2-12 cm thick also develop upward coarsening trends both in grain size and thickness. Stacks of strata within the minor cycles, however, are punctuated by fining upward and thinning layers.

Interfingering of fine-grained deposits and coarse facies dictate formation of occasional lensoidal bed geometry with slight imbrication in detached outcrops. The facies assemblage is haphazardly arranged and grades laterally and vertically into beds of pebbly-, coarse to granule-rich sandstones and granulestones.

Facies analysis of Gulcheru clastics from the best developed stratigraphic sections (Fig.4) near Chitravati Gorge (ca. 520m thick) reveals spatio-vertical development of nine facies which have been classified following Miall (1978), Blair and Mcpherson (1994).

Interplay of factors and cyclicities in the Gulcheru

Sedimentary cycles including coarsening and fining upward sequences as well as complete and incomplete cycles are discernible from the development of different Gulcheru facies. Alternations between beds and interbeds of cyclic sequences were characterized by quasicyclic and dicyclic aspects. The thickness of corresponding bed types did change owing to variations in sedimentation rates in their cyclic disposition.

Sheet-flood couplet facies (Fig.5) in association with channel-fill deposits (Fig.6) characterize the Gulcheru sequence with distinct facies attributes of alluvial fans.

The suite of facies common in both lower and upper sequences indicates deposition by channelised and nonchannelised alluvial systems.

Overall upward-coarsening and thickening cycles within fan bodies are punctuated by thinning and fining upward cycles. The most likely explanations for small scale cyclicity appears to be owing to fan retreat followed by flood basin aggradation and then by gradual fan progradation (cf. Reading, 1996; Selley, 1978; Blair, 1999). In other words, autocyclic episodes are either represented by responses to periods of base level lowering against the marginal fault or they are partly due to

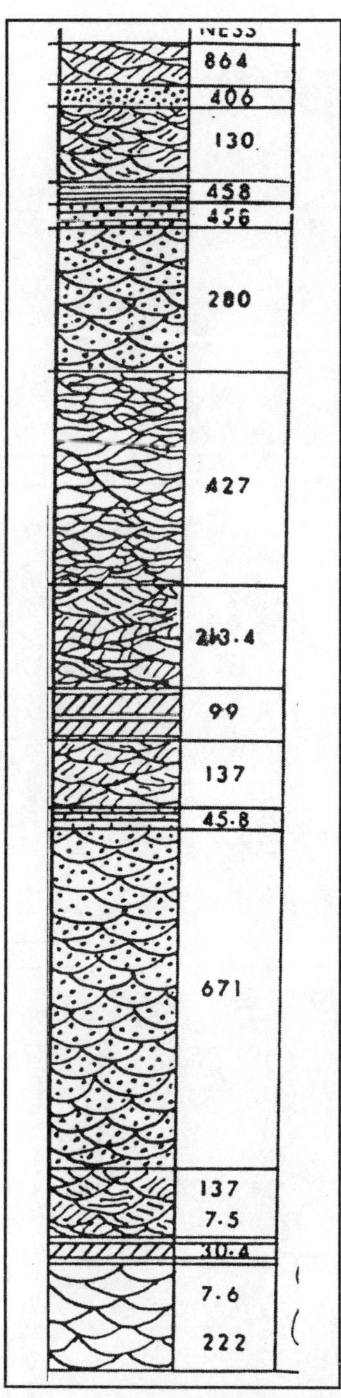

Fig.1 : Representative facies log of Gulcheru Formation near Chitravati Gorge, Cuddapah basin.

Fig.2 : Regional exposure of the Gulcheru Formation near Chitravati Gorge, Cuddapah basin, India.

Fig.3 : Stacked channel facies of Gulcheru Formation, Cuddapah basin, India; Scale : height of the shrub in the right hand corner = 1.2m.

Fig.4 : Stacked cyclic channels near basal Gulcheru Formation over the granitic basement. Couplet facies with cyclicity is discernible; Scale: length of hammer = 40cm.

Fig.5 : Cyclicity within couplet beds, Gulcheru Formation, Cuddapah basin, India; Scale: length of hammer = 40cm.

Fig.6 : Torrential cross-beds with epsilon cross stratification at the top, Gulcheru fluvial clastics, Cuddapah basin, India; length of the bar scale = 10cm.

lateral fan lobe shifting. The latter was important as it clearly happened within a context of basin floor subsidence easily triggered by the tectonic movements. Further supports for direct or indirect tectonic controls on cyclicity in the Gulcheru comes from lateral persistence of boundaries of the fan facies and axially transported alluvium comprising braided and nonchannelised fluvial systems. Similar case histories have also been recorded from rock records (Steel and Gloppen, 1980).

Vertically stacked cyclic sedimentation of alluvial fan and fluvial or aeolian deposits occur in the tectonically most active side of half-graben extensional basin as a result of periodically changing rates of basin subsidence, which is overprinted by allocyclic influence (cf. Denny, 1965; Heward, 1978; Blair, 1987).

In addition, the Gulcheru alluvial fans and braided fluvial systems record sequences of fluvial sediments alternating with deposits of sheet floods and debris flows.

Alternating phases of transportation and deposition of coarser gravel and fine gravel ± sand give rise to facies VIII. This is deposited by distinctive hydraulic conditions (high Froud number, high flow attenuation rate and high deposition rate) related to flow expansion and decreasing slope as well as intrinsic variations in depth and velocity of supercritical flow (Allen, 1983; Blair, 1985a,b; Blair and Mcpherson, 1994). Such couplet facies in association with other three facies variants owing to deposition by fluid gravity flows have been described as characteristic of sandy alluvial sheet-flood deposits and have been reported from modern fans in California (Van de Kamp, 1973; Blair and Mcpherson, 1994), Spain (Harvey, 1984) and England (Wells and Harvey, 1987). Thick (100+m) sequences of sheet flood stratification due to sediment gravity processes generated by the failure of bed rock cliffs have been documented from stratigraphic records. Analogues of Triassic Mount Toby conglomerate of Massachusetts (Hand et al., 1969), Jurrassic Todo Santos Formation of southeastern Mexico (Blair, 1985a; 1987) and the Oligocene-Miocene Pantano Formation of southern Arizona (Balcer, 1984) can be cited.

More specifically sheet flood deposition is a product of migration and washout of submerged antidune bedforms present on the fan surface below the wave base which shoot downslope and then dissipate by mingling with rest of the sheet flood (McGee, 1897; Zielinski, 1982, Blair, 1985a, b). Spatiovertically developed such major sheet flood deposits during geologically significant span of time forms a kind of episodic succession with irregularly spaced thin- and thick event beds. On the other hand, moderately thick fluvial deposits in sequences display some sort of cyclicities that is discernible in areas of substantial subsidence which are continuously filled. Deep-seated fault-dominated terrains under an overall control of extensional regime of Papaghni basinal set-up frequently induced reactivation along fault surfaces producing piedmonts which were subsequently modified through weathering and erosion that supplied huge amount of detritus shed in deposition of terrestrial fan complexes and braided fluvial deposits. Deposition of major floods with periodicities of ten to hundreds of years are not transmitted in the rock record (Dott, 1988; Clifton, 1988). The recurrent interval of large Gulcheru flood episodes preserved in thick flood deposits including those of flash floods, however, are recorded in Chitravati stratigraphic section. This, when compared with similar analogues appear to be in ranges of hundreds to thousands of years. As these cycles of flood deposits are not strictly periodic, they have been designated in Gulcheru scenario as dicyclic in their development.

The upper Gulcheru in continuity with the lower sequence comprises cycles of tabular and overlapping channel facies of sandstones. Trough cross beds in profusion with infills of ferricrete in few cases, are pervasive in the channelised deposits. Red coloured lensoidal sandstone facies, discernible in lower Gulcheru stratal assemblages becomes repetitive in the upper Gulcheru sections. Spatio-vertical facies variations frequently reveal cyclic arrangement of red coloured tabular sand bodies alternating with trough-dominated channel facies. The red coloured lensoidal sand bodies document high angle (20-32) tabular- and wedge-planar cross-strata with super critical and subcritical climb, alternation of grain fall and grain flow layers.

Spatio-vertical assemblage of sedimentary structures suggests that braided fluvial deposition was quite common during the Gulcheru with vertical decrease of flash flood signatures. Cyclically developed red coloured cross-stratified lensoidal bodies with other characteristic facies attributes can be ascribed to aeolian dune-interdune deposition (cf. Hunter, 1977a, b; Rubin and Hunter, 1983; Kocurek and Nielsen, 1986; Kocurek, 1988). Unique repetitive aeolian and fluvial deposition in Chitravati section is a case in hand which substantiates records of allocyclic changes.

The terrestrial fan complex of the lower Gulcheru sequence appears to have been generated in response to successive marginal faulting and subsequently got modified owing to interference by haphazardly arranged facies of different fluvial processes which were controlled by variations in subsidence rates through time. The upper Gulcheru is however dominated by cyclic channelised fluvial facies often interfered by non-channelised coarse fluvial clastic deposits.

Repetitive interference by high energy aeolian environment in the fluvial depositional systems of the Gulcheru stratal assemblages appears to be the result of periodic changes in deep-seated fault-controlled rejuvenated topography and palaeoclimatic variations that affected precipitation, temperature and wind velocities promoting generation of cyclicity between aeolian and fluvial facies.

Switching between arid to humid climatic influence is allocyclic and orbitally forced with resultant cyclic arrangement of aeolian and braided fluvial Gulcheru sediments (cf. Ulicny,1999; Frederiksen et al,1998; Einsele,1992). In addition, such frequencies of climatic variations depicting quasiperiodic signals were recorded in cyclic changes from relatively coarse-grained channel-fill dominated Gulcheru braided systems to transition in the finer-grained flood plain dominated sequences as documented in the Chitravati Gorge section.

The Transition Facies (TF)

Under the influence of oscillatory emergence of the Papaghni subbasin, a 72 m thick Transition facies was deposited over the Gulcheru sediments. Episodic reactivation of faults, repetitive redistribution of 'highs' and 'lows' in physiography and voluminous supply of terrigenous detritus were no longer there. Source area with subdued topography approaching peneplanation and intermittent felsic volcanism fed the basin with feeble supply of fine to very fine sands and mud often intermixed and/or interlayered carbonates and volcaniclastics. Passage from extrabasinal Gulcheru to intrabasinal Vempallae sedimentation is represented by lithosomes of mixed lithology of the TF with some evidences of cyclicity.

Interrelations of limestone and dolomite vary from standpoint of bedforms, their spatio-vertical relationships, continuity of bedding surfaces and internal structural details. Limestones are mainly erosively based, lensoidal and flat-bottomed layers. The lenses are internally cross-bedded. Foresets are asymmetric at their bases. The lensoidal limestones show slight overriding and relatively abrupt decrease in wavelength and amplitude. Profuse ripple-drift laminations, small ripples, interference ripples, tabular cross beds are salient facies attributes. Micro-ripple-drifts in graded laminites with admixture of very fine clastics, carbonates and volcaniclastics have been identified. Thinly laminated felsic ash with pinch and swelling vary laterally and fades out. Ash beds are structureless and possess grading in the coarse tails. Volcaniclastics are discernible from microscopic to exposure levels along with the siliciclastics. Thin-bedded mixed volcaniclastic turbidites have been documented from the basal part of the TF. Fine-grained rhythmites with slumps are common in the basal part of the TF. Typical aeolian bimodality has also been documented from the clastics at four different stratigraphic levels of the formation. Thin-bedded mauve coloured sandstones-siltstones at different stratigraphic levels have been recorded. Shales and glauconitic fine sandstones are thinly laminated, laterally continuous with mild wavy parallelism and horizontal disposition. Chert or cherty bands are intermixed with laminated dolomitic layers showing an overall decrease in the fine clastics. Dolostones with algal laminites interbedded with mudstones and sandstones have been recorded in different stratigraphic levels. The dolostones fills the wavy depressions and ultimately cover the limestone dunes and ripples with horizontal laminae or thin beds. The flat-bottomed limestone lensoidal bodies are found to develop on this substrate. A cyclicity within the limestone-dolostone beds has been identified within the range between 'ripple lamina' and 'dune-thin bed' associations. Each cyclic arrangement of limestone-dolostone begins where chert-dominated interlayering with dolomitic carbonate ends. The association of cherty lithounits and limestone-dolostone couples also show cyclicity.

Cyclicities on the basis of lithic character, their interrelations, gradations from a thin bed to thin lamina, transition from thin-bedded tabular asymptotic foresets to interlaced ripple-drifts and nature of bedforms are discernible throughout the TF. Lithic units commonly are 3-15 cm thick. Their lateral extent ranges from few meters to several hundreds of meters. Beds are separated from one another by thinner or thicker, usually weaker intercalations. Intercalated lithic units form distinct lithosomal assemblages. Intercalations are designated as interbeds with differing composition or structure (cf. Campbell, 1967; Reineck and Singh, 1980; Collinson and Thompson, 1982). Beds with intercalations forming bedding couplets in repetition form bundles of several bedding couplets. Amalgamated thin beds of ash and volcaniclastics define event layers in the TF. Simultaneous supply of sand, mud and volcaniclastics often got mixed up with intrabasinal carbonates and deposited lithosomal assemblages in the facies. Depositional rhythms and cycles in TF resulted from abrupt changes in sedimentation due to depositional events or episodes at random to quasiperiodic time intervals forming episodic and dicyclic bedding. The most prominent of this group are rhythmites with slumps, laterally extended over hundreds of meters and subaqueously settled repetitive pyroclastic ash falls into basins receiving normal background sedimentation (cf. Einsele, 1992; Reading, 1996; Middleton, 2003). Bedding variations in the TF were caused by slow gradual changes in sedimentation generating cyclic bedding even involving formation of three-bed cycles (cf. Einsele, 1992). Some of the laterally continuos thin ash layers serve as marker event beds and their episodic occurrence in stratigraphic record of TF attest to intermittent volcanism in the source area. Interlayering of the ash layers with clastics and carbonates is ascribed to subaqueous settling of the ash. Glauconitic fine sandstones in the sequence is suggestive of deposition in a sea-marginal set-up and their repetition in the stratigraphic column is indicative of oscillatory marine incursions probably owing to eustatic sea level changes. Dolostones with algal laminites, interbedded with mudstones and sandstones lend support to repeated lagoonal impress through time.

Attendant cyclicities in the TF suggest different phases of oscillation in its depositional set-ups. Two extremes are fine clastic dominated lagoonal condition and shallow-marine milieu. Marine incursions only punctuates the cyclic trend in TF. Sands with typical aeolian bimodality at different stratigraphic levels attest to episodic reworking of aeolian sands in shallow-marine and lagoonal environment.

The Vempallae Formation

The TF grades into a sequence (ca.1500m) of well-defined carbonate beds of the Vempallae Formation with sheet-bedded limestones, lenticular dolostone comprising stromatolitic bedding, subordinate fine grained sandstones, shales, mudstones and volcaniclastics, traceable laterally over tens of kilometers. Impress of TF, however, is discernible in a portion of the lower part of the sequence. The Vempallae Formation, in its extents of northern and southern outcrops, depicts variations in depositional style and stromatolitic growth (Fig.7).

Facies analysis of the Vempallae carbonates reveals 21 facies which include :

Fig.7 : Prolific stromatolitic growth from Vempallae Formation, Cuddapah basin, India; bar scale = 10cm.

Intraclastic Dolomitic Packstone, Intraclastic Dolomitic Packstone with sparsely developed stromatolitic laminae, laminated dolosiltite/dololutite with ash pockets, ash beds, laminated mudstone with ash layers, laminated mudstone, tufa, stromatolitic dolomite, laminated dolosiltite/dololutite, oolitic chert, oolitic dolomite, massive dolomite, rippled dolostone, trough cross-stratified turbiditic dololutite and cherty dolostone.

A representative vertical facies log shows some of major facies associations of the Vempallae (Fig.8). The dolostone facies consisting of lenticular-bedded dolostones associated with subordinate wavy and lenticular bedded limestones is predominant in the middle part of the formation. Medium- and small sized channels in dolostones is characteristic of the upper part of Vempallae carbonate facies. Different types of facies attributes discernible are dunes, ripples, ripple-drifts and occasional mud cracked surfaces within the middle dolostone facies. Large-sized, cabbage-shaped, ovoid- to domal stromatolites and highly elongated stromatolitic forms have also been recorded. However, gradational size and shapes of the stromatolites appear and fade out frequently. Impersistent occurrence of different types of stromatolitic bedding is a characteristic feature of the facies.

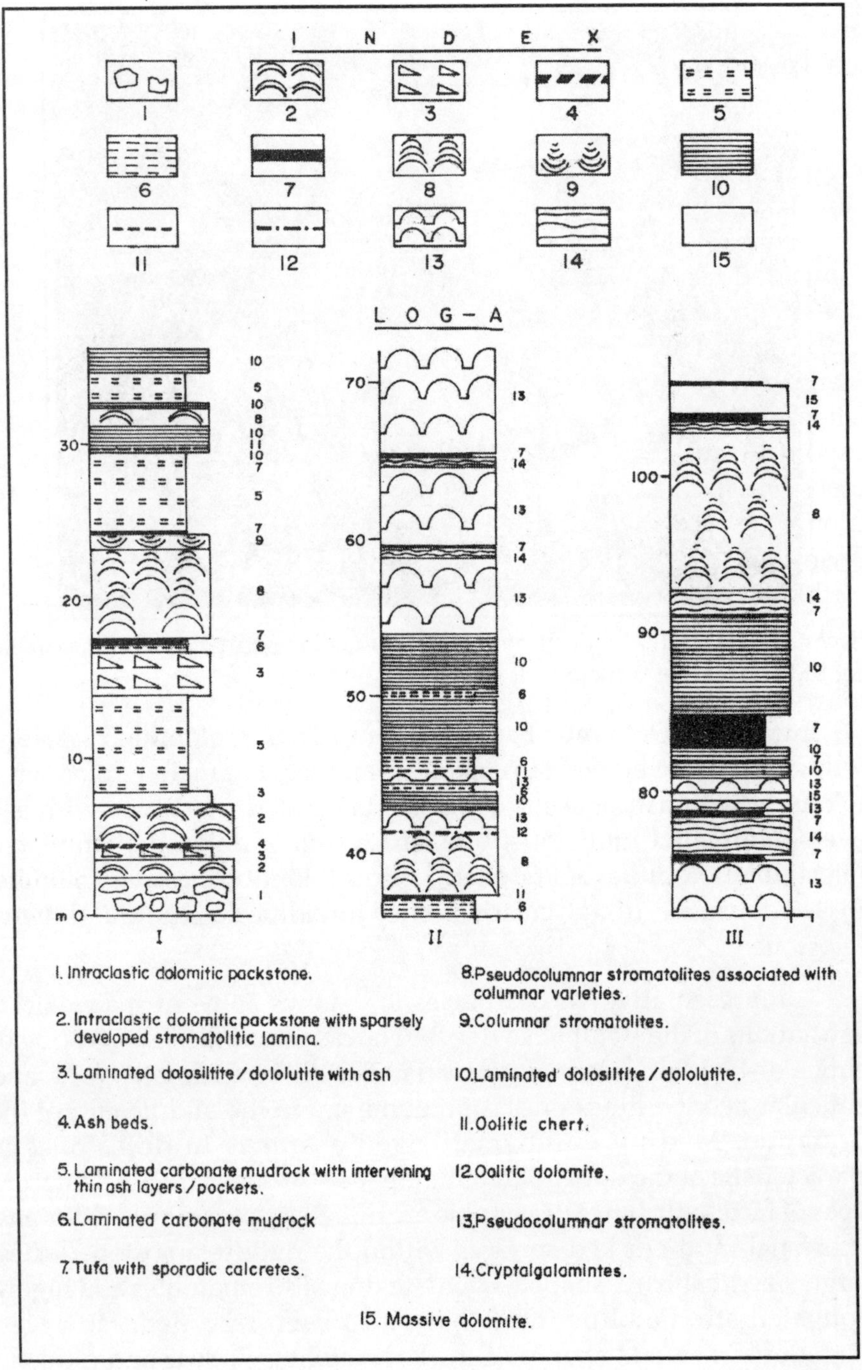

Fig.8 : Representative facies log from Vempallae Formation, Cuddapah basin, India.

Dolomitic element of 'primary' origin develops in carbonate sequence on a very low gradient tidal flats and coastal sabkhas which might be either marginal to shallow-marine or a lagoonal set-up under extreme arid condition through circulation of hot hypersaline brine (Reading, 1996). Transportation, breaking down and deposition of intraclasts signify wave action and frequently generated platformal slope-controlled deposition of debris flows (Fig.9). Muddy limestone-dominated facies with sheet geometry in the lower part characterises low-energy shallow-water, lacustrine- to lagoonal origin (cf. Davies Jr., 1983; Kukal, 1970; Selley, 1978). This might have originated in a low-energy shallow-marine condition below the wave base (cf. Irwin, 1965; Selley, 1978; Reading, 1996). Morphometry of the stromatolites is an important indicator of the energy condition during its growth in depositional set-up (cf. Krylov, 1976; Walther in Reading, 1996). Large colonies of stromatolites could only grow in quiet low-energy conditions. Owing to possible presence of palaeotopographic high towards the north, wave action might have increased promoting development of highly elongated stromatolitic forms. On the other hand, two levels of large cabbage-shaped ovoid and domal stromatolites in the middle order dolostone facies corroborates a low-energy lagoonal-marginal milieu where medium- and small sized fluvial channels redistributed penecontemporaneous dolomitic mud, derived from low-gradient, arid subaqueous flats and got trapped in stromatolitic laminae.

Fig.9 : Debris flow deposits within the Vempallae carbonates of the Cuddapah basin, India; Scale: diameter of the coin = 2cm.

Upper part of the formation is a cyclic carbonate facies with shales, rhythmites, tufas, oolitic cherts/dolomites, volcaniclastic ash and chert layers (Fig.10). The underlying dolostone facies is, however, gradational to the upper calcareous shaly facies. The shaly facies shows gradual increase of terrigenous clastics vertically.

Fig.10 : Rhythmites of volcaniclastic ash beds and impure clastics, Vempalle Formation, Cuddapah basin, India; scale: height of human figure = 170cm.

Successive increment of the terrigenous clastics is a response to the redistribution of topographic highs in the source area. Topographic highs can be added to a peneplained surface either due to oscillatory movements or tectonism related to intracratonic tectonic blocks (Beloussov, 1962; Leeder, 1982,1999). The most intriguing feature of this basinal development is the associated volcanism. Cryptalgalaminites and algal laminites with slight tendency of pseudocolumnar growth float within carbonates intermixed with volcaniclastics which indicates that mixed lithology favoured only the growth of simple forms of algal mats.

Twenty one lithofacies with variable proportion of volcaniclastic elements in a shelfal set-up, often forming asymmetric- to fining upward cycles and punctuated by noncyclic phases both in northern and southern sectors attest to an overall pulsating basin dynamics.

Cyclic and Non-cyclic sequences in the Vempallae Formation

The stratigraphic record of geologic history is characterised by intervals of pronounced rhythmic or cyclic arrangement of facies such that repetitive groups of rock units have component facies which tend to occur in certain order (Grotzinger, 1986). In natural rock record through out the geological time, such cyclic arrangement of facies are rarely complete, due to changes in environmental conditions as well as due to occurrence of noncyclic units. Autocyclicities are responses of internal feedback mechanisms within the sedimentary systems. On the other hand, allocyclicity arises within the sedimentary piles of the formation in response to episodic subsidence, volcanism and eustatic sea-level changes, which are the external controls operative from domains outside the basinal configuration (Grotzinger, 1986 a & b). Milankovitch-forced control on many cyclic systems of the formation can also be deciphered (cf. Grotzinger, 1986 a & b). Upward shallowing platformal carbonate cycles of the Vempallae under variable palaeoenvironmental dynamics, in this context, appear to have been formed in response to low-amplitude (Milankovitch-band) sea-level oscillations. Cyclic carbonate sedimentation of Rocknest Platform, NW Canada can be cited as a suitable palaeoanalogue (Grotzinger, 1986 a & b) for the Vempallae.

Cyclic and non-cyclic units

The Vempallae Formation depicts cyclic sedimentation within the shelfal facies. These are generally asymmetric shallowing upward cycles. The cycles are usually incomplete with intervening noncyclic intervals. The cyclic lithofacies contain tufa-capped sequences, dominated by impure carbonate mud rocks and laminated dolomites with frequent stromatolites towards the southern fringe of the Papaghni basin. This passes northwards into cycles of stromatolitic boundstones and trough-stratified sequences.

The southern cyclic lithofacies is characterised by intraclastic dolomitic packstone, laminated dolosiltite/dololutite, impure carbonate mudstone associated with stromatolitic dolomites and capped by thin tufa layers. The northern cyclic lithofacies consists of infrequent tufa capped sequences of large stromatolitic growth with overlying trough-stratified beds and laminated dolosiltite/ dololutite.

The noncyclic units of the southern cycle comprise carbonates with volcaniclastics and chert as two major dominant elements. Volcaniclastic layers are however thin. The middle noncyclic intervals include thin beds (10-20 cm thick) of oolitic chert/oolitic dolomites. The oolites show radial structures within either a fine-grained chalcedony, chert or dololutite; locally layered with discontinuous chert beds. The oolitic horizons indicate high-energy environment, prevalent for short duration. The noncyclic units in the upper part of the southern cycles are massive structureless dololutite beds, indicative of deposition in a low-energy condition.

The noncyclic units of the northern cycle of the basin are more pervasive and thicker and comparatively more complete than its southern counterpart. The noncyclic interval consists of stromatolitic beds with intervening milky-white chert horizons having interfingering, vertical gradation as well as sharp contacts with the dolomites. This is suggestive of a synsedimentary origin with the dolostones. These chert beds and those associated with the oolites appear to be genetically related to volcaniclastics.

Chitravati subbasin

The Pulivendla Formation of the Chitravati Group unconformably overlies the Vempallae Formation. The Pulivendla in its lower part records cyclic development of trappean-intertrappean sequence (Fig.11) followed up stratigraphically by boulders, bouldery conglomerates and sandstone beds in a cyclic manner.

Trappean-intertrappean Sequence

The trap rocks record mafic and acidic volcanism. Subaerially basaltic lava was emplaced successively with intermittent pause over the unconformity surface of the Vempallae. During the quiescence formation of red soil or red bole due to weathering and erosion of basic lava took place. Basic lava flows with inclusions of unmodified mantle xenoliths (Dasgupta, 1986) constitute a 80m thick sequence which is associated with intervening event layers of red boles indicative of weathering and quiescence.

Associated intertrappeans comprise carbonates, terrigenous clastics, chert, hybridized rocks, impure doloarenites, and dolosiltites. The terrigenous clastics comprising shales, siltstones, fine-grained sandstones with sharp-crested current ripples, tabular cross beds, mud-

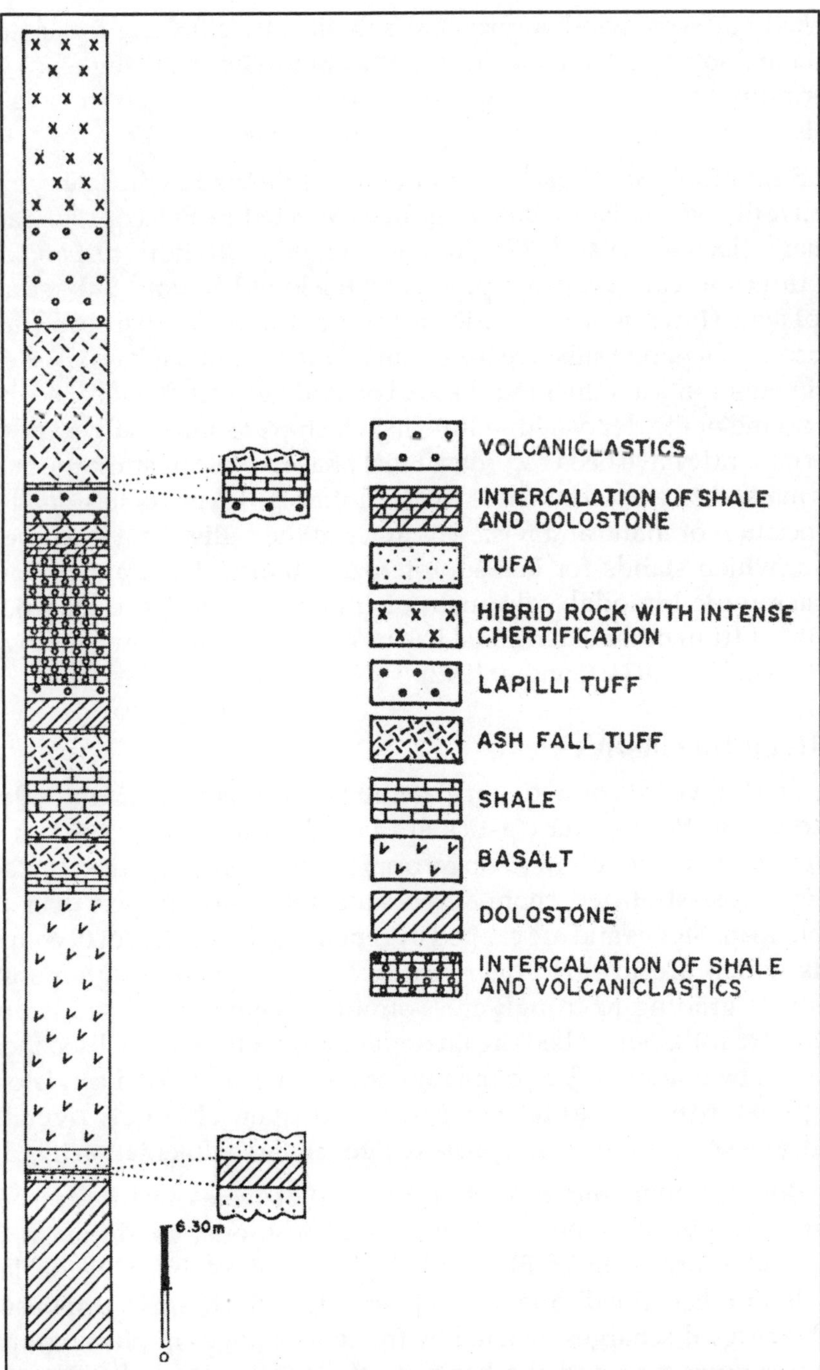

Fig.11 : Representative facies log showing trappean-intertrappean sequence from basal part of Pulivendla Formation, Cuddapah basin, India.

cracked surfaces, wind ripples and subcritically climbing translatent strata are some of the major facies attributes. Facies attributes of the intertrappeans can be ascribed to subaqueous deposition in isolated pools and ponds accompanied by formation of aeolianites.

Stratigraphically basaltic eruption was followed in the later phase by rhyolitic volcanism which might be expected in Palaeoproterozoic scenario (Jacharia et al.,1999). Presence of lapilli tuff, lithic tuff and ash fall tuffs indicate eruptive phases of felsic volcanism. Subaerially emplaced flows were basaltic in the beginning and intermittent quiescence heralded subaqueous sedimentation of limited basinal extent. Sedimentation was either intrabasinal or land derived. Touch of aeolian reworking of the deposited sediments which were subaerially exposed intermittently have been recorded. Final phase of volcanism in the basin was marked by emplacement of welded tuffaceous flows of felsic nature. Association of mafic and felsic volcanics in the Pulivendla is a case in hand, which stands for Palaeoproterozoic bimodal volcanism. Such intracratonic bimodal volcanism of limited extent have also been recorded from other extensional basins of the Precambrian (Goodwin, 1981; Windley, 1984; Condie, 1992).

Pulivendla clastics

The Pulivendla outcrops are laterally continuous over hundreds of meters. The Pulivendla clastics are bouldery and rudaceous in its lowermost part and followed up by rippled sandstones. Spatio-vetrically trough-cross-stratified channelised sandstones are incised over the rippled sandstones and are capped by repetitive development of couplet beds which are in turn, overlain by flat-bottomed conglomeratic channels grading to trough-cross-stratified channelised sandstones (Figs.12 & 13). Couplet beds again dominate over the channelised facies overlain by planar and trough-cross-stratified channelised sandstone. Reappearance of couplet beds in broad open channels over the sandstones marks the closing phase of Pulivendla sedimentation (Fig.14).

Boulder beds and repetition of couplet facies testify episodic interference by extremely high-energy depositional condition in the Pulivendla Formation (cf. Blair and Mcpherson, 1994; Blair, 1999). Flash floods and sheet floods operated repetitively in the basin in response to high-energy discharges dictated by frequent changes in physiography of the source area and the basin itself under an overall control of extensional regime. Significant variations in depth and velocities of supercritical flows and high-energy gravity flows resulted intimate

Fig.12 : Plano-convex channelised cyclic couplet beds, Pulivendla Formation, Cuddapah basin, India; Bar scale = 10 cm.

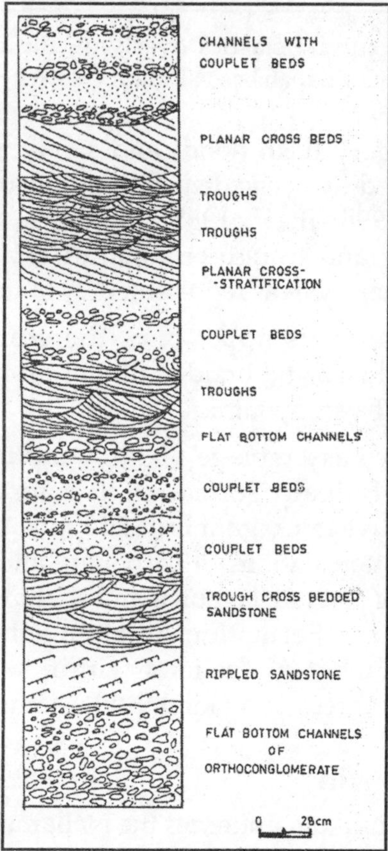

Fig.13 : Representative facies log showing cyclicity in Pulivendla Formation, Cuddapah basin, India.

Fig.14 : Channelised fluvial clastics overlying the conglomerate facies, Pulivendla Formation, Cuddapah basin, India; Scale: length of hammer = 40cm.

association of bouldery flash flood- and sandy alluvial sheet flood deposits characterised by cyclic disposition of couplet facies (cf. Blair and Mcpherson, 1994; Blair, 1985a,b; Allen, 1983). Intervening channelised tabular and trough cross-stratified sandstones signify change in depositional styles as represented by braided fluvial deposits.

The Pulivendla, in its upper part, therefore, attests to cyclic formation of alluvial fans and braided fluvial set-ups, as promoted by frequent changes in basin dynamics.

The cyclic facies assemblage of Pulivendla is inturn overlain unconformably by Tadpatri Formation. It comprises another cyclic association of repetitive thick ignimbrites depicting phreatic volcanism (Figs.15&16) with intercalations of carbonate and terigenous clastics bearing signatures of fluvial-, lagoonal and shallow-marine facies. The uppermost Gandikota Formation overlying the Tadpatri depicts prominent development of cyclicity within sandstone-mudstone sequences bearing impress of braided fluvial set-up.

Nallamalai subbasin

Arenaceous Gandikota underlies the Nallamalai Group. The basal arenaceous and argillaceous Bairenkonda of the Nallamalai Group is of fluvial and/or shallow-marine origin (Nagaraja Rao et al. 1987).

Fig.15 : Repeated ignimbrite facies with shrinkage cracks and drag folds, Tadpatri Formation, Cuddapah basin, India; Bar scale = 5cm.

Fig.16 : Welded tuff with bedded pumice, Tadpatri Formation, Cuddapah basin, India; diameter of the lens cap = 6cm.

The facies assemblage of the overlying Cumbum Formation of the Nallamalai subbasin includes conglomerate-sand facies (now conglomerate-quartzite association), sand-mud alternation facies (now quartzite-phyllite/schist alternation), mud facies (phyllite/schist) and carbonate-mud facies (metamorphosed into crystalline limestone/dolomite-phyllite/schist) at different stratigraphic levels. Mafic and/or felsic to silicic volcanics are often associated with the facies assemblage.

Facies attributes

Channelised coarse- to fine-grained sandstones with massive bedding and thin muddy intercalations, subparallel laminations containing Ta, Tb and Tbc turbidites developed frequently with contourites in spatio-vertical continuity (Fig.17). Dish structures, infrequent sole marks, abundant erosional channels and low angle wedge-shaped cross laminae are also discernible.

Fig.17 : Chertified interbedded sandstone-mudstone turbidites, Cumbum Formation, Cuddapah basin, India; Scale: height of the human figure on the right = 182cm.

Thin interbedded sandstone-mudstone sequences are traceable for hundreds of meters. Laminated sandy beds (6-20 cm thick) are often replaced by muddy units. However, occasional laterally continuous thick massive sandstone lithosomes with muddy contourites (70-90 cm thick) are noticeable. Tb-e, Tde & dominant Tc-e turbidites have been identified.

Thin, irregular, discontinuous, sandy-muddy beds showing wedging and lenticularity are common. Fine sandstone beds are frequently overlain by mudstones with sharp basal contacts. Tde and occasional Te turbidites interbedded with insitu chert characterise the facies.

Sediment deformation such as slumps, mudflow, intraformational folding, contorted strata, breccia and isolated dispersed clasts have been identified as chaotic facies (Fig.18). In some stratigraphic sections, the chaotic facies is repeated with cyclic Tbc turbidites (Fig.18). Strong facies

Fig.18 : A chaotic horizon from cyclic turbidite facies of Cumbum Formation, Cuddapah basin, India. Repetition of Tbc turbidite is discernible; length of the scale = 15cm.

elements such as disorganised to slightly organised lensoidal bodies with basal laminae containing mudstone clasts in muddy matrix, pebbly mudstones, slump folds capped by convolute laminations and clast-rich zones are noticeable.

Facies attributes point to repetitive development of chaotic-, debris flow deposits in association with turbidite-contourite stratal assemblage in different stratigraphic levels within the Cumbum Formation (cf. Stow et al., 1998; Bouma, 1996). Disorganised to slightly organised lensoidal bodies with basal laminae containing mudstone clasts in muddy matrix, pebbly mudstones, slump folds capped by convolute laminations and clast-rich zones characterise the debrites (cf. Shanmugam & Moiola, 1995; Bouma, 1996). Cyclic development of seismites, seismically-affected (reworked) turbidites and tsunamiites at different stratigraphic levels is attested by attributes comprising graded microfaults, fault-bounded graben-like down sagging structures, disturbed bedding in restricted horizons (Fig.19), hydroplastic deformation structures, extensional structures including pinch-and-swells, 'boudinage' dislocation of individual competent beds, brecciation of brittle layers, sand dykes, mudstones with thixotropic deformation and diastasis cracks (cf. Seilacher, 1969, 1984; Shiki & Yamazaki,1996; Sims, 1975; Plazziat & Purser, 1998; Guiraud & Plaziat, 1993). Volcaniclastic detritus mixed-up with terrigenous clastics include chertified pumices and lapilli, clasts

Fig.19: A discrete seismite level from Cumbum Formation, Cuddapah basin, India; bar scale = 10cm.

of chertified welded tuffs and volcanic bombs. Association of insitu chert interbedded with Tde and/or Te turbidites points towards deep-water bathymetry for the Cumbum turbidites.

Presence of volumetrically significant felsic to silicic volcaniclastic elements and repetitive occurrence of pyroclastics enclosing volcanic bombs (Figs.20&21)are suggestive of volcanic activity which brought earthquake waves in its wake by virtue of being highly explosive in nature.

Shallow-marine storm deposits are marked by well preserved laminites, wave ripples, hummocky and swaley cross-stratified beds arranged in several thickening- and thinning-upward sequences. Severe seismic activity in the syn-rift Cumbum scenario is indicated by seismites and seismo-turbidites developed among the storm-induced sediments. They are deposited in transitional offshore (upper bathyal) setting in association with tsunamiites.

Faulting, syndepositional as well as basin marginal might have been accompanied by ground tremors, tsunamis that induced the generation of seismites, seismoturbidites and tsunamiites. Occurrence of seismites is erratic and scattered throughout the Cumbum. Haphazardly arranged seismic signatures within discrete layers throughout the succession have relevance to cyclic development of facies which is very common during Proterozoic sedimentation (Dasgupta, 1998).

Fig.20 : Repeated chertified welded tuffs with volcanic bombs with long axis parallel to the bedding indicating subaqueous deposition, Cumbum Formation, Nallamalai Group. Scale: long axis of the volcanic bomb, centrally placed= 35cm.

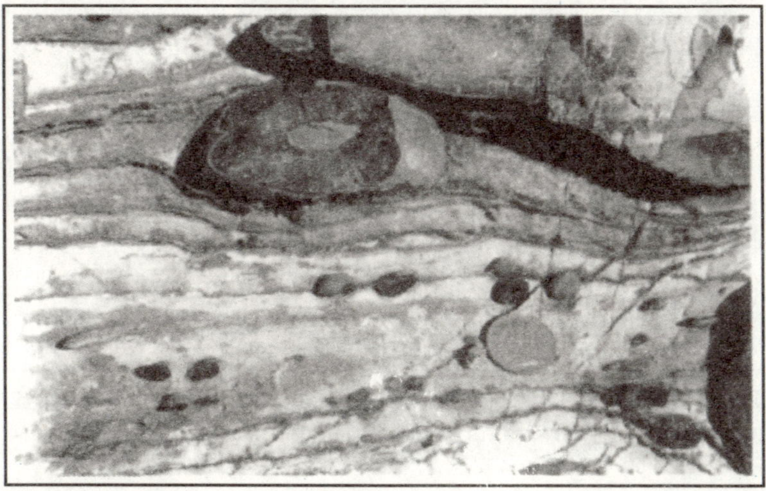

Fig.21: Chertified layered welded tuff with lapilli and volcanic bombs, Cumbum Formation, Nallamalai Group, Cuddapah basin, India; scale: diameter of the coin = 2.2cm.

The Cumbum turbidites with chaotic levels and debris flows in stratigraphic record (Fig.22) supported by recent and ancient analogues seems to have been resedimented owing to variable genetic controls such as overloading, volcanism, diapiric uplift and/or seismicity, common in an extensional regime (cf. Shanmugam and Moiola, 1995; Sixsmith et al., 2004).

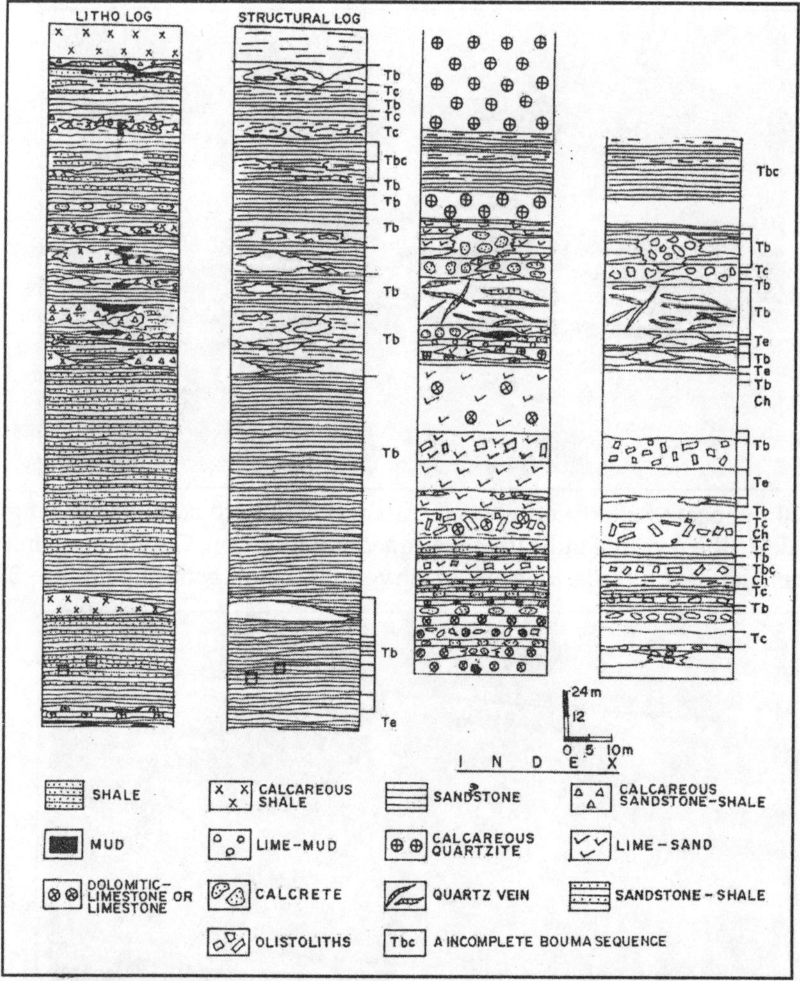

Fig.22 : Representative vertical facies log from Cumbum Formation, Cuddapah basin, India; cyclicity between turbidite and chaotic facies is discernible.

Srisailam Subbasin

In the uppermost part of the Srisailam subbasin, the Egalapenta Member records different types of cyclicities in stratal assemblages. Three different cyclic associations have been documented from the lower-, middle- and the upper part of the Egalapenta Member.

In the lower part of the Egalapenta Member, glauconitic sandstone-mudstones with attributes such as heterolithic sandy-muddy interlaminations, bidirectional cross-beds, rhythmically cross-stratified sandstones comprising sigmoidal cross-bedding, shallow troughs, symmetrical ripples are repetitive with channelised reddish-brown medium- to coarse sandstone facies bearing signatures of profuse trough-cross-stratified beds, ferricrete clasts and mud clast-bearing beds, tabular cross-stratified sandstones, pebble and granule bearing disorganised beds and laminated muddy-sandy beds. The glauconitic sandstones and associated lithic units are of shallow-marine origin (Biswas, 2000). Braided fluvial milieu can be ascribed to the channelised sandstones (Biswas, 2000). The lower part of Egalapenta depicts cyclicity developed owing to repetitive sedimentation of shallow-marine glauconitic sandstones, tidalites and fluvial clastics. Onlaps by glauconitic sandstones and offlaps by fluvial sandstones indicate shifts between offshore and nearshore environments. Cyclicity between shallow-marine and fluvial strata is frequently cited (cf. Reading, 1996; Middleton, 2003; Selley, 1978; Smith, 1974; Cant and Walker, 1978; Long, 1978; Bluck, 1981; Muir and Rust, 1982; Allen, 1983; Simpson and Eriksson, 1993; Miall, 1996).

In the middle part of the member, cycles defined by impure carbonate-sandstone-mudstones and reddish-brown sandstone sheets are documented (Fig.23). Attributes of the impure carbonate-sandstone-mudstones include evaporitic signatures such as anhydrites, enterolithic bedding, chickenwire and bird's eye structures, mudcracked surfaces, shallow isolated troughs. Reddish-brown sandstone sheets are characterized by development of profuse wind ripples, small cross-stratified dunes, horizontal- to low-angle cross-stratified beds, subcritically climbing translatent strata and thin pinches of laminated mudstone beds.

Cyclic occurrence of impure carbonate-sandstone-mudstone with signatures of coastal sabkha and low-angle stratified sandstones representing aeolian sandsheets have been interpreted as episodic marine incursions that promoted growth of such cyclicities in arid climatic zones under eustatic control (Biswas, 2000). Such cyclic associations point towards Milankovitch forcing (cf. Grotzinger, 1986b; Mclane, 1995; Chan, 1989; Wood and Wolfe, 1969; Hardie and Eugster, 1971; Friedman and Sanders, 1978; Fairbridge and Bourgeois, 1978; Hunter, 1981; Kocurek, 1981; Driese and Dott, 1984).

The upper part of the formation records another cyclic association that developed between stacks of climbing lensoidal cross-bedded medium- to fine sandstones and thin beds of laminated, wind-ripple-bearing low-angle cross-stratified sandstone-mudstone beds and wedge-planar cross-beds (Fig.24). Attributes of large climbing cross-stratified

Fig.23 : Flat-bedded impure sandstone-mudstone beds, interbedded with reddish-brown sandstone sheets showing cyclicity between coastal sabkha and aeolian sandsheets, Egalapenta Member, Srisailam Formation, Cuddapah basin, India; scale: length of hammer at lower right hand part = 40cm.

Fig.24 : Low-angle cross-stratifications and dune cross-stratifictaions in reddish-brown sandstones, Egalapenta Member. Height of the shrub : 80cm.

lensoidal sandstones (Fig.25) stand for a token representation of the migrating dunes forming draas (Fig.26) (Biswas, 2000). Wind-ripple bearing, low-angle cross-stratified sandstones and thin mudstones bears the impress of interdune deposits (Biswas, 2000). Such cyclicities between climbing dunes (draa)-and interdune deposits on a regional scale deposit sand seas or ergs. A portion of an erg system between cyclic draa and interdune deposits crops out in upper Egalapenta near

Fig.25 : A dune showing high-angle foresets overlying the mudstone-sandstone interdune deposits, Egalapenta Member, Srisailam Formation, Cuddapah basin, India; scale: length of hammer = 40cm.

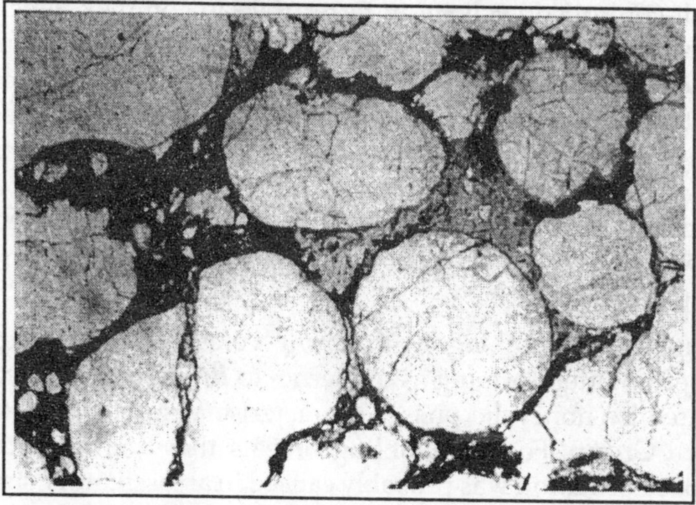

Fig.26 : Typical aeolian bimodality among quartz sand and silt from Egalapenta Member, Srisailam Formation, Cuddapah basin, India; the diameter of largest quartz grain = 0.74mm.

the Krishna gorge. Derivation of huge volume of sand from granitic-gneissic basement and episodic reworking of the same in a high-energy aeolian environment with distinct imprints of aeolian bimodality generated sand seas. Cyclicities in aeolian and associated facies have been frequently reported (McKee, 1966; Inman et al., 1966; Kocurek and Dott, 1981; Kocurek and Fielder, 1982; Fryberger et al., 1983).

The Egalapenta displays a complex interplay of fluvial-, shallow-marine, arid sea-marginal sabkha and aeolian processes. The Egalapenta complex represent a spectrum of cyclic aeolian system types in which end members are dry, wet and stabilised systems with records of interference by fluvial-aeolian interaction, sabkha phases and allocyclic shallow-marine milieus. Gradual change from wet fluvial condition to dry arid climate might point to mechanisms such as fluctuations in groundwater volume in continents or thermal expansion and contraction, of sea-water (Ulicny, 1999) which could have controlled to some extent the eustatic fluctuations. Association of fluvial- and cyclic aeolian dune-interdune and coastal sabkha facies corroborates climatic fluctuations from humid- to arid and possible controls of the climatic variations on the dynamics of the desert systems. The cyclicity also appears to be the response of climatic fluctuations which might have been orbitally controlled (Frederiksen et al., 1998).

Role of cyclicity in basinal evolution

Fault-controlled cyclicity in terrestrial fan complex which marks the initiation of Cuddapah Supergroup demonstrates significant role of extensional tectonics in basin evolution. Fining upward cycles in alluvial fans that punctuates coarsening upward trends indicates recession of piedmont-like source from earlier site of sedimentation and subsequent shift of proximal fan set-up away from the basin. Thus gradual enlargement of the protobasinal phase might have occurred in the Cuddapah under the influence of extensional regime during the Gulcheru sedimentation.

Lithofacies assemblage with strong volcaniclastic elements in a shelfal set-up often forming asymmetric- to fining upward cycles and punctuated by noncyclic phases characterise Vempallae strata of the Papaghni Group. Facies modeling reveals that the cyclicity in the Vempallae Formation was probably caused in an oscillatory shelfal set-up with inteference by periodic pulses of volcanism which accompanied faulting in an extensional regime rather than in response to global sea level change.

The Naramrajupallae lithology of the Nallamali Group reveals cyclic association of chaotic facies with bounding turbidite sequences. From structural and vertical facies assemblage, fining upward sequences can also be observed. Olistostromes and associated facies around Naramrajupallae with fining upward trends can be attested to the influence of allocyclic controls such as variations in the local rate of sea level change and/or intensity of tectonic activity in shelf/ near shore/ hinterland areas. The Naramrajupallae tectofacies comprising chaotic facies gradational to clastic-carbonate couple in each cycle probably originated under the overall control of syndepositional faulting in an extensional regime.

Cyclicity in the Egalapenta composite complex shows accumulation of cycles of draas often interfered by shallow-marine facies. Transgressive-regressive cycles of shallow marine facies might have developed eustatic oscillatory changes in sea level. Such cycles probably exerted an overall control on the genesis of cyclic sedimentation of draas complex.

Outward progradation of fault systems in the protobasinal stage from the initial rift in the brittle upper crust probably indicates that the upper crustal extension was spatially separated from the lower crustal/ mantle lithosphere extension. Depth-dependent extension based on variations in lithospheric rheology appears to have significant bearing on the evolution of Cuddapah basin.

REFERENCES

Allen,J.R.L.,1982. Developments in sedimentology,: Sedimentary structures. Their character and physical basis. Elsevier (Amsterdam). v.30A and B.

Allen,J.R.L.,1983. Studies in fluviatile sedimentation : bars, bar complexes and sandstone sheets (low-sinuosity braided streams) in the Brownstone (L.Devonian), Welsh Borders. Sed.Geol., v.33, pp.237-293.

Balcer, R.A., 1984. Stratigraphy and depositional history of the Pantano Formation (Oligocene-early Miocene), Pima county, Arizona, Unpublished MS thesis: Tucson Univ. Arizona, 107p.

Banerjee,I.,1974. Barrier coastline sedimentation model and the Vindhyan example. In: Contributions to the Earth and Planetary Sciences, Golden Jubilee Volume (A.Dey ed.). Quart.J.Geol.Min.Met.Soc.India, pp.101-127.

Beloussov, V., 1962. Basic problems in geotectonics. Mcgraw Hill, New York.

Bhattacharji,S. and Singh,R.N.,1984. Thermo-mechanical structure of the southern part of the Indian Shield and its relevance to Precambrian Basin evolution. Tectonophysics, v.105, pp.103-120.

Bhattacharji,S.,1986. Asthenospheric upwelling, lineament reactivation, magmatic episodes and ore mineral localization in Proterozoic Basin Evolution on the Archaean Indian Shield. International Basement Tectonic Association, Publ. No. S. pp. 187-200.

Biswas, A, 2000. Facies analysis and depositional milieu of the Proterozoic Egalapenta Member, Srisailam Formation, in parts of the Krishna Gorge West, Cuddapah basin, India. Unpublished Ph.D. thesis, University of Calcutta, India, 119p.

Blair, T.C. and Mcpherson, J.G., 1994. Alluvial fan processes and forms in Abrahams, A.D. and Parsons, A.J. (Eds), Geomorphology of desert environments, London, Chapman and Hall, pp. 354-402.

Blair, T.C., 1985a. Depositional chronology, sedimentary processes and the resulting vertical stratification sequences in the Roaring river alluvial fan, Rocky Mountain National Park, Colorado. In: Flores, R.M. and Harvey, M.D. (Eds.), Field guide book to modern and ancient fluvial systems in the United States, Fort Collins, Colorado, Proceed. 3rd Int. Fluvial Conf., pp. 96-101

Blair, T.C., 1985b. Nonmarine sedimentological response to tectonism in rift basins : a comparison of the Jurassic Todos Santos Formation, Chiapas, Mexico, with Quaternary deposits in Death Valley in California (Abstract), SEPM Annl. Mid Year Meeting Abstracts, v. 2, pp. 11-12

Blair, T.C., 1987. Tectonic and hydrologic controls on cyclic alluvial fan, fluvial and lacustrine rift-basin sedimentation, Jurassic- lowermost Cretaceous Todos Santos Formation, Chiapas, Mexico, Jour. Sed. Pet., v. 57, pp. 845-862.

Blair, T.C., 1999. Cause of dominance by sheet flood vs. debris flow processes on two adjoining alluvial fans, Death Valley, California, Sedimentology, v. 46, pp. 1015-1028.

Bluck,B.J.,1981. Upper Old Red Sandstone, Firt of Clyde. In: Field Guides to Modern and Ancient Fluvial Systems in Britain and Spain (T.Elliot, ed.). Proc.2nd Cong.Fluvial Seds.Univ.Keele, U.K., Int.Ass.Sediment., 5.14-5.20.

Bouma, A.H., 1996. Review of fine-grained submarine fans and turbidite systems. Lecture notes for the U.S. Department of Energy, 73p.

Campbell, C.V.,1967. Lamina, lamina set, bed and bedset. Sedimentology, v.8, pp.7-26.

Cant,D.J. and Walker,R.G.,1978. Fluvial processes and facies sequences in the sandy braided South Saskatchewan River, Canada. Sedimentology, v.25, pp.625-648.

Chakraborty, C. and Bhattacharyya, A., 1996. Fan-delta sedimentation in a Foreland Moat: Deoland Formation, Vindhyan Supergroup, Son Valley.

In : Bhattacharyya, A. (Ed.), Recent advances in Vindhyan Geology, Geol Soc. India, Bangalore Mem. 36, pp. 27-48.

Chan,M.A.,1989. Erg margin of the Permian White Rim Sandstone, SE Utah. Sedimentology, v.36, pp.235-251.

Clifton,H.E. (ed.),1988. Sedimentologic consequences of convulsive geological events. Geol.Soc.Am.Spec.Pap., v.229, 157p.

Coe, A. J.(ed.), 2003. The sedimentary record of sea-level change, The open University Press, Camrudge, 288p.

Collinson,J.D. and Thompson,D.B.,1982. Sedimentary Structures. George Allen and Unwin Publ., London. 194p.

Condie, K. C., 1992. Plate tectonicsand crustal evolution, Butterwarth Heinmann, 282p.

Costa, J.B.S., Hasui, Y. and Pinheiro,R.V.L. 1992. Bacias sedimentaries, Edit Univ.,Univ.Fed.Para,Belem (PA),116pp.

Crawford,A.R. and Compston,W.,1973. The age of the Cuddapah and Kurnool Systems, South India. J.Geol.Soc.Aust., v.19(4), pp.453-464.

Dasgupta,P.K.,1996. Cyclicity in the Proterozoic Cuddapah Supergroup and its bearing on Basin Evolution, India. Session, 3-6, (Sedimentation in Intracratonic basins) 3-6-5, 01214, 0925 p.198, Abs. Vol. 2 of 3, 30th IGC, Beijing, P.R.China.

Dasgupta, P.K. & Sengupta, A. 1997. Cyclicity in the transition facies of the Cuddapah Supergroup and role of oscillatory movement in basinal evolution. Unpublished project report submitted to Dept. of Sc. And Tech., Govt. of India, 140p.

Dasgupta,P.K.,1986. Picrite borne mantle xenoliths from Cuddapah Basin, South India. Ind. J. Earth Sciences 13(4), 333-338.

Dasgupta,P.K.,1998. Seismic signatures from thin bedded olistostromes and associated turbidites : Cumbum Shale, Cuddapah Basin, India. Extended abstract, abstract volume, 15th International Sedimentological Congress, Alicante, Spain. pp.210-212.

Davies,A. Jr., 1983. Depositional systems: A genetic approach to sedimentary geology. Prentice Hall , 669p.

Denny,C.S., 1965. Alluvial Fans in the Death Valley region, California, Prof. Pap. USGS, No. 466, 62p.

Dhoundial, D. P. 1987. Purana basins of PeninsularIndia. In : Purana Basins of Peninsular India, Geol. Soc. India, Bangalore Mem. 6, pp. 1-4.

Dott, R.H., Jr., 1988. An episodic view of shallow-marine clastic sedimentation. In: De Boer, P.L., Van Gelder, A. and Nio, S.D. (Eds.), Tidal-influenced sedimentary environments and facies, Reidel, Dordrecht, pp. 3-12.

Driese,S.G. and Dott, R.H.Jr., 1984. Model for sandstone-carbonate "cyclothems" based on upper member of Morgan Formation (Middle Pennsylvanian) of northern Utah and Colorado. Am.Assoc.Petrol.Geol.Bull., v.68(5), pp.574-597.

Drury,S.A and Holt,R.W., 1983. The tectonic framework of the South Indian Craton: a reconnaissance involving Landsat imagery. In: Structure and Tectonics of Precambrian Rocks in India (S.Sinha-Roy, ed.). Hindusthan Publishing Corp. Press, New Delhi, pp.178-185.

Einsele, G., 1992. Sedimentary basins, Springer-Verlag, 627p.

Fairbridge,R.W. and Bourgeois,J., 1978. The encyclopedia of sedimentology, encyclopedia of earth sciences. Dowden,Hutchinson and Ross Inc., Stroudsburg. v.6, 901p.

Frederiksen,K.S., Clemmensen,L.B. and Lawaetz,H.S.,1998. Sequential architecture and cyclicity in Permian desert deposits, Brodick Beds, Arran, Scotland. J.Geol.Soc.Lond., v.155(4), pp.677-683.

Friedman,G.M. and Sanders,J.E.,1978. Principles of sedimentology. John Wiley and Sons. 792p.

Fryberger,S.G., Al-Sari,A.M. and Clisham,T.J.,1983. Eolian dune, interdune, sand sheet and siliciclastics sabkha sediments of an offshore prograding sand sea, Dhahran area, Saudi Arabia. Bull.Amer.Petroleum Geologists, v.63, pp.280-312.

Goodwin, A.M., 1981. Archaean Plates and greenstone sets.In: A. Kroner (ed) Precambrian Plate Tectonics, Elsevier, Amsterdam, pp.105-135.

Goodwin, P.W.A. and Anderson, E.J., 1985. Punctuated aggradational cycles: A general hypothesis of episodic stratigraphic accumulations. Jour. of Geol., v. 93, pp. 515-533.

Grotzinger,J.P., 1986a. Cyclicity and palaeoenvironmental dynamics, Rocknest Platform, Northwest Canada, Geol. Soc. of Amer. Bull., v.97, pp.1208-1231.

Grotzinger,J.P., 1986b. Upward shallowing platformal cycles : a response to 2.2 billion years of low amplitude, high frequency (Milankovitch Band) sea level oscillations. Palaeoceanography, v. 1 (4), pp. 403-416.

Guiraud, M. & Plaziat, J.C., 1993: Seismites in the fluviatile Bima Sandstones: identification of paleoseisms and discussion of their magnitude in a Cretaceous Synsedimentary strike-slip basin (Upper Benue, Nigeria). Tectonophysics, 225, 493-522.

Hand, B.M., Wessel, J.M. and Hayes, M.O., 1969. Antidunes in the Mount Toby Conglomerate (Triassic), Massachusetts, Jour. Sed. Pet., v.39, pp. 1310-1316.

Hardie,L.A. and Eugster,H.P., 1971. The depositional environment of marine evaporites : a case for shallow, clastic accumulation. Sedimentology, v.16, pp.187-220.

Harvey, A.M., 1984. Debris flows and fluvial deposits in Spannish Quaternary Alluvial Fans: Implications for fan morphology. In: Koster, E.H. and Steel, R.J. (Eds.), Sedimentology of Gravels and Conglomertae, Can. Soc. Petrol. Geol., Mem. 10, pp. 123-132.

Heckel, P.H., 1986. Sea-level curve for Pennsylvanian eustatic marine transgressive-regressive depositional cycles along mid-continent outcrop belt, North America, Geology, v. 14 pp. 330-334.

Heward, A.P., 1978. Alluvial fan sequence and megasequence models with examples from the Westphalian D-Stephanian B Coalfields, northern Spain. In: Miall, A.D. (ed), Fluvial Sedimetology, Mem. Can. Soc. Petrol. Geol., v.5, pp. 669-702.

Hunter,R.E., 1977a. Basic types of stratification in small aeolian dunes. Sedimentology, v.24, pp.361-387.

Hunter,R.E., 1977b. Terminology of cross-stratified sedimentary layers and climbing-ripple structures. J.Sedim.Petrol., v.47 (2), pp.697-706.

Hunter,R.E., 1981. Stratification styles in eolian sandstones : some Pennsylvanian to Jurassic examples from the western interior USA. In : Recent and Ancient nonmarine depositional environments : Models for exploration (F.G.Ethridge and R.M.Flores, eds.), Spec.Publ.Soc.Econ.Paleont.Miner. Tulsa, v.31, pp.315-329.

Ingersoll,R.V. and C.J.Busby.1995. Tectonics of sedimentary basins. In: C.J. Busby and R.V.Ingersoll (Eds.) Tectonics of sedimentary basins, Blackwell Science, Oxford,pp.1-51(Chapter-1).

Inman,D.L., Ewing,G.C. and Corliss,J.B.,1966. Coastal sand dunes of Guerrero Negro, Baja California, Mexico.Bull.Geol.Soc.Amer., v.77, pp.787-802.

Irwin, M.L., 1965. General theory of epeiric clear water sedimentation. AAPG Bull, v. 49, pp. 445-459

Jachariah,J.K., Bhaskar Rao,Y.G., Srinivasan,R. and Gopalan,K.,1999. Rb-Sr-Nd isotopic systematics of Uranium mineralised stromatolitic dolomite from Proterozoic Cuddapah Supergroup, South India and constraints on age and provenance. Chem.Geol., v.162, pp.49-64.

Kaila,K.L. and Tewari,H.C.,1985. Structural trends in the Cuddapah Basin from deep seismic soundings (DSS) and their tectonic implications. Tectonophysics, v.115, pp.69-86.

Kaila,K.L., Roy Chowdhury, K., Reddy,P.R., Krishna,V.G., Narain,H., Subbotin, S.I., Sallogub, B.B.,Chekunov,A.V., Kharetchko,G.E., Lazarenko,M.A. and

Ilchenko,T.V., 1979. Crustal structure along the Kavali-Udipi Profile in the Indian Peninsular Shield from Deep Seismic Sounding : J.Geol.Soc.India, v.20, pp.307-333.

Kaila,K.L., Tewari,H.C., Roy Chowdhury,K., Rao,V.K., Sridhar,A.R. and Mall,D.M., 1987. Crustal structure of the Proterozoic Cuddapah Basin of India from deep seismi soundings and gravity data. In : S. Asano and W.D.Mooney (editors), Seismic studies of the continental lithosphere. Tectonophysics, v.140, pp.1-12.

King,W.,1872. The Kadapah and Karnul Formations in the Madras Presidency. Mem.Geol.Serv.India, v.8, Pt.1, p.346.

Klein, G, deV, 1989. Origin of coal-bearing cyclothems of North America. Geology, v. 17, pp. 152-155.

Kocurek,G. and Fielder,G., 1982. Adhesion structures. J.Sedim.Petrol., v.52, pp.1229-1241.

Kocurek,G. and Dott,R.H.,Jr., 1981. Distinction and uses of stratification types in the interpretation of aeolian sand. J.Sedim.Petrol., v.51, pp.579-595.

Kocurek,G. and Neilson,J., 1986. Conditions favourable for the formation of warm-climate aeolian sand sheets. Sedimentology, v.33, pp.795-816.

Kocurek,G., 1981. Significance of interdune deposits and bounding surfaces in aeolian dune sands, Sedimentology, v.28, pp.753-780.

Kocurek,G., 1988. Late Palaeozoic and Mesozoic eolian deposits of the Western Interior of the United States. Sed.Geol., v.56, pp.413.

Krylov, I.N., 1976. Approaches to the classification of stromatolites. In: Walter, M.R. (Ed.), Stromatolites, Elsevier, Amsterdam, pp. 31-43.

Kukal, Z., 1970. Geology of recent sediments. Prague, Academic Publishing House of the Czechoslovak Academy of Sciences, 490p

Leeder, M.R., 1982. Sedimentology, process and product, George Allen and Unwin, London, 344p.

Leeder, M.R., 1999. Sedimentology and sedimentary basins, Blackwell, 592p.

Long ,D.G.F., 1978. Proterozoic stream deposits : some problems of recognition and interpretation of ancient sandy fluvial systems. In. Fluvial Sedimentology (A.D.Miall, ed.). Can.Soc.Pet.Geol.Mem., v.5, pp.313-342.

McGee, W. J., 1897. Sheet flood erosion, Geol. Soc. Am. Bull., v. 8, pp. 87-112

McKee,E.D., 1966. Structures of dunes at White Sand National Monument, New Mexico (and a comparison with structures of dunes from other selected areas). Sedimentology, v.7, pp.1-69.

McKenzie, D.P., 1978. Active tectonics of the Alpine-Himalyan Belt: the Aegean sea and surrounding regions, Geophys. Jour. Royal Astronomical Soc., v.55, pp. 217-254

McLane,M.,1995. Sedimentology. Oxford University Press, 423p.

Miall, A.D.,1978. Fluvial sedimentology, Can. Soc. Pet. Geol. Mem 5, 859p.

Miall,A.D.,1996. The Geology of Fluvial Deposits: Sedimentary Facies, Basin Analysis and Petroleum Geology. Springer-Verlag, Berlin, 582p.

Middleton, G.V., 2003. Encyclopedia of sediments and sedimentary rocks, Kluwer, 821p.

Muir,I.D. and Rust,B.R., 1982. Sedimentology of a Lower Devonian coastal alluvial fan complex: the Snowblind Bery Formation of Cornwallis Island, Northwest Territories, Canada. Bull.Can.Petrol.Geol., v.30, pp.245-263.

Murthy,Y.G.K., 1979. Salient features of the revised geological map of the Cuddapah basin. 3rd Workshop on Status, Problems and Programmes in Indian Peninsular Shield. Institute of Indian Peninsular Geology, Hyderabad, pp.1-21.

Murthy,Y.G.K., 1981. The Cuddapah Basin — A review of basin development and basement framework relations: 4th workshop on status problem in the Cuddapah Basin. Institute of Indian Peninsular Geology, Hyderabad, pp.51-72.

Nagaraja Rao,B.K. and Ramlingaswamy,G.,1976. Some new thoughts on the stratigraphy of Cuddapah Supergroup. Seminar on Kaladgi-Badami, Bhima and Cuddapah Supergroup, Mysore. 17-20 abstracts.

Nagaraja Rao,B.K., Rajurkar,S.T., Ramalingaswamy,G. and Ravindra Babu,B.,1987. Stratigraphy, structure and evolution of Cuddapah Basin. In: Purana Basins of Peninsular India (B.P.Radhakrishna, ed.). Bull.Geol.Soc.Ind. Bangalore, Mem.6, pp.33-86.

Narain, H., 1987. Geophysical constraints on the evolution of Purana basins of India with special reference to Cuddapah, Godavari and Vindhyan basins. In : Purana Basins of Peninsular India, Geol. Soc. India, Bangalore Mem. 6, pp. 5-32.

Plaziat,J.C. & Purser,B.H.,1998: The tectonic significance of seismic sedimentary deformations within the syn- and post-rift deposits of the north-western (Egyptian) Red Sea coast and Gulf of Suez. In : Sedimentation and Tectonics of Rift Basins : Red Sea Gulf of Aden. (B.H.Purser and D.W.Bosence, eds.), Chapman & Hall, London. pp.347-652.

Radhakrishnan, B.P., 1987. Purana basins of Peninsular India, Geol. Soc. India Bangalore, Mem. 6, 518p.

Rajurkar,S.T., 1979. Pre-Kurnool age of the Cuddapah Orogeny. Indian Mineralogist, v.18, pp.32-39.

Rajurkar,S.T. and Ramlingaswamy,G., 1975. Facies variation within the upper Cuddapah strata in the northern part of Cuddapah basin. Geol.Surv.India Misc.Publ.No.23, Part 1, pp.157-164.

Rajurkar,S.T. and Ramlingaswamy,G., 1978. Atmakur fault and its bearing on the evolution of Kurnool basin. Proceedings of the Symposium Purana Formations of Peninsular India. Univ.Saugar, Sagar, pp.242-245.

Reading,H.G., 1996. Sedimentary Environments : Processes, Facies and Stratigraphy. 3rd edition. Blackwell Science Ltd., 688p.

Reineck,H.E. and Singh,I.B., 1980. Depositional sedimentary environments – with reference to terrigenous clastics. Springer-Verlag, Berlin, 2nd ed. 549p.

Riegel,W., 1991. Coal cyclothems and some models of their origin. In: Einsele, G., Ricken, W., Seilacher, A. (eds), Cycles and events in stratigraphy, Springer- Verlag, pp. 733-750

Rubin,D.M. and Hunter,R.E., 1983. Reconstructing bedform assemblages from compound cross-bedding In: Brookfield,M.E. and Ahlbrandt,T.S. (Eds.) Eolian Sediments and Processes. Elsevier, Amsterdam, pp.407-427.

Schwarzacher, W. and Fischer, A.G., 1982. Limestone-shale bedding and perturbationsof the earths orbit. In: Einsele, G.and Seilacher, A. (eds.). Cyclic and event stratification, Springer-Verlag, pp. 72-95

Seilacher,A.,1969: Fault graded beds interpreted as seismites. Sedimentology, v.13, pp.155-159.

Seilacher,A.,1984: Sedimentary structures tentatively attributed to seismic events. Marine Geology, v.55, pp.1-12.

Selley, R.C., 1978. Ancient sedimentary environments and their subsurface diagnosis, Chapman and Hall, 317p.

Sen,S.N. and Narasima Rao,CH., 1967. Igneous activity in Cuddapah Basin and adjacent areas and suggestion on the palaeogeography of the basin. In: Proc.Symp.UMP GRB and NGRI.Publ. Hyderabad, A.P., India, 8-261-285.

Shanmugam, G. and Moiola, R.J., 1995. Reinterpretation of depositional processes in a classic flysch sequence in the Pennsylvanian Jackfork Group, Ouachita Mountains: AAPG Bull., v. 79, p. 672-695.

Shiki,T. & Yamazaki,T.,1996: Tsunami-induced conglomerates in Miocene upper bathyal deposits, Chita Peninsula, Central Japan. Sedimentary Geology, v.104, pp.175-188.

Simpson,E.L. and Eriksson,K.A.,1993. Thin eolianites interbedded within a fluvial and marine succession: early Proterozoic Whitworth Formation, Mt. Isa Inlier, Australia. Sed.Geol., v.87, pp.39-62.

Sims,J.D.,1975. Determining earthquake recurrence intervals from deformational sructures in young lacustrine sediments. Tectonophysics, v.29; pp.141-152.

Singh,I.B.,1980. Precambrian Sedimentary sequences of India: their peculiarities and comparison with modern sediments. Precambrian Res., v.12, pp.411-436.

Sixsmith, P.J., Flint, S.S, Wickens, H.DeV. and Johnson, S.D., 2004. Anatomy and stratigraphic development of a basin floor turbidite system in the Lainsburg Formation, main Karoo Basin, South Africa. Jour. Sed. Res., v. 74(2), pp. 239-254

Sloss,L.L., Krumbein,W.C. and Dapples, E.C., 1949. Integrated facies analysis. Geol.Soc.Amer.Mem.39.,pp.91-124.

Smith,N.D.,1974. Sedimentology and bar formation in the upper Kicking Horse River, a braided outwash stream. J.Geol., v.81, pp.205-223.

Steel, R.J. and Gloppen, T. G., 1980. Late Caledonian (Devonian) basin formation, wetern Norway- Science of strike-slip tectonics during infilling. In: Reading , H.G. and Ballance, P.F. (eds.), Sedimentation in oblique-slip mobile zones, Spec. Publ, Int. Assoc. Sedim., v.4, pp. 79-103.

Stow D.A.V.; Faugeres J.-C.; Viana A.; Gonthier E., 1998. Fossil contourites: a critical review. Sedimentary Geol., v. 115(1), pp. 3-39.

Turcotte, D.L. and Oxburgh,E.R.,1973. Midplate tectonics. Nature, v.244, pp.337-339.

Ulicny,D.,1999. Sequence stratigraphy of the Dakota Formation (Cenomanian), southern Utah: interplay of eustasy and tectonics in a foreland basin. Sedimentology, v.46, pp.807-836.

Van de Kamp, P.C., 1973. Holocene Continental sedimentation in the Salton basin, California:a reconnaissance. Bull. Geol. Soc..Am., v.84, pp. 827-848

Van Wagoner, J.C., Mitchum, R.M., Posamentier, H.W.and Vail, P.R.,1987. Seismic stratigraphy using sequence stratigraphy, PartII: Key definitions of sequence stratigraphy. In: Bally, A.W (ed.), Atlas of seismic stratigraphy, AAPG Stud. In Geol., v. 27, pp. 11-14.

Vail, P.R., Audemard, F., Bowman, S.A., Eisner, P.N. and Perez-Cruz, C., 1991. The stratigraphic signatures of tectonics, eustacy and sedimentology, In: Einsele, G., Ricken, W., Seilacher, A. (eds), Cycles and events in stratigraphy, Spriger-Verlag, pp. 617-659.

Vail, P.R., 1987. Seismic stratigraphy interpretation using sequence stratigraphy. Part I: Seismic stratigraphy interpretation procedure. In: Bally, A.W (ed.), Atlas of seismic stratigraphy, AAPG Stud. In Geol., v. 27, pp. 1-10.

Wells, S.G. and Harvey, A.M., 1987. Sedimentologic and geomorphic variations in storm-generated alluvial fans, Howgill Fells, northwest England, Bull. Geol. Soc. Am., v. 98, pp. 182-198.

Windley, B.F., 1984. The evolving continents, 2nd edn., Wiley and Sons Ltd., 399p.

Wood,G.V. and Wolfe,M.J.,1969. Sabkha cycles in the Arab/Darb Formation off the Trucial coast of Arabia. Sedimentology, v.12, pp.165-191.

Zielinski, T., 1982. Contemporary high-energy flows, their deposits and reference to the outwash depositional model, Prace Naukoe Uniwersytetn Slaskiego, NR 519 Geologia, v.6, pp. 98-110.